Synthese Library

Studies in Epistemology, Logic, Methodology, and Philosophy of Science

Volume 401

The aim of *Synthese Library* is to provide a forum for the best current work in the methodology and philosophy of science and in epistemology. A wide variety of different approaches have traditionally been represented in the Library, and every effort is made to maintain this variety, not for its own sake, but because we believe that there are many fruitful and illuminating approaches to the philosophy of science and related disciplines.

Special attention is paid to methodological studies which illustrate the interplay of empirical and philosophical viewpoints and to contributions to the formal (logical, set-theoretical, mathematical, information-theoretical, decision-theoretical, etc.) methodology of empirical sciences. Likewise, the applications of logical methods to epistemology as well as philosophically and methodologically relevant studies in logic are strongly encouraged. The emphasis on logic will be tempered by interest in the psychological, historical, and sociological aspects of science.

Besides monographs *Synthese Library* publishes thematically unified anthologies and edited volumes with a well-defined topical focus inside the aim and scope of the book series. The contributions in the volumes are expected to be focused and structurally organized in accordance with the central theme(s), and should be tied together by an extensive editorial introduction or set of introductions if the volume is divided into parts. An extensive bibliography and index are mandatory.

More information about this series at http://www.springer.com/series/6607

Joseph Agassi

Ludwig Wittgenstein's Philosophical Investigations

An Attempt at a Critical Rationalist Appraisal

Springer

Joseph Agassi
Department of Philosophy
Tel Aviv University
Tel Aviv, Israel

Synthese Library
ISBN 978-3-030-00116-2   ISBN 978-3-030-00117-9 (eBook)
https://doi.org/10.1007/978-3-030-00117-9

Library of Congress Control Number: 2018953692

This Springer imprint is published by the registered company Springer Nature Switzerland AG
The registered company address is: Gewerbestrasse 11, 6330 Cham, Switzerland

As for modern methods of philosophizing ... Substantial forms, Occult Qualities, are exploded; ... they are only empty sounds.
William Wotton, Reflections upon Ancients and Moderns, 1694, Ch. 27

The questions that we forbid you to investigate ... are not only insoluble; they are illusory and devoid of meaning.
Henri Poincaré, Science and Hypothesis, Ch. X

And everything descriptive of a language-game is part of logic.
Wittgenstein, On Certainty, §56

Getting to know Wittgenstein was one of the most exciting intellectual adventures in my life.
Russell, obituary for Wittgenstein, Mind, 1951

I regarded him from the start as a mystic and metaphysician of the refined type, as an antiscientific person through and through and I dared to say so as the admiration of Wittgenstein was the fashion... Schlick enjoyed ... attacks on ... my violent remarks on this fellow Wittgenstein.
Neurath to Carnap, June 16, 1945

I have not found in Wittgenstein's Philosophical Investigations *anything that seemed to me interesting and I do not understand why a whole school finds important wisdom in its pages.*
Russell, My Philosophical Development, 230

"To say that metaphysics is nonsense is nonsense."
Friedrich Waismann, "How I see Philosophy"

"the cult of the piecemeal approach"
Ernest Gellner, Words and Things, Ch. 16

What is the use of studying philosophy . . . if it does not improve your thinking of everyday life, if it does not make you . . . more conscientious . . . in the use of . . . dangerous phrases. . .?
Wittgenstein to Malcolm, 16 Nov. 1944

Bach wrote on the title page of his Orgelbüchlein, "To the glory of the most high God, and that my neighbor may be benefited thereby". That is what I would have liked to say about my work.
Rush Rhees, Ludwig Wittgenstein, Personal Recollections, 181

I have carefully endeavored not to deride, or deplore, or detest but to understand.
Spinoza, Tractatus Politicus, 1 §4

Preface

This book is a rewrite (plus annotations) of the lecture notes of a course delivered in 1992 at the Department of Philosophy of York University, Toronto, on Wittgenstein's *Philosophical Investigations.* Of these 13 lectures, 7 are on the background to the philosophy of Wittgenstein and 5 are on it: 1 on the philosophy of young Wittgenstein, 1 on his transitional period, and the final 3 on the philosophy of mature Wittgenstein, chiefly on his *Philosophical Investigations*. The last and concluding lecture concerns the analytical school of philosophy that grew chiefly under its influence.

I first heard of Wittgenstein in 1953, when I was a graduate student at the London School of Economics and Political Science. The philosophical literature then hardly mentioned Wittgenstein.[1] I was then ignorant of him, as well as of Karl Popper. A paper of his (Popper 1952) on the value of metaphysics for science caught my attention, and I became his student (Agassi 2008, 72). Hearing Popper on Wittgenstein made me read his *Tractatus Logico-Philosophicus* (*TLP*). Trained as a physicist, I misread[2] it, viewing it as positivist old style. As it referred to Newton and not

[1]The case of Carnap is typical. As his views diverged from those of Wittgenstein, he found it increasingly difficult to express his debt to him (Schilpp 1963, 25); Wittgenstein considered him a plagiarist (Hintikka 1991). Carnap (1937) tacitly endorses Russell's deadly critique of Wittgenstein, with no mention of it, much less as deadly. He finally deemed Wittgenstein an irrationalist (Schilpp 1963, 26; Witherspoon 2000).

[2]The subtitle of Cook's 2005 book is *The Twentieth Century's Most Misunderstood Philosopher.* Black 1971, 1 says, "No philosophical classic is harder to master. According to Wittgenstein himself, it was misunderstood by Russell, Moore and Frege; and even Ramsey…"; see also Engelmann 1967, 94; Cavell, 1976, 151; Sluga and Stern 1996, Preface; Biletzki 2003, Introduction: "Rarely has a philosopher been so widely interpreted"; "a deluge"; a "massive corpus of interpretations." Landini 2007, 1–2: "we find diametric opposition among even the most prominent philosophical interpretations"; "It is difficult to avoid the pessimistic … conclusion that no satisfactory account … will be found." Hintikka (2006, 43) goes furthest: "nobody I know (with the exception of von Wright and perhaps a couple of other philosophers) has anything like a firm overall grasp of Wittgenstein's philosophy, including its development." They all dodge

to, it looked to me obsolete.[3] When Wittgenstein's first posthumous book, his *Philosophical Investigations* (*PI*), appeared, I tried to read it. I failed. Most of the leading commentaries on it I found rather unpleasant: they are hagiographic and obscure and thus, perhaps also unawares, metaphysical. Even his leading biographer, Ray Monk (1990, 2012), who made efforts to be open, has omitted too many unpleasant details from his tremendously detailed life of Wittgenstein. Few scholarly commentators like Paul Horwich (2012, Preface) show that it is possible to write an appreciative, informed, and balanced commentary on Wittgenstein's philosophy without concealing the inadequacy of his anti-metaphysics (*op. cit.*, xiii & note 7); Horwich judged Wittgenstein's view erroneous but still valuable. I try here to support this.

Nowhere in Wittgenstein's works did I find any restatement of the traditional characterization of metaphysics, reflected in G. E. Moore's description of it as "a general description of the whole universe" and the view of it of Moritz Schlick (1979, 215) as a "synoptic view," [4] both true to tradition. They naturally viewed realism as a part of the true metaphysics. Both ascribed it to Wittgenstein; Moore, but not Schlick, considered this stance a rejection of Wittgenstein's anti-metaphysical stance. Perhaps this reflects no more than Schlick's famous veneration of Wittgenstein as a sort of demigod.[5] It is hard to judge, as already then the literature on Wittgenstein suffered excessively from hagiography.[6] It still does. Bertrand Russell dismissed it whole, together with the whole output of mature Wittgenstein.[7] In disdainful response to it, he said ([1959] 1993, 217–218), Wittgenstein "threw away his talent and debased himself" because he had "grown tired of serious thinking." Still, he appreciated Wittgenstein's sincerity, comparing him to Blaise

---

the first rule of interpretation: first discuss the question why does the text at hand deserve attention in the first place. Particularly Hintikka is puzzling. He spoke of him in admiration, never explaining why, while rejecting his sole claim (*PI*, §133) that clarification will oust all metaphysics and criticizing him severely. The ideas that he advocated, he admitted, Wittgenstein "would in all likelihood have rejected" (2006, 41).

[3] *TLP*, §6.53-7; the view of metaphysics as meaningless was fashionable then. Eddington (1928, 91) found it impossible to avoid all metaphysics, but he declared his private. (So did Popper (1935, §29).) Schrödinger declared it unavoidable: one cannot open one's mouth and speak, he said, without metaphysics pouring out of it.

[4] Diverse, often vague, definitions of metaphysics appear in Rorty 1967 classic collection. None of them is traditional, and none of them is Wittgenstein's, although most (not all) of their originators are his fans.

[5] The expression is of Schlick's wife (Blanch Guy), recorded by Popper (Hahn 1995, 17); Stern (2007, 312) cites her to refer to Schlick's "reverential attitude of the pilgrim."

[6] The first public reaction to the Wittgenstein cult was Geoffrey Warnock's 1972 book that presented *English Philosophy since 1900* as indebted to G. E. Moore, not to Wittgenstein.

[7] Russell [1959] 1995, 159. He respected Tolstoy as a writer and as a moralist; "his theories are of course worthless" (letter to Goldie Dickinson, July 20, 1904). Wittgenstein's Tolstoyan mysticism marred his close friendship with Russell. Monk (1990, 211) blames Russell for this. He notes Russell's view of Wittgenstein's religiosity as tiresome but dismisses it as due to Russell's atheism. This forces him to ignore Wittgenstein's atheism (*TLP*, §5.4733) put grammatically ("God" has no reference). See McCutcheon 2001, Preface.

Pascal and to Leo Tolstoy. Wittgenstein's antics did not impress him, but he indulged him, in deep empathy with Wittgenstein's anguish if with nothing else.[8]

Current analytic literature is largely commentaries, largely on texts of Wittgenstein. All his texts repeat the traditional[9] positivist total dismissal of metaphysics as meaningless[10], and most if not all of his current followers reject this total dismissal. This requires reflection.[11] His admirers value his demand to write unpretentiously. Possibly, they hope to restore the precritical *naïveté* that Pascal, Tolstoy, and Wittgenstein yearned for. This suggests that his endorsement of the hackneyed mystic idea of the ineffability of the true metaphysics is a corollary of sorts to his demand for unpretentiousness. Mystics of all sorts say, initiates cannot teach; they can only point the way. Perhaps this explains Wittgenstein's straightforward and yet undeniably cryptic style.[12] Yet, whereas little disagreements exist within traditional mystic traditions, for example, the Zen canon, disagreement about Wittgenstein's texts abounds and varies. This disagreement is reducible by the adoption of some commonsense rules of interpretation (Agassi & Meidan 2016, 8, 91). First, explain the need for hermeneutics: try to read a text as interesting, i.e., as suggesting discussion of a problem or as assessing an interesting answer to it. Second, avoid

[8]Excess sincerity can serve as a tool for dogmatism. (Sartre identified seriousness with commitment and commitment with unreserved dogmatism.) Notoriously, Wittgenstein either dismissed criticism with contempt or admitted it, invariably adding to the admission that he should commit suicide. That his disposition to commit suicide was sincere only worsened the situation. Talented as he assuredly was, he never tried to avoid causing misunderstanding, his declared passion for clarity notwithstanding. Generally, any repeated unexplained dismissal of criticism of any text renders it increasingly opaque and so better ignored.

[9]Gomperz (1941, 173) followed tradition and listed some familiar terms that he deemed meaningless (terms, not statements).

[10]Ignoring the habitual ambiguity in the analytic literature about the distinctness of the two analytic schools, Dummett (1960, 76) dismisses Gellner's 1959 critique of analytic philosophy: "As a serious piece of philosophical criticism, Gellner's book is totally vitiated by his failure to distinguish between the targets of his attack." Dummett could illustrate this by citing some criticism of Gellner that applies to only one school. He did not trouble himself to do that.

[11]*The Stanford Encyclopedia of Philosophy,* Art. Metaphysics does not mention Wittgenstein!

[12]Russell rejected the philosophy of young Wittgenstein as mystical yet as saying much that is allegedly unsayable. Anscombe (1995, conclusion) explains Wittgenstein's vagueness as due to his continuous inquisitiveness. This excuse is always valid and raises the question regarding the criteria for distinguishing the still vague from the hopelessly muddled. Now the difficulties that Wittgenstein faced relate not to problems but to his task of freeing people from metaphysics (Monk 1991, 291). Ambrose and Lazerowitz ([1972] 2014, 16) say this clearly: (as the aspiration of metaphysics is "to get a general comprehensive picture of the universe") his aspiration was to "not satisfy the craving but make you cease to have it." He denounced "our craving for generality" as it rests on "certain craving of the metaphysician which our ordinary language does not fulfil" (*Blue Book* 17; Monk 1991, 527). True, metaphysical reflections did pain him; to others they brought joy. Are they addicts? If so, then Wittgenstein is a therapist who imposes his services. His therapy is thus unusable: the use of general terms is unavoidable and so is the curiosity that propels the learning that expands "our ordinary language." This shows how penetrating was Pole's 1958 characterization of Wittgenstein as a conservative. The hostility that Pole met at the time was extraordinary.

defensiveness. In particular, avoid reading new ideas into old texts.[13] Third, if comments of authors on their texts are available, consider them seriously but in a critical mood – as if written by others (except when information about the authors helps contextualize their texts).[14]

The concern here is with Wittgenstein's (posthumous) *Philosophical Investigations* (1953). He advised there (*PI,* x) his reader to contrast it with his first work (1922) as all too narrow.[15] Whatever part of his early views he later withdrew, he retained his hostility to all effort to articulate metaphysics.[16] Viewing his first book as too narrow, he later argued that the vernacular is free of metaphysics. He allowed for common religious expressions, but as having no truth-values; that is to say, as meaningless; that is to say, as defective. Some of his followers reject this as a violation of his Tolstoyan respect for sincere religion, but they say so only *sotto voce*.[17]

The early commentators on Wittgenstein portrayed him rightly as one who had advanced one and only one thesis. They worded it clearly: metaphysics is meaningless. They did that with great admiration and alacrity. From the start, the thesis wanted specification: what locution is meaningless? Wittgenstein refused to answer it. A panel discussion took place in London in the 1940s, soon after the war, on the mind-body problem, with three leading Wittgenstein disciples as participants. The ensuing public discussion must have been quite a spectacle, since echoes of it reached me a decade later. In that discussion, speakers denied that they were in metaphysical disagreement, since metaphysics is meaningless. This frustrated the audience. Wittgenstein had hoped to dissolve all philosophical disputes; this story presents him as a generator of a new kind of dispute.[18] His claim to have improved

---

[13]Cavell 1979, 20; Hintikka 1996, Introduction; Goldstein 1999, 8.

[14]Hintikka (1991) presents Wittgenstein as "An Impatient Man" in order to explain his having (wrongly) charged Carnap with plagiarism. Hintikka (1996, Ch. 1) also explains this way Wittgenstein's enigmatic style. He never discussed Wittgenstein's refusal to control his (all too sincere) impatience in an effort to write clearly.

[15]This is not the whole truth; there was an exception: early Wittgenstein admitted tautologies, even though only reluctantly (*TLP,* §§4.461; 5.633–5.641, 6.53); mature Wittgenstein did not (Pascal 1973, 25). See Wright 2006 on Wittgenstein's view of mathematics as senseless ("sinnlos"). Quine derided this "meaning by courtesy." Wittgenstein viewed tautologies as uninformative; Quine (1988) notoriously insisted that there is no sharp dividing line between the informative and the uninformative.

[16]Katz 1990, Introduction. Conant (2005, 175) limited Wittgenstein's condemnation of metaphysics to the past. Hintikka (2003, 7, 174 and Ch. 13) limited it to meaningless assertions translatable to every language. This makes no sense: there are no rules for translating nonsense. Wittgenstein demanded of Popper an instance of a philosophical problem. Popper named the problem of induction; Wittgenstein dismissed it as merely scientific. Wittgenstein (1929, §6.31) says, the problem of induction is very hard and scarcely begun a task.

[17]Thus, Ryle (1963, 143–144) said cryptically, "When he said, 'Do not ask for the meaning, ask for the use', he was imparting a lesson which he had to teach himself after he had finished with the *Tractatus*."

[18]Kripke (1972, 42, 155) has a new solution to the mind-body problem (Agassi 1995b, 271), with no mention of Wittgenstein's disapproval of such a move. Hacker (1997, last paragraph) goes

the philosophy of life is awkward just because he prevented philosophical discussion from the ability to achieve that: he cut the branch on which he was sitting. This is the most serious critique of Wittgenstein. Although leading thinkers Russell and Popper have advanced it, almost all commentators ignore it. My effort is to bring it into center stage. Yet admitting this we may see in Wittgenstein's very arguments against metaphysics some metaphysical ideas unfold that relate to language and that we may find refreshing – especially his anti-reductionism and his dissent from Russell's staunch opposition to *das Drittes Reich* (the third realm) or *das Reich der Gedanken* (the realm of thought) of Frege.

A recent essay that has won the prize of the American Philosophical Association (Malmgren 2011) discusses the possibility of critical examination and assessment of philosophical theories. The essay has no mention of earlier works on this question, of Russell, Wittgenstein, or Popper. This wants a remedy. Let my presentation of the philosophy of Wittgenstein against its background assist in the return to a modicum of scholarship. No matter how we judge and appraise it, and whatever we declare the right philosophy, we should try to have our discourse on it somewhat informed.

Herzlia, Israel, Summer 2018

---

further and ascribes to Wittgenstein the view that computers do not think. By contrast, the contributors to *Philosophy of the Social Sciences*, Vol. 46, No. 6, 2016, explicitly eschew Wittgenstein's anti-metaphysics. This is a welcome change.

Two Terminological Notes

Technical Terms The distinction between technical and ordinary terms reflects awareness of changes of meanings of words due to new theories (beginning with Newton). The claim that Wittgenstein has introduced technical terms suggests that he offered a new theory. He denied this repeatedly and emphatically. His allegedly technical terms, "life-forms," "family resemblance," "rule following," and "private language," are not technical. His term "meaningless" is[1]: he always used it in Russell's sense (of violating grammar and thus having no truth-value). Ordinarily a meaningless move in a discussion is unhelpful; a meaningless expression is one not in the common vocabulary (e.g., "Jabberwocky"). Ordinarily, seemingly meaningless terms invite interpretation. Some analytic texts confuse the technical sense of a term with its ordinary sense. This invites careful reading of these texts and possibly some criticism.

Translation Some of my translations of Wittgenstein's texts slightly deviate from their familiar ones. An example is "Bekräften und Entkräften" that means increase and decrease of strength. The published translation is "confirmation and disconfirmation" that is Carl G. Hempel's objectionable, stilted neologism for "confirmation and infirmation."

[1] *PI*, §116; Read 2010. Ayer (1991, 3) reports that Morris Lazerowitz rejected the ordinary sense of the term as used by Moore and by Ayer. This is a half-truth; Ayer did so only late in his career: in the first and most famous of his books (1936), he used the term in Russell's sense. To complicate things further, Toulmin (1975) pronounces Wittgenstein's ordinary terms Austrian rather than English. This, he adds, made Wittgenstein incomprehensible to his students, much to his frustration. No: frustrated and isolated as Wittgenstein surely was, he knew English well enough. Malcolm ([1967] 2001, 4) says, "It is probably true that he lived on the border of mental illness."

Acknowledgment

My discussions of natural deduction rest on lecture notes of Karl Popper's lectures at the London School of Economics of the 1950s and 1960s. (For more see the bibliography at the end of this work.)

Nimrod Bar-Am of Sapir Academic College, Israel; Gilead Bar-Elli of the Hebrew University of Jerusalem; Daniel Cohen of Maccabee Seed Company, Davis CA; and Ian C. Jarvie of York University, Toronto, have read early versions of this study and made many corrections and many useful suggestions. My profound gratitude to them all for their extraordinary generosity and sagacity.

Abstract

The positivist hostility to metaphysics altered with Frege's refutation of the traditional theory of meaning as reference, whose main appeal was that it seemed naïvely realistic, and thus as metaphysics-free. Russell rejected Frege's new theory of meaning as metaphysical, offering a partial alternative to it. He rendered logical paradoxes meaningless within his formal language by creating rules to block them. This was the birth of the analytic method.

Young Wittgenstein viewed metaphysics meaningless in Russell's sense (of rendering it ungrammatical). His early work is a translation of neutral monism to modern logic; it is a monumental achievement yet a failure – since his language is too limited.

Wittgenstein's posthumous *Philosophical Investigations* ignores formal languages while retaining the view of metaphysics as meaningless. He said that he advocated only this idea. It was very popular in the middle of the twentieth century and is now *passé*.

Wittgenstein is very popular and almost incomparably influential. This requires reflection and explanation. His concern with language is a part of the Fregean shift of concern from truth to truth-or-falsity, namely, to truth-value identified with meaning in Frege's system. It is also the decision to reinterpret any odd sentence that may look metaphysical so as to render it a part of commonsense.

Analytic philosophers seek a theory of meaning, considering the concern with language a major aspect of philosophy proper. What else characterizes this school is still under dispute. The default reading of Wittgenstein is that he considered genuine philosophical problems impossible to word. It is not quite the consensus. Jolley (2014, 24), for instance, argues ably for the opposite view: he sees in Wittgenstein (*PI*, §109) reference to problems essential for rational discussion. I wish I could share his reading. Alas, Wittgenstein's place would thereby be explained by bluntly ignoring his philosophy. This implies disrespect for him, even if tacitly. His followers view his concern with language a new and invigorating source of interest in philosophy. His critics still find in it a regrettable indifference to genuine philosophical problems.

The Sapir-Whorf hypothesis says, language hides traces of diverse systems. They are often defunct. Unawares, analysts unearth these systems and deem metaphorical all magical expressions. This reflects a metaphysical anti-magic prejudice. It is much better to discuss magic directly and openly and reject it explicitly – in preference for better metaphysical systems.

Contents

Chapter 1
Background

For decades now two philosophers are the most popular within departments of philosophy worldwide, Martin Heidegger and Ludwig Wittgenstein. Each of them is a leader of a major school of philosophy – the phenomenologists (or the Continental) and the analytical (or the Anglo-American) – with a marginal group of philosophes who make efforts to synthesize them. There is little or no dialogue between schools. Very crudely, the analysts are pro-science and the phenomenologists are pro-poetry. (Poetic truth, Heidegger said, is superior to scientific truth.) Regrettably, this is not quite right: what characterizes these schools is hostility – to science or to poetry respectively. Being pro-science and pro-poetry, the present study shows little sympathy for any hostility.[1]

Since leading analytic philosophes are largely familiar with logic, they are also familiar with the defect of the method of essentialist definition that was the traditional standard tool of philosophers. Leading phenomenologists, on the contrary, employ this tool extensively. (Perhaps this is an excessive concession, since these applications are often sham. Thus, the most famous (mock-) essential definition is Heidegger's: Man as Afraid and Bored. Jean-Paul Sartre has offered a famous variant of it that is equally pejorative: Man as Afraid and Disgusted. That Man is Truth-lover and Justice-lover they ignore.) This is a major advantage of the analysts over the phenomenologists. Another defect of phenomenology is its flirtation with barbarism, if not its open endorsement: Heidegger was a member of the Nazi party; his philosophy appeals to violence. By comparison, the defect of the cult of personality that the analytic school suffers from (Bartley 1982–1983) is mild: its treatment of Wittgenstein as a super-philosopher is boring, as it keeps on its research agenda a dreary item: what was his teaching and what is its significance?

---

[1]This ignores almost all the smaller philosophical schools with the exception of the critical rationalists. It is the only school to have taken the cudgels against Wittgenstein's heritage (Popper [1935] 1959, Preface to the English Translation). Likewise, this pays little attention to the existentialist aspect of analytic philosophy and the Kafkaesque ideas to find in Wittgenstein and even in Quine (Stroud 1995; Williams 2004).

J. Agassi, *Ludwig Wittgenstein's Philosophical Investigations*, Synthese Library 401,
https://doi.org/10.1007/978-3-030-00117-9_1

What did Wittgenstein teach? Why is he the leader of a school? He said, his only message is that all metaphysical expressions are hopelessly ungrammatical, and since this message has no current advocates (as we know too little about grammar). It seems then that from the point of view of those who aspire for a rational reconstruction of the history of philosophy (Goldstein 2004, 171), he has no major role in it. Critical rationalists share this view. This study is an attempt to improve upon it in a friendly manner and see what we can learn from the seven decades of study of Wittgenstein's heritage.

1.1 The Tragedy of Young Wittgenstein

Frege, Peano, Russell, and a handful of other philosophers developed logic in splendid isolation. It is hard to imagine the gulf between the ignorance of, and indifference to, modern logic at the time and its being so popular and so self-understood today. Two prodigies entered this highly exclusive club like old pros: the two friends Ludwig Wittgenstein and Frank P. Ramsey. The latter was the younger of the two – by more than a decade. He was perhaps the only critic of the *Tractatus Logico-Philosophicus* whose criticism Wittgenstein accepted with no qualm. He died young, before Wittgenstein embarked on a new philosophical career. The tragedy of Wittgenstein was the disparity between the tremendous success that he had as a young undergraduate student, when both pioneers, Frege and Russell, befriended him and treated him as an equal, and his immense sense of failure as they rejected his work that he had written under most incredibly trying conditions. Their great respect for it was no consolation to him. He felt he was a failure and left the scene. He returned about a decade later to start a new philosophical career.

The achievement of young Wittgenstein was his version of scientism. What was his later attitude to science? Did it change? Some commentators considered him hostile to science or to reason all the way. These commentators distorted his attitude to science or to reason by making the following mistake. They may have found that he rejected their own views of science or of rationality and concluded that he rejected science or rationality *tout court*; this is a common error.[2] It is possible to consider any loss of interest in science as a loss of respect for it; this does not hold in the present case. Russell, who rejected Wittgenstein's philosophy as mystical, avoided this confusion. Whether one agrees with his reading of Wittgenstein's early views or with his dismissal of his later views, one may notice that he was familiar with it and cautious in his wording of his disagreement. Also, having dismissed the later

[2]Hostility to scientism is the last reason for the ascription of any anti-science attitude – regardless of any attitude to scientism.

writings of Wittgenstein he still spoke of him with sympathy. He possibly never felt free of his protective attitude to him.[3]

Russell's appreciation of the philosophy of young Wittgenstein (despite his dissent from it) as valuable differs sharply from his view of the philosophy of the mature Wittgenstein as trite. It is the use of the best achievements of modern logic to present scientism at its best; it convinced Russell more than anything else did that scientism is erroneous: philosophy is more than the limiting human knowledge to science and to science alone. This is a tribute to Wittgenstein, not different from the tribute historians of physics pay to the failure of Laplace to square Newtonian mechanics with the metaphysics of Descartes or the failure of Maxwell to develop a mechanical model for the aether. Wittgenstein was not ready to accept a compliment of this sort. This is regrettable. He sadly insisted on equating comprehension with agreement and appreciation,[4] concluding that there is no value to Russell's magnificent Introduction to his book. He even left open his own valuation of his book while he was forging his new outlook. He thus lived in constant inner conflict.

Wittgenstein's philosophical career divides into early, interim and mature. Most of the interim period he spent as a schoolteacher in small, remote places. When Ramsey visited him with a request to clarify a passage in his book, he said he did not remember. He then put that book behind and embarked on a new philosophical career. Naturally, his philosophy changed only in some respects. Such changes are of two kinds: the new may but need not contradict the old. Did Wittgenstein change his mind [significantly] or did he only add to his early work? He did not say, and debate on this still rages. His fans avoided discussing all options, perhaps because they suspect that he considered it good public relations to shun this issue. In any case, does it matter? He said, it does: he recommended contrasting his mature and early texts in order to improve comprehension (*PI*, Preface).

The chief question here is, did he change his view about science?[5] His early work radiated scientism. This is the culmination of his early work (*TLP*, 6.53):

> The correct method in philosophy would really be the following: to say nothing except what can be said, i.e. propositions of natural science – i.e. something that has nothing to do with philosophy – and then, whenever someone else wanted to say something metaphysical, to demonstrate to him that he had failed to give a meaning to certain signs in his propositions...

Those who view this passage as the culmination of the achievement of young Wittgenstein stand apart from all other commentators on it. In particular, they totally dismiss the major extant criticism of it as a total misunderstanding. That criticism is that he had failed to defend his view, asserting it dogmatically. Those who voice this

[3]Russell's letter to Popper of 12 November 1946 (Grattan-Guinness 1992, 15) may support this reading.

[4]This is hard to uphold. Indeed, on occasion Wittgenstein deviated from it too (Monk 1990; 2012, 313).

[5]Howes 2007, 23. Kindi 2017; Ryle 1957, 1996: "The Work of an Influential but Little-Known Philosopher of Science; Ludwig Wittgenstein". Proctor 1959, "Scientific Laws and Scientific Objects in the *Tractatus*".

criticism of his early work overlook its cardinal aspect: he was following a well-established, traditional attitude – traditional positivism or scientism – and his achievement was the completion of Russell's program: to develop logic into a system in which philosophy and logic are coextensive.[6] Russell failed to achieve it and finally he gave it up (Russell 1940, last two pages); Wittgenstein did achieve it – during the First World War and as a prisoner of war soon after. It was a relatively clear task and the question was not how each move in the process of its completion is justifiable but rather, is the task performable? Wittgenstein answered it by performing the task. This way he also showed that the price for achieving it was much too high; thus, it radically reduces the unattraction of the project. Russell explained all this in his Introduction to the book.

Now, according to the consensus,[7] this great achievement of Wittgenstein is a failure. This is unfortunate, and quite avoidable, as the fate of logicism shows: logicism is the idea that mathematics is a part of logic; Russell worked it out in detail and thus proved it a failure. (He claimed that the axiom of infinity is essential for mathematics yet is no part of logic proper.) Russell's work is an achievement nonetheless. The same holds for the earlier failure of James Clerk Maxwell's effort to develop a model for the aether: Thanks to this failure, it became reasonable for Einstein to examine the option that there is no aether. Now, my reading of Wittgenstein's *Tractatus Logico-Philosophicus* as an achievement and as a failure (as it shows how costly it is to achieve its declared aim) is not new: it is the declared thesis of Russell's Introduction to that work. Wittgenstein dismissed that Introduction as a reflection of Russell's failure to comprehend the book. The worst aspect of the story is that Wittgenstein never tried to clarify that misunderstanding; he only lamented the failure of Frege, Moore, and Russell to comprehend him. This is a bit ironic, considering that his life-long task was clarification. It even raises the suspicion that Wittgenstein, renowned for most uncompromising sincerity, was in this case rather disingenuous, or at least self-deceiving.[8] This suspicion may explain the paucity of discussion of this story and particularly the similarity between the ideas of Russell and of young Wittgenstein.[9]

[6]Black 1971, 24. The title of Chapter 2 of Russell's 1914 text, "Logic as the Essence of Philosophy", expressed the new program. He gave it up; young Wittgenstein executed it as best he could.

[7]The "Vienna Circle" deemed reasonable the cost of getting rid of metaphysics. Carnap 1937 pretended that no cost was involved: he altered Wittgenstein's system by admitting Russell's three criticisms of it. (He accepted the quantifiers, infinity and the meta-language.) He thus tacitly denied that this invalidates the achievement. At least he realized that his alterations required of him to alter the concept of metaphysics. This he did: all traditional metaphysical systems confuse words and things, he said. Even proverbially pedantic Quine who considered this idea admirable admitted that it is a failure. Carnap's characterization holds for Pythagoreanism but not for classical atomism (Popper ([1935] 1959, §85). Carnap dismissed Popper's criticism by declaring that he had no objection to atomism, overlooking its status as a metaphysical system contrary to his view.

[8]Wittgenstein showed bad faith here. Since he and Russell went over the manuscript of the book together, his attitude to Russell's Introduction was odd and required explanation. It deserves mention though that he would have preferrred to have his book published without Russell's Introduction (but not enough to use his wealth to procure publication).

[9]Black 1971, 23*ff.*

The case of young Wittgenstein is by far not the only case of taking criticism amiss. Russell admired Frege for his not having taken amiss his (Russell's) demolition of his (Frege's) system just because such a detached conduct is still exceptional. Wittgenstein was hostile to criticism and even to mere dissent (although he dissented from Otto Weininger whom still he admired for his sincerity). Before his first visit to the "Vienna Circle", Schlick warned its members not to ask him too many questions. His sensitivity, the oddity of his ideas and his precision made it very difficult to understand him. Notoriously, young Wittgenstein's assertions are enigmatic, whereas their thrust is clear; by contrast, mature Wittgenstein's assertions are very clear but not their thrust. What this thrust is comprises a major problem of this study. *Grosso modo*, mature Wittgenstein clung to the view of young Wittgenstein in declaring all metaphysics meaningless. Even this is under controversy.

1.2 Metaphysics Needs No Defense

Disagreement rages about the contents of Wittgenstein's output. However far this disagreement goes, and however much it may color our discussion, it is hard to disagree about its chief characteristic. His early output is the presentation of a new version of the traditional denunciation of metaphysics as meaningless by a new use of the powerful ultra-modern logic; his late (posthumous) output does the same by the use of the powerful logical analysis of language (or language analysis or simply analysis).[10]

Before discussing the denunciation of metaphysics, we may examine the very idea of the denunciation in the kingdom of ideas. The simplest denunciation of any intellectual activity is by the view of it as inciting immoral conduct. Whatever we think of morality or of intellectual activity, there is no dissent from the view that this is problematic, since any idea can be misused and no idea has to be. In the late nineteenth century nihilism was deemed demoralizing and thus as inviting immoral activities. Yet critics of it did not recommend a taboo on discussing it. Similarly, no one before Wittgenstein has denounced metaphysics as such, even though (obviously and trivially) some metaphysics is culpable this way. Still, freedom of speech is a basic tenet of democracy, and yet the law limits it. The paradigm cases are two: disturbing the peace (shouting "Fire!" in public) and inciting for riots. Traditional opponents of metaphysics never suggested that there should be a law against its study, or even that it is immoral. The nearest to it was Wittgenstein's denunciation of metaphysicians as slum landlords. He spoke as a police officer,[11] causing discomfort even to some of his close followers.

---

[10]Russell 1940, 327: "The operation by which, from the examination of a whole W, we arrive at 'P is a part of W' is called 'analysis'."

[11]"A bad philosopher is like a slum landlord. It is my job to put him out of business". Fann 1967, 69; Rhees, 1984, xv, 117; Canfield 1986, 102; Hallett 1991, 125; Edmonds and Eidinow 2001, xx;

Like other anti-metaphysicians, Wittgenstein considered metaphysics deceit and its success as due to its appeal to human weakness. This is not peculiar to him: already the Bible condemned false prophets as deceitful. His new argument against metaphysics has strengthened and altered anti-metaphysics: he said all metaphysical propositions are confused, and inherently so: their being pseudo-propositions is irreparable, except by depriving them of their metaphysical character, by rendering them commonsensical or scientific. Otherwise, they display efforts to transcend the limits of language. This, contrary to a common understanding of Wittgenstein, is a denunciation not of metaphysical ideas but merely of efforts to articulate them.

Some texts are very confused; of these, some may qualify as metaphysical with no need to discuss the nature of metaphysics. What is objectionable is Wittgenstein's denunciation of all metaphysicians as slum-landlords. If we ignore this and center on assertions, then what is objectionable here is Wittgenstein's declaration that all that is sayable is sayable clearly and that one should not try to articulate what is not sayable. This is both undemocratic and detrimental to research: this usually starts with stuttering, at times ending with clear statements and at times not. Wittgenstein took back his demand for utter clarity, although unfortunately he did so only implicitly – when he admitted that utter clarity is impossible.

One may view all this as quibbling: what signifies is that, as tradition has it, Wittgenstein's output at its apex denounces metaphysics as a matter of course. Why then is this so obvious? It is hard to believe that liberal Immanuel Kant asserted that obviously metaphysics is obscure and incomprehensible and that to study it is simply to waste time. (He admitted that it is very hard to stay away from it.) He valued liberalism as a moral and a political doctrine; he did not consider it metaphysics, as metaphysics is speculative and he considered liberalism certain: he took his categorical imperative as both demonstrated and as the very basis of liberalism. Moreover, he expected that in the future metaphysics would have the right to claim scientific status. Wittgenstein and his followers could deny that liberalism is metaphysical only by considering it practical and commonsense. This denial may face the objection that quite possibly some assertions are both commonsensical and metaphysical. The disagreement then concerns the question, is commonsense metaphysics possible? Moore and Russell said, yes, Wittgenstein said, no.[12] All three admitted that liberalism is commonsensical. A criterion to demarcate metaphysics from commonsense is necessary for the resolution of this dispute. If this dispute is

---

Nielsen and Phillips 2005, 15; McGuinness 2012, 314. This way Wittgenstein explained his hostility to certain metaphysicians. Yet his demarcation of metaphysics did not gain general approval. He slightly ameliorated his hostility when he allowed for (sincere and unpretentious) metaphysical poetry. This made Carnap replace "meaning" with "cognitive meaning".

[12]This is not to claim understanding of Wittgenstein on commonsense. He seems to have included in it Spengler's historicism (Klagge 2010); alternatively, he might have included in it Tolstoy's historicism. Neither version is commonsensical. Nor is it scientific. Hence, it is metaphysical (Berlin 1955).

unresolvable, then Wittgenstein was in error: he said unanswerable questions are meaningless. This raises another question: what question is unanswerable?[13]

As analytic philosophy distinguishes between science, commonsense and metaphysics, a discussion of the demarcation of these will make it clearer. Without discussing it, however, we all tend to agree about the metaphysical character of both the doctrine of the inborn evil of humans (*Genesis* 8:21) and the Enlightenment doctrine of human natural goodness. Both had tremendous impact. Hence, if Wittgenstein was right and they are inherently confused and thus to be exorcized, then we may ask, how does his philosophy account for this. For, both of these doctrines have common-sense consequences. The suggestion that these doctrines demand improvement deviates from modern anti-metaphysics. For, even if metaphysics is in constant need of improvement, it is still in no need for defence against anti-metaphysics: there is room for improvement of everything human, both in matters of clarity and in matter of contents and more.

This is not to dismiss the enormous output of Wittgenstein, nor of the multitude of his fans, not even the hagiographers among them. They comprise the majority of pro-science philosophes today and, if possible, their influential ideas and techniques want examination and improvement, not dismissal.

1.3 The Analytic School and Politics

Language analysis, analysis for short, names (1) a doctrine, (2) an activity and (3) an approach. As a doctrine, its status is under dispute. As an activity, it is traditional. As an approach, it is new, and it characterizes a leading school of philosophy. The first thing to notice is that its special attention to the everyday furnishes it with the tinge of a prosaic philosophy of life, one that is very different from that of Wittgenstein (who was Tolstoyan and who thus tended towards asceticism).

The analytic school associated with the name of Wittgenstein rejects the technique of essential definitions and replaces it with the technique of philosophical analysis, or language analysis, or linguistic analysis. It is supposed to construct linguistic philosophy or the philosophy of language anew. It is the opinion that the concern with language is a major aspect of philosophy proper. The nearest that this study will approach a discussion of the Continental or the phenomenological school will be the critique of the classical doctrine of essential definitions. Since this critique is devastating, little remains to say about phenomenology. Indeed, the sweeping dismissal of metaphysics that characterizes early analytic philosophy applies with

[13]Scientific researchers would love to know what question is answerable. This is context-dependent, and even obviously so. This and the idea that science can grow by criticizing commonsense, Russell observed, demand a revision of analytic philosophy.

ease to much phenomenology. Otherwise, that sweeping dismissal of metaphysics that characterizes early analytic philosophy is the chief target of criticism here.[14]

To begin with, the analytic school identified itself with the view that analysis is the only proper tool for philosophy[15]: it had only one declared task, to unmask metaphysical expressions as ungrammatical.[16] This view of the status of metaphysics in general was the one and only characteristic of the analytic school.[17] It is now *passé*. Nevertheless, it still applies on occasion to particular cases or in particular ways. The latest output of this school is largely commentaries on works of Frege, Russell and Wittgenstein. These are largely analyses of diverse philosophically significant expressions, present or past. Since the first clearly analytic work is Russell's path-breaking essay "On Denoting" (1905), much of the current analytic discussion concerns this paper. This is impressive, as it reveals the tremendous variety of options that Russell ignored. This is no criticism, since his concern with meaning was largely an offshoot of his logicism. Indeed, his theory does not resolve all of the puzzles that he posed in that essay (Linsky 1967, 127): he solved the problem as far as his study of the foundations of mathematics require. The question then is, why do analytic philosophers study the use of the definite article in further detail? Perhaps linguists have use for it for one or another concern they have. What justifies calling this activity philosophy rather than linguistics is the Wittgenstein-style tradition. Some question it, but it is better to let it ride and discuss the Wittgenstein-style tradition as it is.

Wittgenstein took care to distinguish between truth and clarity; he enjoined his followers tirelessly not to contradict metaphysicians: he wanted them to expose the confusions inherent in metaphysics and do no more. (He abhorred confusion, commentators have observed, as it leads to self-deception.) This may suggest a clear distinction between curable and incurable grammatical errors. Traditional grammarian will reject this idea of Wittgenstein. He did not care: he was engaged in a special kind of confusion, the one that emerged out of efforts to transcend the [permanent] limits of language. He had no rule to distinguish these confusions from any other, nor could he possibly know that these limits are permanent.[18]

Before delving into any philosophical discussion, it behooves us to ask, what is sinful in being somewhat confused, perhaps even destined to be somewhat confused

---

[14]This is not very accurate, since anti-metaphysics is ubiquitous and prevalent.

[15]Russell's view of commonsense was scientifically informed and fallibilist. He was thus a terrific analyst.

[16]*TLP*, §6.53: "The right method of philosophy would really be to say nothing . . . except natural science . . . and then, whenever someone else wants to say something metaphysical . . ." it should be debunked.

[17]Ryle said repeatedly, there is too much discussion about philosophy and not enough doing philosophy. This amounts to the imposition of his agenda. Fortunately, as editor of *Mind* he was liberal.

[18]Jack London reports in his 1908 dystopia *The Iron Heel* the (future) refutation of Berkeley's idealism. This is a conjecture, of course. Conjectures about future conjectures are absurd (Popper 1957, Preface).

by the very context of our human condition? In any case, what is the good of dismissing one confusion after another? Is this task terminable? This repeatedly-asked question points at the expectation that philosophy should be useful in some way. This may be an error. We venerate high art although it is largely useless. If Socrates was right in asserting that the unexamined life is not worth living, then philosophy, namely, the critical examination of whatever we value, is as good as possible for making life worthwhile. Was this Wittgenstein's goal? Maimonides and Spinoza viewed philosophical contemplation the only thing that makes life attractive. Analytic philosophers found all this troubling. In 1958 a very famous and influential book by Geoffrey Warnock, *English Philosophy Since 1900*, concluded by registering a complaint against the demand from philosophers to adjudicate on matters practical in disregard for their recently attained "professional status" that frees them of it. In 1990 Elizabeth Anscombe, one of Wittgenstein's leading exponents, said he was not an "ordinary man's philosopher". Wittgenstein's biographer Ray Monk viewed him as a guru and a poet (1990; 2012, 20, 291). Hilary Putnam (2008) presented him as a philosopher of life. What is this? What is its link to analytic philosophy if there is any?

Analytic philosophy[19] is the study language, sentences, and words: their use, usage, or ordinary use.[20] Early in the day, Russell gave a lovely example for an obvious run-of-the-mill analysis (1905, 489):

> I have heard of a touchy owner of a yacht to whom a guest, on first seeing it, remarked, 'I thought your yacht was larger than it is'; and the owner replied, 'No, my yacht is not larger than it is'. What the guest meant was ...

This is Russell's example for his view that quite often we all analyze what we hear. He was right, of course. He did not intend to validate all analysis this way, however, much less to say that all philosophy is analysis. Hence the question, what is the limit to useful analysis? Wittgenstein has answered this question (*PI*, §133): analysis is to continue until success is achieved and metaphysics causes pain no longer (*PI*, §133). Those who have no need for his cure[21] may want an alternative aim for their philosophy; those who do not admire his therapeutic attitude[22] want an alternative to it.

[19]The notoriously superficial rationalism of early-stage analytic philosophy masks the mysticism popular at the time that served as a replacement of the lost faith and as means for secularizing the philosophy of life. This is the root of the popularity among intellectuals then of superficial philosophies like those of Nietzsche and of Tolstoy. Compare on it critical Schwartz 2006 and apologetic Putnam 2008.

[20]Ryle [1953] 1963, 108; Hanfling 2013, Introduction.

[21]Evidently, Wittgenstein's view of therapy is irrefutable. Viewed as compulsory, it is deplorable.

[22]Many a language analyst is captive of Bacon's immature view that there is no admiration in dissent.

First, the analysts were zealous[23]; their zeal soon tapered off.[24] They were first busy debunking metaphysics but their doctrine softened and some of them dared engage in metaphysics. They still agree, however, to keep this activity very limited. Russell wanted to reform philosophy by efforts to make do without Plato's heavens.[25] Wittgenstein went much further: he tried to make do with no metaphysical theories at all, and he demanded to silence all metaphysicians: he compared them to slum landlords and declared his job to silence them. Suppose he was right. Now, prohibition comes in different degrees of severity – legal, moral and rational: the law allows some immoral conduct (abiding by the law while violating moral law is regrettably common, but the opposite conduct is at least as seldom as breaking the law); honesty allows some irrational conduct; the edicts of reason are then the strictest. Analytic philosophers usually evade discussing the rules of rational discourse,[26] preferring to center on one rule: avoid talking metaphysics/nonsense. This is too vague. Indeed, this very rule was relaxed in stages (Hempel 1950). Since we tend to interpret vague assertions or ask for their elucidation, we hardly ever meet irredeemable nonsense, such as nonsense verse. This kind of nonsense is indeed unusual (Parsons 1994, 41).

Analysts often dismissed the views of their adversaries as illogical or as nonsensical. This dismissal needs some proof or at least a better idea of logic and of sense. Only one possible proof of this kind exists; it lies within the language that young Wittgenstein had developed. At most, it holds in that language. Russell and Carnap rejected that language as too narrow.[27] Ordinary-language analysts may have done so too. It is hard to say, since they declared easy victory over the metaphysicians with hardly a discussion (perhaps because they considered self-evident an incidental remark of young Wittgenstein[28]). Wittgenstein admitted that his first book is too

[23]Fans were particularly zealous in the early days. Neurath's peers had no choice but to treat as a joke in poor taste his view of Wittgenstein as a metaphysician (Pietarinen 2011, 71; Rorty 2000, 129; Cook 1994).

[24]This view of Wittgenstein as a metaphysician of sorts is now quasi-official.

[25]Russell began the trend in his 1905 "On Denoting", where he opposed Frege's theory of meaning as he (rightly or not, opinions differ) considered it metaphysical, the admission of Plato's Heaven. Wittgenstein explained his own rejection of this move of Frege (*PI*, §58) but he also cast his net much wider (*PI*, §116), declaring his program thus: "we bring back the words from their metaphysical to their everyday use".

[26]Whether Wittgenstein was a rationalist is under dispute (Glock 2001, 195–7). Glock deemed Wittgenstein only partly a rationalist. Some view his philosophy as concerned with reason. Bennett 1989 opens his discussion with the suggestion that Wittgenstein's project is very ambitious as its subject is rationality. Grice 2001, 5 says, Wittgenstein "aims at the clarification of the notion of reason or rationality". Føllesdal 1996 conclusion subjects analytic philosophy to rational debates, which is very nice, of course.

[27]Carnap won praise for his having changed his views openly while admitting error. Not here, though.

[28]*TLP*, §4.111: "(The word 'philosophy' must mean something whose place is above or below the natural sciences, not beside them.)" This argument is question begging.

narrow, but he never stated an overall judgment on it. Possibly he never lost hope to have it vindicated.[29]

Analysts still dismiss opponents who do not swear by commonsense. The absence of discussion of it permits oversight of the prevalence of magic and prejudice and superstition. We may perhaps ignore this, since tradition allows limiting discussion to science and ignoring superstition. When traditional philosophers, especially David Hume, yielded to commonsense, they did so in order to rescue science from desperate skepticism.[30] This move does not work since commonsense has no authority and is open to improvement.[31] Consider the problem that troubled Wittgenstein most: how to reinstate certainty. Commonsense allows for what Robert Boyle called "moral certainty".[32] The demand to achieve certainty is quite common; the recognized procedures for it are imperfect and hopefully improvable. They are not good enough for philosophical or logical certainty. For that, diverse pro-science philosophies are usually equated with the view of commonsense certainty – moral certainty – as rational belief and thus as probability. Popper has refuted this popular idea repeatedly. Earlier, Bernard Shaw (1912, Preface) ridiculed it: make a list of assertions from Scriptures, he suggested, and ask different people to say which of them they deem commonsense. Responses to a sufficiently varied list will not converge.[33]

The most ironic fact here is that it is the return to sense-perception as the rock bottom of all experience. Moore considered the theory of sense *data* commonsensical.[34] He said so in defiance of the obvious fact that it is very far from being common: in western culture, it is controversial; elsewhere it is almost totally absent. Aristotle advocated it despite its familiar refutations; John Locke altered it in the

[29]This my conjecture explains Wittgenstein's return to philosophy upon having heard Brouwer lecture on finitist mathematics, as well as his determined refusal to express a detailed opinion on his early views. It also explains the surreptitious nature of his changes of opinions. Rodych 1999, 174 discussed his altered view of sense, no less.

[30]Moore, Wittgenstein, Austin and Kripke endorsed commonsense as authoritative; not Shaw (1903), Russell (1953) or Ayer (1969, 13).

[31]Agassi and Wettersten 1980; Agassi 2013, 211 (§14.14). Stroll (1994) defends Wittgenstein's reliance on commonsense, as advocating ideas too vague to be refutable. Popper 1945, Ch. 25, note 26 responded to this, saying "Irrefutability is not a virtue". Thus, modern commonsense is superior to traditional commonsense.

[32]Agassi 1975, Ch. 15: Partial Knowledge, 346–9.

[33]Austin 1956, "Plea for Excuses" defended the authority of all commonsense ideas thus:

> our common stock of words embodies all the distinctions men have found worth drawing, and the connexions they have found worth marking, in the lifetimes of many generations: these surely are likely to be more numerous, more sound, since they have stood up to the long test of the survival of the fittest, and more subtle, at least in all ordinary and reasonably practical matters, than any that you or I are likely to think up in our armchairs of an afternoon-the most favoured alternative method.

This is standard, insipid conservatism in a new garb: it justifies every conceivable long-accepted traditional atrocity.

[34]Moore 1953. Ch. II, 28–40; O'Connor 2012, 95.

wish to support the Copernican hypothesis despite its conflict with common observations: he declared that we do not see motion but infer it. This is contrary to both *naïve* observations and careful up-to-date empirical studies of perception. Thus, although in some sense some sense-perception theory is decidedly commonsensical, traditional, strict sense-*data* theory is not: "We primarily observe things, not observations" (Dewey 1929, 12). The refutations of sense data theory abound.[35]

We learn from experience; this is an observed fact. How we do that is under dispute.

Problems are lumped here as belonging to the philosophy of life; roughly, these problems engage not only philosophers. Wittgenstein's condemnation of philosophy as crazy refers only to academic philosophy. To illustrate this he told a charming anecdote. Someone watches in bewilderment a conversation I have with a philosopher. I explain it to him: "this fellow isn't insane: we are only doing philosophy" (Rhees 2003, 55). In Oxbridge then, the expression "this sounds odd" was nasty: they kindly gave permission only to scientists to sound odd, not to philosophers. How odd!

Wittgenstein's interest in the philosophy of life thus placed him apart from his followers (Conant 2001, 23–4). Whereas he played guru (Monk 1990; 2012, 20, 291), they were professional academics (Warnock 1958, conclusion). His writings are perplexing, but he forbade others to emulate him. He also allowed expressing odd views on religion and on politics only to himself (Nielsen 2011). He was not alone in this. Carnap mentioned in passing in his autobiography that he was a socialist[36]; his followers ignore it (Agassi 1995a). This again presents the attraction of Wittgenstein as the escape from tough moral and political problems. Possibly the attraction of Wittgenstein was in his postponement of the very frightening debates on the consequences of Auschwitz and of Hiroshima for enlightened liberal philosophy. There is a literature about the postponement of this discussion. Incidentally, Hempel opposed it.[37] I will skip all this here.

To consider Wittgenstein a philosopher of life is to give up (tacitly) his role as a leading thinker. This is due (not to any deep idea about philosophy but) to the logic of the situation: to say that Wittgenstein offered only a philosophy of life is an admission of defeat. Most commentators refuse to do that. Thus, Avrum Stroll (1994, 3) says, Wittgenstein was the greatest philosopher since Kant because of his having invented a new way of philosophizing. Stanley Cavell read this as the use of available everyday knowledge as grist for the philosopher's mill.[38] He responded to the criticism of language analysis as wanting empirical base[39]: native speakers

[35]Chapman 2009, 200.

[36]Schilpp 1963, 23. Carnap stood up valiantly against the tide of McCarthyism.

[37]Hempel depicted the hostility of the "Vienna Circle" to religion as an asset only orally: in the session in his honor at the American Philosophical Association Eastern Division 1983 Boston meeting.

[38]Cavell 1962, 1976, Chapter 2: "The availability of Wittgenstein's Later Philosophy".

[39]Mates 1964. Gellner [1959] 2005, Schlagel 1974 and Uschanov 2002 offered the same criticism in a less complementary assessments. Freeman Dyson went furthest: Wittgenstein's refusal to

possess all the empirical knowledge that analytic philosophical research needs. This was a step towards considering Wittgenstein a guru. Beforehand, Carnap ([1928] 1967) viewed the philosophy of life favorably, but as irrelevant to philosophy proper. His follower Wolfgang Stegmüller (2012, 10, 106) declared common scientific optimism mere hope, and thus an attitude, not a statement, and thus not in need of verification (Agassi and Wettersten 1980). Wittgenstein was a living tribute to this idea: he claimed to have improved the philosophy of life by his mere disgracing of academic philosophy.[40]

Leading analytic philosopher Hilary Putnam declared Wittgenstein a great philosopher of life, comparing him to Socrates.[41] To test this we have to decide whether the appeal of Wittgenstein was in his postponement of the debates on Auschwitz and Hiroshima today. Alas, the literature about the postponement of this debate by decades is problematic. Let us skip it here with no decision. For myself, let me say in passing, I find many thinkers compete for the position of the choice philosopher of life, and I doubt Wittgenstein is the lead candidate for it.

1.4 Analysis; Paradoxes of Analysis

The two Greek terms "analysis" and "synthesis" are antonyms ("detachment" and "attachment"). They appear intermittently in various works on scientific method of Aristotle and in some famous ancient mathematical works, particularly in a few scarcely comprehensible short passages in a text of mathematician Pappus of Alexandria. Twentieth-century historians of Renaissance science have brought back from oblivion the anti-Copernican Aristotelian Giacomo Zabarella, who wrote brief and obscure passages that echo Pappus. So did Newton in a famous passage (*Opticks*, last Query). The discussion of all this, though clearly pertaining to scientific method, is much too vague to signify.[42]

The difficulty regarding the attitude to science of mature Wittgenstein (and of his followers) concerns analysis, not synthesis. The paradigm for synthesis is Euclidean geometry: tradition considered its axioms the simplest; its text moves through series

---

discuss his first book made Dyson pronounce him a charlatan: https://www.youtube.com/watch?v=byi3vOnVodQ

[40]Joad 1950 censured "logical" positivism as moral indifference, using as evidence Ayer's emotivism (Ayer 1936, 109). The ensuing angry response presented Ayer as a concerned citizen – as if Joad was attacking Ayer's conduct rather than his views (Royall 1952). Ayer's philosophy of life is very nice; his project was to reconcile it with analytic philosophy. He finally admitted defeat by despairing of analytic philosophy and returning to classical empiricism in the eve of his life.

[41]As Monk 1990, 471–2 notes, Wittgenstein disapproved of Russell's philosophy of life but avoided arguing with him. Monk 1990, 243 reports that Wittgenstein disappointed members of the "Vienna Circle" when they expected him to give a lecture: he faced the wall and read to them a text of Rabindranath Tagore.

[42]See Hintikka and Remes 1974; see also Marchi et al. 1982.

of theorems that are increasingly complex and surprising.[43] Analysis is the motion in the opposite way, from the complex theorems to the simplest axioms (Robinson 1936). The paradigm is the field of mathematical analysis. In analysis, a mathematical theorem was traditionally supposed to lead logically to increasingly simpler theorems, ending with axioms – as these are allegedly not open to further analysis. Traditionally, ever since Newton presented the words for analysis and synthesis as synonyms with the words for induction and deduction, scholars took it for granted that remarks about analysis and synthesis apply to all research. The excitement about the passages on it rests just on this: the understanding of these passages is the acquisition of ability to contribute to human knowledge. The efforts to overcome the obscurity of these passages is thus justified: the matters that they handle are inherently difficult but worthy of investment of great efforts to understand them, since this just might open vast new opportunities. Newton put induction on a par with some mathematical procedures – even though in science it does not lead to utter certitude, he admitted; it is the best we can have, he added. Here Newton's methodology is problematic, as Hume has argued: contrary to expectations, it does not sit well with Locke's theory of perception.[44]

Philosophical analysis differs from (Newton-style) scientific analysis. Wishing to introduce scientific method to philosophy, Russell said (1905), the main task of philosophical analysis is to remove paradoxes. No, said Wittgenstein: it should remove confusion; moreover, all philosophical (including theological) doctrines are cases of confusion for analysis to clear. Young Wittgenstein took for granted that all this has firm grounding in science and perfectly clearly so (*TLP*, Preface):

> what can be said at all can be said clearly; and whereof one cannot speak thereof one must be silent.

This rendered his work rather limited, as he noticed later (*PI*, Preface). To broaden the scope of his investigations he had to be less strict and allow some

---

[43]The paradigm case is that of Thomas Hobbes who at age 40 noticed Euclid for the first time and expressed incredulity before he was convinced (Grant 1990, 148).

[44]Mature Wittgenstein did not say what statement is scientific. Proops 2001, 392 cites him to say that his not having given any example for a scientific statement is a defect. Tejedor 2011, 22 disagrees: "That we do not have access to any examples of these elementary propositions should not hinder our ability to acknowledge that they are essential to the concept of proposition to work." Young Wittgenstein's elementary propositions do not exist, she adds. Not so: he said his elementary propositions are scientific observation reports; and these abound. Like other conventionalists, he viewed scientific theory as means for arranging them simply (*TLP*, §§6.341–2). Like Mach and Russell, he viewed them as reporting sensations. Anscombe 1957, 25–8 used his refusal to explain as an obvious refutation of Popper's reading of him. Her comment amounts to debunking Wittgenstein (Hintikka 1996, 127, 174) and it should therefore be ignored as a slip. Popper's reading now prevails (Hintikka 1996, 126–7) for want of an alternative. Biletzki 2003, 53*ff.* admits this (reluctantly and tacitly). Neurath's protocol sentence are intentionally vague about all this. The relatively atomic propositions of Popper [1935] 1959 §28 are distinctly realistic, e.g., the 4-co-ordinates of a planet plus its luminosity, or as the trail of an electron-path in a bubble chamber. Kraft [1952] 2015 rightly reports that Popper introduces "observation" as an undefined concept.

inaccuracies. It is not clear which, except that those that invite metaphysics are to shun at all cost.

Both philosophical analysis and anti-metaphysics are traditional since the scientific revolution; so is also the view that the two boost each other. Wittgenstein's innovation was his alleged proof that analysis ousts all metaphysics. This he never retracted, although he did change his view about analysis. His mature philosophy thus rests on his early philosophy: he could not retract it to the full.

Consider traditional philosophical analysis in some detail. A well-known simple example is the inapplicability of some adjectives to some nouns. One classical example for this is the adjoining of the adjective "sweet" to any noun from the geometrical vocabulary: "this triangle is sweet". Analytic philosopher Leonard Linsky considers the size of the national flag in such a context. Every flag has a size, yet *the* national flag does not – since the law does not specify size. The law of the land bestows the status of a flag on any object that displays certain clear, legally specified characteristics. The proper use of the word is thus contextual.[45]

In traditional texts on grammar, examples illustrate rules. In traditional texts on philosophy, analytic philosophy recognizes both examples and rules but nothing else.[46] In what sense of the word "analytic" is philosophy analytic?[47] This is not clear at all. Quite a few analytic studies comprised nothing but examples.[48] Yet the little that they did say was often problematic and in dire need of explanation. Gilbert Ryle (1949, 7) called "category mistakes" and "systematically misleading expressions" errors that he deemed metaphysical.[49] Carnap differed: he declared them a

[45]Linsky 1967, 49–56, 122. French et al. 1979, 14, 37, 109. This is the perennial discussion on verbal determination. This discussion began in antiquity; it is still going strong. The example cited here is from Solomon Maimon. He threw doubt on Kant's doctrine of the categories. He was thus a pioneer in the study of ranges or scopes of variables, to use current terminology. The last word on this is the opening sentence of Hintikka 1997 of which he was very proud: "The scopes of quantifiers and of other logically active expressions are one of the most important determinants of the logical form of the sentences of both formal and natural languages." Hintikka says, this item still suffers neglect. This renders incomplete all proofs in the analytic literature. The reason for this is the Tarski-Quine non-synonymy. In the traditional example, "foot" as in descriptions of animals and in geography (foothill) have different ranges. Even the same word may have different context-dependent ranges: the same quality considered sinful in Christian texts and laudable in modern texts have different ranges, not to mention the name of the color red and the many other uses of the same word. Hence, the theory of possible worlds is plainly deceptive (Hintikka 1967; Linsky 1977, Ch. 7).

[46]Even were the analytic school of philosophy right in all of its claims, its conceit that it has won the competition with all other schools, a conceit that began with *TLP*, is somewhat questionable. The most obvious examples of this conceit is the way analytic philosophers dismiss critics. See below.

[47]Langford, Carnap (1949) and Arthur Pap (1950) discussed verbal determinations (like "red is a color"): do they comprise synthetic *a priori* knowledge? The answer depends on context.

[48]Waismann 1956, §§VI, VII: "What do you find in reading Ryle or Wittgenstein? Lots of examples with little or no logical bone in between. Why so many examples?" "There is something deeply exciting about philosophy, a fact not intelligible on such a negative account." Wittgenstein answered this question: he had only examples to offer, he said, not any articulated theory.

[49]It is not clear whether these two categories are coextensive. For more confusions about Ryle's views, see Julia Tanney, *Stanford Internet Encyclopedia of Philosophy*, Art. Ryle. For more

special kind of confusion: between the meta-language and the object language. Ayer said they were a special kind of illusion, in a variant of Wittgenstein's view of them as verbal illusions of a special kind. Is this diversity about metaphysics not metaphysical? To develop this debate, analytic philosophy needs a philosophical opinion; all leading historians of philosophy should agree on it, or else it will beg the question. So far there still is nothing remotely like this.[50]

Things are not that bad. Behind Wittgenstein's anti-metaphysics stands a guideline: try to avoid confusing your public! Alas, analytic philosophers confused their publics repeatedly, by merely skipping specifying the contexts of their discourses. Very few introductory presentations of the outlines of the ideas of this school of thought are available, and most of them are utterly unsuitable as introductions, being idiosyncratic, difficult to read, and much too defensive.[51] Authors determined to avoid being defensive will find it easy to improve upon the extant literature if they begin by discussing some key objections that provoke a defensive attitude without being so provoked. Whether these objections are satisfactorily answerable or not is often under dispute; discussing it may relieve the unease of uninitiated readers. Let me mention one that is still with no rejoinder. It is so famous it has a name: the paradox of analysis or the paradox of language analysis: the analysis of a verbal item necessarily ends up with a statement of synonymy, and as such these statements (be they true or false) are trite.

The paradox may suggest that all verbal analysis is redundant; it may suggest the opposite: since there is no perfect synonymy, no analysis is ever complete. The choice between these extremes rests on the question of the purpose of the analysis in question. Here let us ask, what is the profit accrued by any successful analysis? This kind of question is obviously always wise. The aim of analysis always depends on circumstances, since analysis is a mere tool.

Consider scientific analysis first. The analysis that Newton spoke of was induction, that is to say, the discovery of laws of nature, given the gathered facts on which it allegedly rests. The paradox here, then, is this: either induction leads to new results and is thus an invalid inference, or else it is trite. As the aim of induction is to discover laws of nature, the first option is the right one. This is paradoxical indeed:

---

confusions about Carnap's view see Bar-Hillel 1950. If philosophy is therapy, then it has generated iatrogenic diseases galore.

[50]A survey is wanting of the views about the demarcation of metaphysics or of philosophical illusions. The disagreement between Wittgenstein and Carnap stands out. Yet even that is unclear. A passage from Wittgenstein's *Philosophical Investigations* (§392) seems in accord with Carnap: "analysis oscillates between natural science and grammar". Still, whereas Wittgenstein deemed this marginal, Carnap deemed it standard. A comprehensive study on the contrast between Wittgenstein and Carnap is wanting.

[51]For the removal of defensiveness from a text, a simple technique is readily available: find the criticism that its author is trying to overcome and explain the situation regardless of agreement or disagreement with its author. Example: Imre Lakatos took the latest definition of the polyhedron and showed that each of the abstruse qualifications in that definition comes to block a criticism of the major theorem about it (the Descartes-Euler theorem). The result is his exciting 1970 *Proofs and refutations*.

how can we derive new knowledge from old? Either the new is implicit in the old, and then, as already Francis Bacon has noted, it is not quite new, or it is not logically derivable from it, and then it is new but merely speculative. This paradox is not answerable.[52] The application of analysis in physics may be very useful, even though as inductive inference it is logically invalid. The question is, does it have to be valid? Why? Because we want to avoid error, said Bacon. This is impossible. As Popper has observed, jumping to conclusions is inevitable; hence, he added, we should distrust our conclusions and test them severely and repeatedly.

Not so with mathematical analysis. There the obviously right strategy is to insist that inferences be valid. In the process of developing axioms, only valid inferences serve. When analysis is not logically valid, as in physics, we have one kind of trouble; when analysis is a valid inference, as in mathematics, we have another kind of trouble; but trouble is there anyway. To advance this matter, we may look at the goal in question. What is the goal of mathematical analysis? It is to find the simple building blocks from which to synthesize whole edifices. How do we find the simple items in mathematics? How do we separate the simple from the complex? What secures the conclusion that the axioms discovered are simple? A general answer to these questions is impossible, and we should admit that those who declared that there is no universal criterion of simplicity in mathematics gave up the traditional search for the simplest mathematical axioms. Indeed, quite a few edifices have equivalent sets of different axioms; the choice between them then may be arbitrary; it may also be context-dependent.

This is significant as it has an analogue in the philosophy of language: the idea that some sentences are the simplest building blocks of a language is the idea that they are atoms of that language. This view is a philosophical system; it is the famous philosophical system of logical atomism, the idea that some sentences are the simplest building blocks and others are compounds of atoms of two sorts, the mental and the physical. Its originator is Wittgenstein, with or without the aid of his teacher Russell.[53] Russell disproved its basic tenets.[54]

This does not dispose of the paradox of analysis even within mathematics: what is an axiom is neither determined nor arbitrary; an axiom-system is any mathematical theory with a finite set of theorems – its axioms – that are necessary and sufficient for entailing all of its theorems. As there are logically equivalent alternative sets of

---

[52]Bacon 1620, II, Aph. 19 suggest that induction is not a valid inference but an act of a new intuitive insight, a mystic union (in mediaeval parlance, "putting the understanding and nature on a par").

[53]Russell began his research towards logical atomism before Wittgenstein appeared. That doctrine has a few versions (Russell [1959] 1996, 84; Ayer 1971, 54). Beaney 1998, 465 says, "Wittgenstein's logical atomism is . . . subtler than Russell's", namely, the one that Russell 1914 ascribed to Wittgenstein. See Pears 1985, esp. xv. Popper [1935] 1959, §38 relativized version of it has rendered them all obsolete – since they aimed at certitude and he aimed at describing extant scientific practice.

[54]Russell 1918, 520: "all the names that it [logical atomism] would use would be private to that speaker and could not enter into the language of another speaker." Wittgenstein agreed: private language is impossible.

axioms, they are all legitimate, each of them exhibiting its own merits and limitations.

In science, the paradox is the problem of induction. Some philosophers say, there is an *a priori* valid principle of induction. Leibniz identified it with the principle of parsimony or of simplicity. Kant identified it with the principle of causality. Others, the conventionalists (who view the truth of scientific theories as truths by conventions) or instrumentalists (who view scientific theories not as descriptions but as mere instruments for prediction), viewed scientific theory as uninformative. Young Wittgenstein agreed: he denied generalizations their literal reading. He also followed Russell in endorsing the law of simplicity as the principle of induction (*TLP*, §§6.31, 6.36, 6.363), in addition to his endorsement of conventionalism (*TLP*, §§6.34–6.3611, 6.373).[55] Finally, he discussed Newton's mechanics as if he had not heard of Einstein. Alas, most commentators shun this (Penco 2010, 360).

As to the philosophical paradox of analysis, Wittgenstein never tired of stressing, it leaves things as they are – meaning, it does not add information. It is not redundant, though, as it helps relieve people of their confusion and of the discomfort that comes in its wake. Removing the discomfort due to philosophy was the sole aim that mature Wittgenstein recognized.

Proper analysis in any field does not increase information but serves other functions. Young Wittgenstein discussed logic and mature Wittgenstein discussed language games. He declared that these are parts of logic (Wittgenstein [1969] 1975, §56); his disciples replaced logic with common sense: analysis does not increase the stock of commonsense knowledge. This is the solution of the paradox: logic and mathematics do not increase informative knowledge, but they do increase other kinds of knowledge.

Take the example of philosophical analysis mentioned here. "This triangle is sweet" looks grammatical but is not.[56] This example serves a purpose within logic. The more modern example mentioned here, the national flag having no size, serves in legal cases. To call them both ungrammatical rather than false is not a matter of philosophy but of common knowledge. Conversing with children who are acquiring their mother tongue elicits this phenomenon. When we first make grammatical errors, we may have trouble understanding grammatical corrections – mistaking them for factual corrections. What is strange is that the decision that Russell made as he decided to declare his paradox ungrammatical was such a great breakthrough. Similarly, Wittgenstein's view – metaphysical errors are grammatical rather than factual – sounded odd. Now that thanks to the widespread use of computers,

[55]This seems a redundancy, except that Wittgenstein offered an interesting insight into conventionalism: "But what *does* characterize the picture is that it can be described *completely* by a particular net with a *particular* size of mesh" (*TLP*, 6.342).

[56]The view of the sweet triangle as ungrammatical depends on the meaning of "grammar" to which I come soon. It does not appear in traditional grammar texts, as these are patched-up Aristotelian messes. Already Frege changed grammar and inaugurated the distinction between semantics and syntax. Yet his prime student Carnap ignored semantics until Tarski appeared on the scene (Coffa 1991, 283).

knowledge about formal languages is widespread, the idea is much more familiar. Consequently, Wittgenstein's view of all traditional metaphysics as ungrammatical is now much less intriguing than it was then.

Perhaps the greatest contribution of modern analysis to the improvement of ordinary conduct is this. It debunks – it should help debunk – a kind of defensive conduct: when the wording of an error is improved, it is likely to remain an error. Moreover, in such cases, the defensive may try to deny their having expressed an erroneous opinion, admitting instead having made an error in articulation; it is, thus, an attempt to look wise or knowledgeable; that attempt achieves the opposite effect, since articulation is an essential, basic skill. Wittgenstein said, philosophers are prone to utter seemingly proper statements, even profound ones. He proscribed discussing the question, are they true or false[57]: the meaningless has no truth-value (Frege).[58]

1.5 Grammatical Errors

How do we decide whether we should correct a vague expression or view it as hopelessly ungrammatical?

Wait; does it matter?[59] Admittedly, the analytic literature is full of cases that may occur in everyday experience. Yet most of it is redundant: in such cases, people ask for clarification or take their chances and clarify as best they can. Did any analytic discussion of such a case bear fruit? It is hard to know since grammars of ordinary languages are fluid.[60] It is simpler for vague expressions that appear in computer programs. In such cases, one may give up and one may alter the vague expression and hope for the best. One artificial language is of great importance to logicians and to historians of computer science: PMia (pronounced pea-em-ia), the language of *Principia Mathematica*, the 1910–1913 three-volume book on mathematical logic by Russell in collaboration with his mathematics teacher, Alfred North Whitehead.

---

[57] *PI*, §110: "this proves to be a superstition (*not* a mistake!), itself produced by grammatical illusions." Hence, superstitions are meaningless. Lerner 1994 and Richter 2004, 168 note that elsewhere Wittgenstein diagnosed superstitions differently. This is not surprising: as superstitions come in all shapes and colors, all of Wittgenstein's ideas of them are too narrow. See next note.

[58] Are myths erroneous? Are superstitions? Are myths superstitions? As anthropologists disagree on this (Jarvie and Agassi 1967), what should philosophers do? On the status of magic systems, Peter Winch read Wittgenstein as a relativist. Quine's Foreword to Hallen and Sodipo [1986] 1997 offers a different view on the place of languages in magic systems. The paucity of analytic discussion on magic is intriguing.

[59] Of course, this matters for one who hopes to use grammar as means for barring metaphysics. This requires speaking of syntax and of semantics separately, not of the (Aristotelian) traditional grammar that conflates them (Bar-Am 2008, 103).

[60] Should grammarians begin discussing language synchronically or diachronically, ontogenetically or phylogenetically? All options are agenda, each with some measure of success, and all need unification.

Despite objections, and despite its being out-of-date, it still is the cornerstone of modern logic[61] and of the foundations of mathematics and of computer science.[62] In PMia, examples of everyday vague expressions do not appear since it is limited to mathematics. This is easily changeable, since one can add to PMia descriptive terms. Discussing young Wittgenstein's language, Russell mentioned the paucity of alternative formal languages and raised the question of their viability. The current proliferation of formal languages offers new freedoms. They have altered the status of Wittgenstein's mode of thinking (van Bendegem n.d.).

Wittgenstein advised his readers repeatedly to observe language in action. This is reasonable, since communication is of supreme value and adding to its clarity adds to its efficiency. The vagueness of metaphysics may be quite burdensome. Another burden on communication is speakers' mistaken assumption about hearers' background knowledge. In such cases, speakers may have to elaborate in order to communicate successfully.[63] This option is available particularly in face-to-face discussions. Researchers often speak opaquely and hope to clarify their discussion after some progress; young Wittgenstein's demand that they keep silent is thus possibly right but impractical. This happens particularly in heuristic that rests blatantly on some metaphysical suppositions (since the enrichment of the contents of a metaphysical theory may render it empirically testable; Popper [1935] 1959, §85). The paradigm case still is atomism. It has untestable (metaphysical) versions as well as testable (empirical) ones; adding assertions, say about the size of atoms or about their mode of bonding, may render an ancient metaphysics empirically testable.[64]

Empirical evidence indicates that people exposed to an unfamiliar word are prone to give it some familiar meaning.[65] Thus, the utterance of some conspicuously ungrammatical sentence is likely to sound as corrected or as telescopic or as metaphoric. Philosophers often protested against this as an obstacle to perfect comprehension. Young Wittgenstein is one of them; mature Wittgenstein deemed the ideas of young Wittgenstein too limited (*PI*, Preface, §§12, 23, 38, 44, 46, 48, 81, 97, 114). He treated ordinary language as if it were formal and he kept his injunction: speak clearly or shut up.[66] He based this claim on the supposition that whatever can be said can be said clearly. There is no objection to (meaningful or meaningless)

---

[61]No doubt, this success is largely due to Quine's superb 1940 *Mathematical Logic*.

[62]There are objections to this assertion, of course, I do not know how weighty (Agassi 2006, 264).

[63]There may be a tacit agreement for pretense: a reference may be vague on the tacit agreement to leave it vague in (self-) deceit. The easiest is the misuse of the definite article (its use with no specification). All one can do against it is to advise friends to ignore systematically vague discourse (Bar-Hillel 1962).

[64]The variety of versions of realism (Niiniluoto 1999, 13) troubles some and (rightly) delights others.

[65]Ordinary members of Shakespeare's original audience could not fully comprehend his rich language, yet they did so well enough to keep him employed. The same goes for nonsense verse. Today linguists agree that natural languages are flexible enough to consider grammatical all sufficiently comprehensible expressions.

[66]Stroud 1965, 504 observes that Wittgenstein's examples seldom illustrate what he said they do.

expressions open to different readings unless they mislead. Wittgenstein claimed that all metaphysics mislead. Is this assertion not metaphysical? Deception is ubiquitous. The deception that Wittgenstein made his business to combat is particular because it causes the pain that he had suffered from in his adolescence. When Russell or Tarski declared paradoxes meaningless, they were concerned with formal systems, where a paradox is deadly. Young Wittgenstein shared their concern; mature Wittgenstein did not.

Formal languages have explicit rules for sentence formation, for making a string of words a comprehensible sentence. Natural languages in literate societies are semi-formal, especially when discussing some expertise: a speaker addressing expert readers may ignore most other potential readers. Listeners may pretend that they comprehend what they do not comprehend, and speakers may rely on this pretense and only pretend to be saying something; all of them (with the exception of the science-fiction that shares much with nonsense poetry) are deceitful slum-landlords; only some of them are metaphysicians.[67]

When Wittgenstein declared that all metaphysical sentences are sham, he did not have to declare individual metaphysicians deceitful slum-landlords. (He did so anyway.) Pseudo-sentences, like pseudo-vectors and pseudo-enzymes, have misleading appearances with no one to blame. Accusations of pretense are redundant; it is regrettable that Wittgenstein insisted on accusation.

To render the analysis of informal texts at all possible, Wittgenstein offered specific techniques. The choice of techniques naturally depends on some ideas and on aims. Wittgenstein said he offered therapy by exposing metaphysical discourse as inherently pretentious bogus. The philosophy behind his technique, then, should be metaphysics-free. This is impossible. Since not all muddle and pretense is metaphysical, and since mature Wittgenstein offered his method of analysis as therapeutic, his technique rests on two (questionable) generalizations. Some commentators discuss it in oversight of his characterization of the illness that his therapy should cure as the craving for metaphysics.[68] Let us ignore them for a while and look first at his technique as an antidote to metaphysical pretense. It is, watch language in action! Juliet Floyd clarifies[69]: "The job of philosophical clarification consists in investigating the contexts in which an expression might appear to be apt" and, of course, to condemn them as metaphysical when used too freely. Wittgenstein's *Philosophical Investigations* is mostly examples (Schroeder 2006, 121 calls it "only an album")

[67]This is true of much sociology of the Frankfurt School. Metaphysician Hegel asserted that no one had understood him. Heinrich Heine found this ludicrous. Wittgenstein came close. Niels Bohr used to repeat the Jewish anti-cabbalist joke: what the audience do not understand are secrets and what even the speaker does not understand are secrets of secrets.

[68]Anscombe rightly presented Wittgenstein's idea as his demand to limit metaphysical thinking to wordless thinking. In *TLP*, §5.61 he spoke of logic as the limit of the world. In *PI*, §§22, 341, 523, and esp. §527 he sought to look at sentences not for their content but for their music. This may throw new light on his picture theory of language.

[69]Floyd 2007, 191, 179. She notices a few generations of scholars engaged in Wittgenstein hermeneutics. See also Fischer 2011, Peterman 1992 and Savickey 2017, Ch. 6.

and comments on them; they all serve as arguments against the view that metaphysical assertions are possible. To the question, which metaphysical texts was he after, he seems to have answered, all the usual suspects. His friend Engelmann cited him to approve of some metaphysical poetry as expressing the prosaically inexpressible in some poetic fashion – provided it is unpretentious. Again, we see, it is not the metaphysics but its (alleged) pretense that raised Wittgenstein's ire – the pretense of having managed to express the inexpressible.

To distinguish between proper and improper sentences Wittgenstein had to offer rules of sentence-construction and sentence-transformation,[70] with well-formed strings of words being sentences; their having or not having truth-values is their having or not having meanings.[71] Some sentences are true/false by virtue of the rules of formation (syntax) alone. These are tautologies/contradictions, or logically true/false sentences. Thus, words and sentences have meanings in different ways or have different kinds of meanings. (Consider a one-word sentence and notice the difference between that word and that sentence. Example: "hello" and "hello!") There is a huge linguistic literature on this matter. Even the simplest and oldest of these ideas – of naming as giving meanings to names, or of names as designations – has a huge literature on it. Wittgenstein began there but ended with Wittgenstein's maxim, Watch language in action![72]

The two strands of Wittgenstein's school offer two kinds of analysis: one applying to formal languages, the other to natural ones; one linked to Vienna, the other to Oxford or Cambridge (Oxbridge); one of his first published book (1922), the other of his first posthumous book (1953) that is at the center of the present study. He first endorsed the received view – of Aristotle, Frege and Russell – of logic as shaping the ideal language, the essence of language. He then shared the received (post-Tarski) retraction of it as an error. How much this retraction matters is still under dispute. It hinges on the following question. Did mature Wittgenstein disagree with young Wittgenstein significantly?[73] If so, on what, and how exactly?

---

[70]This is the chief activity in introductory modern-logic classes, in disregard for Quine's doubts.

[71]Not quite. A well-formed sentence with an empty name ("Pegasus"), Wittgenstein seems to have decreed meaningless (*PI*, §§131–3 and p. 217). A well-formed formula with an empty name display partial meaning (or a variety of kinds of nonsense: beware of the Jabberwocky!). No *a priori* considerations can make a name *show* whether it is vacuous or not; this may but need not refute the *show itself* system of young Wittgenstein.

[72]Russell's theory of definite descriptions is different, as it renders false any assertion about a non-object. Science agrees: it declares refuted a theory that uses significantly a seemingly empty noun ("the center of the universe"; "phlogiston"). The most famous book in the analytic tradition on naming, Kripke 1972, Abstract (*cf. PI*, §453; Agassi 1995b, 245), is evasive:

> Some topics essential to a full presentation of the viewpoint argued here, especially that of existence statements and empty names, had to be omitted altogether.

"For philosophical problems arise when language *idles*" (*PI*, §38). Why Wittgenstein found this objectionable he did not say: he too used metaphors to evade tough questions (Weiler 1961, 207).

[73]This debate has lost its punch, however, as the following observation of Hintikka 1996, 94, won popularity: Wittgenstein adhered only to the letter of his first book, as he altered his reading of it.

Let us approach the question differently, then. Why does the difference between the schools matter? Isaiah Berlin answered (Ignatieff 2011, 83–4): the one is hard, the other is soft. Now, as logic and commonsense are not in conflict (Agassi 1985) are the two kinds of analysis in conflict? Do they lead to conflicting conclusions? The obvious detail of the distinction between the two concerns nonsense. What exactly is it and why care about it? There is no formal way to demarcate it sharply: it is context-dependent.[74] This renders Wittgenstein's anti-metaphysics too sharp to hold strictly even within commonsense alone. Practitioners of Wittgenstein-style ordinary-language analysis, then, cannot let go of young Wittgenstein. They discuss at length the difference between his early and later views.[75] The weakest aspect of his *Philosophical Investigations* is its loose sense of logic: it used "logic" and "grammar" as (near) synonyms, although the rules of logic are much clearer and significantly more explicit. The answer to this is, for some tasks, even the best system of logic is still not sufficiently sharp. True. Still, in some philosophically significant contexts, the use of the words "logic" and of "grammar" as (near) synonyms is an eye-opener; elsewhere it is simply erroneous.[76]

Even the sharpest system of logic is not sharp enough to prove anti-metaphysics.[77] As no single formal system is generally preferable, a string of words may be kosher in one system and non-kosher in another. Young Carnap declared anti-metaphysics proven with no qualification; old Carnap expressed the hope that a metaphysics-free language will be available soon.[78]

---

[74]Russell's stratification of sets blocks his paradox. His theory of types introduced it axiomatically rather than as rules of syntax. The same holds for the standard presentation of the Zermelo-Fraenkel system. The difference was not clear between the two ways of formal presentation; Skolem-Lövenheim theorem illustrates its significance (Carnap 1937, §71 d–e). Some intuitionists insist on the priority of rules. Most mathematicians are reluctant to discuss them. Gödel expected mathematicians to subdue this reluctance to allow for non-standard arithmetic. He was disappointed: two axiomatic proper alternatives appeared soon to replace Robinson's rules. Paul Cohen's earlier technique of forcing that he used for proving the independence of Cantor's continuum hypothesis had a similar fate (Weaver 2014). All this requires additions to classical symbolic logic; even the *Principia* then is a mere historical monument.

[75]Young Wittgenstein declared rules meaningless since it is possible to disobey them (*TLP*, §§6.42–6.423); mature Wittgenstein disagreed (*PI*, §345): a never-obeyed rule does not deserve the name of a rule.

[76]Wittgenstein gave up the picture of language of his *Tractatus*. His *Investigations* included "more under logic than most people" allow: it includes a "world picture", a "common conceptual framework" (Wang 1991, 233–4), thus smuggling commonsense metaphysics into logic. To allow for this, his *Investigations* "does not develop any alternative picture of language" (Williams 2004, 5).

[77]This is not to endorse the demand to prove anti-metaphysics. Knowing the reason for it is helpful, though, since they were reasonable and significant although they are so no longer. For, the traditional hostility to metaphysics rested on Bacon's doctrine of prejudice. Philosophers from Kant to Mach and Husserl did not bother to justify their hostility to it. Wittgenstein justified it, saying, metaphysics deprives thinkers of their peace of mind. No one takes this seriously, not even the advocates of the New Wittgenstein theory that (rightly) considers therapy his central concern.

[78]Carnap 1956, last two sentences:

> Views vary a great deal as to the probability and even the possibility of such a development; and many will especially oppose, with either scientific or metaphysical arguments ... the

Perhaps there is no need to argue against anti-metaphysics any longer, but merely to observe that this is how things looked in the early days of analytic philosophy: at the time, many analytic philosophers argued against metaphysics from logic alone. Some saw no problem in the application of logic to natural languages; others admitted that they had to construct artificial languages and argue that they are metaphysics-free. Their effort was doomed to failure (Popper 1959, Preface to the English Translation). Here again we come to details of the use of the analytic tools, of which too little was said here thus far. So perhaps now is the time to say a few words on it.

Early analytic philosophers have claimed resolutely, and in an extremist fashion, that their doctrine imposes itself: it is obligatory for anyone who does not dismiss logic. This extremism characterizes young Wittgenstein and the early stages of analytic philosophy; it is gone for good.[79] Leading second-generation analytic philosophers do not claim any more that logic makes their philosophy incontestable; they have also changed their view: second-generation analytic philosophers disagree with their predecessors; they are now tolerant of philosophical tradition and of traditional metaphysical assertions that first-generation analytic philosophers abhorred. Acknowledged leading heirs[80] of Wittgenstein like Saul Kripke and Hilary Putnam have returned to some version of naturalism Quine-style, though in some analytic garb, having reopened questions once considered forever closed. Both Russell's and Wittgenstein's versions of naturalism are outdated. Russell and Einstein destroyed the dichotomy between truths by nature and truths by convention, considering received opinions putative truths by nature, hopefully the best available approximations to it.[81]

---

possibility of the last step, the assertion of physicalism.... My personal impression, in view of the progress made within the last decades ... is that the whole development ... seems today much more probable and much less remote in time than it appeared even thirty years ago.

[79]Warnock 2013, 5–7, recognized philosophical problems as "general questions" on the "disputed frontiers between language and formal logic", 155; he rejected the dismissal of all metaphysics, 160–1. Early in the day, he discusses (Warnock 1950) the status of singular abstract terms like the North Pole (that Quine 1947 found uninteresting). This raises the task of correlating the old-style conceptual analysis with the new-style analysis of statements and their corresponding ideas about metaphysics.

[80]This ascription is more social than intellectual. It illustrates the difficulty involved in the task of discussing ideas in isolation from their social habitat – a task that tradition since Descartes considered imperative. It is still advisable to perform it as best possible, even though this is never fully performable and hence it is always somewhat questionable and given to alternative readings. For example, it is advisable to read Heidegger on the supposition that his Nazi affiliation is irrelevant to it, and then to refute this supposition, which is not too difficult. To refuse to acknowledge this refutation is to raise suspicion of bad faith.

[81]Approximation to the truth poses very interesting metaphysical problems. Wittgenstein took it as utterly unproblematic (*PI*, §§38, 81, 130). Incidentally, he took usage to approximate at times sense and at times intended sense – in accord with Russell's initial (1905) analysis. That analysis

Analytic discussions center round the meaning of sentences, namely, their truth-or-falsity (does x possess a truth-value?), leaving for later occasions the discussion of truth (is x true?). This already Frege did as he postponed discussing induction and natural science. This is the procedure of removing confusion before removing error. It justifies Wittgenstein's quest for clarity. This quest he later modified. The idea that fixing meaning and clarity of an assertion precede the question of its truth has its reason and is very attractive. It is not new, however, as it is the basic idea of Galileo (see the opening of his first dialogue). Wittgenstein tried to adapt this idea to modern logic, thereby raising its significance. (He viewed himself a new Galileo; see below.)

Thus, Berlin was perceptive when he associated the difference in style and output of the logical positivists and the language analysts with the difference in their interests: the former cared about the natural sciences and the latter about the humanities. Indeed, young Wittgenstein took seriously ideas of leading physicists Hertz and Boltzmann, unlike mature Wittgenstein, whose idols were Tolstoy and Tagore. Even though the output of members of the two Wittgenstein schools did not explicitly side with the natural sciences or with the humanities, Berlin's observation stands. Yet it is limited. After all, Russell and his followers, for a conspicuous example, exhibited passionate interest in both. Yet they did acknowledge the difference between these schools. With all of his sympathy with the concern for the humanities and his distaste for the scientism of the logical positivists, Russell did side with the logical positivists, dismissing the language analysts as he did, for their refusal to notice that science impinges on the humanities by its disposition to correct the worst errors of prevalent commonsense. All his life he considered the early output of Wittgenstein significant (although erroneous) yet his posthumous output too trite to deserve notice.

Russell's opinion still is the best available, even though we should try to ameliorate it if we can and try to see the positive contribution of his mature output and of his followers in the school of analytic philosophy. For that, two tasks seem to me essential to perform first. The first is social: we must clean it all. We have to consider a great impediment to scholarship the aggressive tone of the later philosophy of Wittgenstein and the defensiveness of his disciples in the analytics school of philosophy, not to mention the hagiography that is simply distasteful. Bartley accused Wittgenstein's heirs of improper conduct. His complaint won corroboration from some inside sources, such as that of Arne Naess. Now we cannot expect that scholars will fully adhere to the rules of conduct of the commonwealth of learning (there is no utopia). Yet the damage due to violation of etiquette is usually temporary: it causes the postponement of improvements that some intellectual elites deem undesirable (Agassi 1981a, b, 119). This causes a time lag that is becoming an increasing burden with the general progress of the modern world. It also has to do

---

approximates the concept of approximation to the ideal comprehensive truth that Einstein developed and that Russell soon adopted wholeheartedly. Popper later developed the theory – his (qualitative) theory of verisimilitude.

with contemporary philosophy, not with that of Wittgenstein and of his immediate heirs. The time has come to take stock of the early days of Wittgenstein's dominance. This study is hopefully a part of this process.

The second task is the main defect in Russell's philosophy, his sensationalism Locke-style. That it has met with ample empirical refutation is no big deal: it is always possible to amend a theory to accommodate for empirical innovations. When one undertakes such a process, one is concerned with keeping the characteristic of the theory that renders it significant. In this case, this characteristic is of supreme importance: the maintenance of commonsense realism. This is why Russell said, if sensationalism will turn out to conflict with commonsense, then he wished posterity to ascribe to him commonsense rather than sensationalism. Now this conflict is obvious: sensationalism leads to neutral monism that conflicts with commonsense. This conflict was allegedly resolved by rendering neutral monism sheer methodology. It turns out to be psychology rather than methodology. Moreover, taking theories at their face value renders them realist in the sense of Frege's idea of meaning: they have a truth-value. This is a major revolution: tradition deemed scientific theories true (verified or tautological) or meaningless (*façon de parler*); Popper ([1935] 1959, Ch. 1) was the first to follow Frege here, thereby raising the problem of demarcation of scientific theories in a new manner.

Wittgenstein did not go into all this: his concern was philosophy, and he soon ignored science as much as he could. His chosen philosophical task was then merely exposing metaphysics as sham. Now, the wording of a sentence that seems possibly metaphysical and that for some reason we cannot ignore, invites critical examination. Wittgenstein's *Philosophical Investigations* offers many discussions and Wittgenstein's followers offered different kind of these. What do all these amount to? How does an analytic philosopher diagnose a piece of seemingly meaningful nonsense? Wittgenstein had one answer: follow the use of words: they are useful if and only if they are non-metaphysical.

Chapter 2
A History of Anti-metaphysics

Francis Bacon (a contemporary of Shakespeare) inaugurated the scientific revolution. His success was largely due to his preaching of hostility to metaphysics. He objected to it for one and only one reason: its method is speculative. He asserted repeatedly that speculations are obstacles to scientific progress and promised steady scientific progress after the removal of that obstacle. His breath-taking, most influential opposition to speculation is his doctrine of prejudice. Preconceived notions, he observed, influence our perceptions and then naturally our experience supports our preconceptions; this way observations tend to turn our speculations into prejudices and theses block scientific progress. Thus, he concluded, it is most advisable to avoid at all cost making any hypothesis, as this is dangerous: hypotheses are prone to be false and the observations made in their light are prone to mislead; those who entertain hypotheses, he explained, become committed to them in their inability to perceive refutations or else in their reluctance to admit error. Metaphysics is thus objectionable since it is hypothetical; if the urge to speculate is under check and gives way to small, humble experiments undertaken patiently, with no guidance of any theory whatsoever, then all science-based metaphysical truths will arrive by themselves in good time.

Bacon was in error: no one can ever be free of all prejudice and of all metaphysics (Russell 1956b, 77; 1968, 223). Theory-free observation is then likewise impossible.[1] The way to reduce prejudice and observation-errors is to develop a critical

[1]Bacon wisely allowed to view proper observation reports either as theory-free or as theory-laden where the theory behind them is the true one, the one that will emerge from them (like wine out of grapes): in either case, they are unbiased. Carnap ignored this as he studied empirical verification. He introduced a set of pure observation statements and he called it L_0; it is a language since it is the extension of the language of the functional calculus by the addition of pure observation terms. This should render the statements of that language theory-free. They are supposedly verifiable and so they are allegedly certain. Popper ([1935] 1959, §38) proved this erroneous, since universal names of observed objects are dispositional. Thus, contrary to Carnap's intention, L_0 is no genuine extension of the functional calculus.

J. Agassi, *Ludwig Wittgenstein's Philosophical Investigations*, Synthese Library 401,
https://doi.org/10.1007/978-3-030-00117-9_2

stance and examine as many options as possible; when metaphysical systems are concern, developing competing metaphysical systems may help. To that end, discussions of metaphysical questions are possibly helpful.[2]

2.1 The Doctrine of Prejudice

Bacon's hostility to speculation in general (as leading to dogmatism) makes sense; it does justify his hostility to metaphysics. The importance of his theory of induction is that it provided incentives to perform small experiments and thus to the development of the modern scientific societies (Agassi 2013). As Russell (1956b, 77) considered any claim to be free of prejudice sheer humbug, he was hostile to a metaphysics for different reasons: first, he rejected Plato's metaphysics, or, even more specifically, Plato's Heaven; second, he wanted his philosophy scientific: his view of rationality, his endorsement of the requirement for grounding beliefs, rendered his default attitude anti-speculative in general. By contrast, Wittgenstein explained his anti-metaphysics by observing that concern with it is futile and greatly disturbing.[3] This suggests that he understood it; he wanted it exorcised the way Russell exorcised his paradox: Russell rendered his paradox meaningless in PMia. Whether Russell's paradox disturbs anyone is irrelevant to his exclusion of it; as it is intolerable in any formal system, the wish to exclude it was understandable. As to Wittgenstein's finding metaphysics disturbing, it is no argument: some people find love disturbing and some people find metaphysics greatly enjoyable. Among them were the great physicists James Clerk Maxwell and Albert Einstein. Wittgenstein ignored this.[4]

The sense of "meaningless" in all of the output of Wittgenstein is that of Russell,[5] namely, being ungrammatical, having no truth-value. Mature Wittgenstein added

---

[2]Shaw 1908, Agassi 2013.

[3]Surprisingly, what confusions are metaphysical has never been clearly discussed (Ayer 1985, 137). Wittgenstein always took it for granted that there is no problem to recognize a metaphysical statement when bumping into one. He described metaphysics as "the bewitchment of our intelligence by means of language" (*PI*, §110) not as a demarcation but as a diagnosis. Stirring as this description may be, it is both too narrow and too wide. Alternatively, it is too vague. Moreover, metaphysics can be a delightful challenge. The "Vienna Circle" identified metaphysics with theology and with obscurantism. (They were not familiar with mature Wittgenstein.) Quine 1988 laid great stress on the continuity between science and metaphysics as well as between the meaningful and the meaningless. This throws interesting light on Popper's demarcation between science and metaphysics.

[4]TLP, §6.53: "This method would not satisfy the other – he would not feel that we teach him philosophy – but it would be the only strictly correct one."

[5]Popper 1986, 116 noted that his criticism of Wittgenstein's use of "meaningless" was unanswered. Popkin and Stroll 1996, 309 said, when facing an ungrammatical sentence people correct them one way or another with no fuss (and if need be they check their corrections). Mature Wittgenstein allowed for this, but claimed ability to show that some expressions are inherently meaningless; this is his idea that meaning is use: metaphysical assertions have no use. Now the slogan "meaning is

that they are idle. Gellner responded with the anthropological observation that meaningless expressions need not be idle. The response to him is that they are idle as information. This is obvious: each piece of information has to have truth-value. What Wittgenstein meant then is that if you wonder whether an expression has a truth-value, examine its use. Now, notoriously, some people pretend that empty verbiage is significant. Thus, Wittgenstein's suggestion to watch language in action amounts to the advice to listen critically to what people say, to notice that some people appeal to the gullible. This is not much of a message, practically valuable though it surely is.[6] As far as his *Philosophical Investigations* is concerned, the force of the same criticism lies elsewhere: replacing the alleged applicability of Wittgenstein's thesis to the ideal language with its applicability to ordinary language renders the whole affair somewhat dreary. I will try to avoid returning to all this. It is difficult to do so, since many of his followers mistakenly take his technical term "meaningless" as the ordinary word and they mistakenly take his use the ordinary words "clarification",[7] "family resemblance", "private language" and "rule following" as technical terms that he had invented and forgot to explain. They seem unable to imagine that this distortion of his texts is an insult to him. It is.

Analysis is the tool of analytic philosophy, the very means for proving the truth of its central tenet. It is not clear what that tenet is. Initially, it was Wittgenstein's: there are no proper metaphysical assertions. Traditionally, the word "metaphysics" means the articulation of an aspect of an image of the universe as a whole. This way it appears, say, in Aristotle's *Metaphysics Alpha* and in Bacon's *Novum Organum*. The previous Lecture did not discuss the value of the purge of metaphysics from our discourse, and so, as a whole, it seems pointless. Here, then, are grounds for the hostility to metaphysics that renders slaying it valuable (be it by analysis or by any other means).

Bacon's doctrine of prejudice (1600) is at the background to all modern philosophy. It was most influential in the Age of Reason, in seventeenth century and eighteenth century scientific philosophy, though at the time it was scarcely stated or discussed. It is still very popular and influential. Locke developed his dominant perception theory under its impact. Schlick was an ardent supporter of it (Coffa 1991, 421). So was Otto Neurath, his heir as the head of the Vienna Circle after his tragic demise.[8] Some of the writers who enter our discussion were less hostile to metaphysics than this doctrine demands.

Bacon viewed science as the salvation and redemption of humanity, as the means by which we will soon reach total comprehension of Nature and full control over

use" is a part of behaviorism yet Wittgenstein did not support behaviorism. (Commentators have ascribed it to Ryle, but they were in error.)

[6]Wittgenstein's response to Popper in their famous face-to-face exchange (Popper 1986, 122–5; Edmonds and Eidinow 2001, beginning of Ch. 1) went much further. This, however, can hardly count as a considered opinion. The report that Wittgenstein wrote the next day, in which he called Popper an ass (*op. cit.*, 287), supports this.

[7]See Diamond 2004, 214.

[8]Neurath 1983 declares Popper's demarcation of science a prejudice. See Hacohen 2002, 272, 280.

Her. He took this as natural, since every human possesses the marvelous natural capacity for contributing to the growth of human knowledge: the simple collection of easily available information will give rise to true theories of great value, both intellectual and practical. This marvelous capacity will soon establish Heaven on Earth. What obstacle, then, is there on the road to salvation that has prevented us from achieving salvation and redemption thus far? Bacon's answer is, prejudice: it blocks the natural, otherwise relentless growth of science. He explained how this obstacle prevents progress. It is the disposition to be lazy and guess, the temptation to suggest hypotheses, he said, that is the obstacle to scientific progress. Making hypotheses looks innocuous, but it is not. A hypothesis is unlikely to be true and its empirical refutation will not eliminate it, as its advocates will not admit to holding a false view. People are loath to give up views and admit having erred even in the face of effective criticism; we prefer to stick to our hypotheses at any cost. When forced to yield, we try to minimize cost: we make the smallest corrections to our hypotheses, ones that are required as means of neutralizing any admittedly valid criticism. It is always possible to rescue a hypothesis from any criticism by making some small adjustment, at times by limiting the denotation of a word.[9]

William Whewell (1857) criticized Bacon on two points. First, hypotheses need not be prejudices; otherwise, it would be impossible to give up Newton's theory of light. The renunciation of this theory as the result of empirical criticism (in 1818) is a strong criticism of Bacon's doctrine of prejudice. In addition, said Whewell, hypotheses are indispensable. Without them science cannot develop at all. A hypothesis, he suggested, should not be given up prior to its having been effectively criticized, subjected to severe tests. Upon passing severe tests, it gains scientific character and deserves endorsement. Popper has advocated the strategy that already William Whewell had described a century earlier but with a reverse judgment: agreeing with Whewell on method but not on the status of theories, he considered empirically testable hypotheses forever scientific and forever doubtful. In between, during the period of the crisis in physics (around 1900), two great thinkers, Pierre Duhem and Henri Poincaré, suggested that (in opposition to Bacon's caution) the rescue of a valuable theory by a minute correction is advisable. Propriety requires that the rescue should be open, not defensive, and produce profit that exceeds cost. All the alternatives to Bacon's view of scientific method were great improvements. They all modified his doctrine of prejudice, but only Popper criticized it as the root of radicalism. Thus, the reason Duhem and Poincaré declared scientific theory as without truth-value or as uninformative (tautological) is that they refused to declare Newton's theory false, since this sounds an insult to the greatest scientist of all time. It took Einstein and Popper to declare that viewing it as an approximation to later

[9]All this is ancient knowledge. Bacon's excessive condemnation of *ad hoc* amendments was beneficial as it undermined the scholasticism that was dominant then. Disdain for *ad hoc* hypotheses increased under the joint influence of Copernicus and of Bacon. Duhem and his followers returned it to fashion. Reasonable practice allowed the reluctant use of it as a merely temporary measure (as Copernicus has displayed regarding epicycles). Duhem exaggerated this to support his denial of the realist reading of scientific hypotheses.

theories is maintaining its import. According to Popper, the history of ideas is the history of great errors that are parts of our heritage. In a way, this should hold for Wittgenstein too.

2.2 Radicalism

Arch-inductivist Bacon said, unless we know what had impeded the growth of science in the darkness of the Middle Ages, the danger is very likely that a new darkness will soon prevail. The (alleged) realization that Aristotle's metaphysics is the cause of the medieval darkness is also the (alleged) realization that replacing one metaphysics with another is useless, as the process of degeneration and stagnation will then repeat itself.

Going to the root of the trouble is radicalism. ("*Radix*" is the Latin for root.) It sounds appealing as it invites thoroughness. Sigmund Freud said, it is no use curing a symptom, as the illness that gives rise to it will express itself again, perhaps in different symptoms. In particular, curing a phobia forces the anxiety that expresses itself in it to express itself in another: it is thus the anxiety, not the phobia that invites treatment. Freud's theory, incidentally, is in conflict with the established practice of treating symptoms when they are worse than the illness, or when treatment of the illness itself is too costly or unavailable. It is also in principle impossible, since it requires that we keep ourselves in perfect health. The same goes for Bacon's doctrine, as it is the demand to avoid all hypotheses at all costs. This demand rests on the (Pyrrhonist) hypothesis that it is possible to avoid entertaining hypotheses: all one has to do is to look at the disadvantage of an appealing theory and to look at the advantage of an unappealing theory. This is a complex set of hypotheses; its advocates rescue it by small alterations.

There are likewise many hypotheses that serve as the basis of science. By Aristotle, science begins with developing a proper metaphysics, a proper foundation for science. Bacon, on the contrary, said that science begins with information; to start with metaphysics rather than with observations, he said, is to start with a hypothesis that becomes a prejudice that corrupts all research.

Bacon's doctrine of prejudice began the hard-nosed tradition of hostility to metaphysics. It is obviously erroneous: though it describes a danger correctly, there are many other worse dangers. We do what we can and hope for the best. The evils of dogma and prejudice are plainly unavoidable – although we hope to reduce them somewhat. The most fruitful efforts to reduce the danger, Bernard Shaw suggested, are by playing the devil's advocate, namely, by the use of competing systems. The hostility that Bacon advocated was not to the content of metaphysics but to its scientific status: it was speculative yet it posed as scientific: it was pseudo-scientific *par excellence*. It was Bacon who introduced the discussion of pseudo-science, and because he considered it a competitor and an obstacle to the advancement of the real thing.[10] He characterized metaphysics by reference to status not to contents: it is first principles; he also characterized science by reference to status not to contents. Hence, a scientific metaphysics is conceivable.[11] Granting this allows for the possibility of a metaphysics that in the future may claim scientific status. Obviously, then, Bacon's view of metaphysics as a pseudo-science, radical as it is, is less radical than that of Wittgenstein. Indeed, Kenny describes Wittgenstein admiringly as the chief crusader against pseudo-science (1984, 125–136, "The Homunculus Fallacy").

Bacon's radicalism is much more appealing than the later, modified versions of it. Its target was the tremendous authority of Aristotle and the scholastics. He discussed *epagoge* ("a bringing in") not only in the sense of dialectics (*elenchus*, "argument of disproof or refutation; cross-examining, testing, scrutiny esp. for purposes of refutation"), but also in the sense of induction, of concluding from particular assertions to some universal ones. What Aristotle said about this is unclear[12] – perhaps because no set of rules articulate analysis well enough; it is not a logic (Kneale and Kneale 1962, 36, 307). Bacon, too, said that one should first clean one's mind of its error and confusion – in order to open up to true contemplation. Yet, unlike Aristotle, Bacon declared dialogue useless (as it did not prevent scholastic dogmatism). He said, clearing one's mind is a matter of good will and decision. Only when one's mind is clean, he said, it is ready for enlightenment. What is this enlightenment? If Aristotle and Bacon said what it is, they agreed with Plato on it (Agassi 2013, 25). Plato was not clear about it, except that he said it is the proper mystic experience. To this Bacon added that prior to the enlightening experience one must observe – as homage to Nature.

Since anti-metaphysics was initially a version of scientism, there is room for the observation that not all scientists endorse it. Leading physicists Max Planck and Einstein rejected it, though Niels Bohr probably adhered to it and Max Born and Werner Heisenberg supported it enthusiastically. For a while, received opinion was that Heisenberg was a supporter of the "logical" positivism of the "Vienna Circle". Current received opinion is different. In any case, this is a marginal point: science is no beauty contest. Still, since the tradition of science is radical, and radicalism justifies the hostility to metaphysics, the support of metaphysics within the ranks of science is rare and refreshing. It conflicts then both with Bacon's anti-metaphysics and with the traditional views for metaphysics that Bacon opposed. Traditionally,

---

[10]Popper returned to the problem merely in order to offer an alternative to Wittgenstein's unintended preference for metaphysics over science. Popper's solution is the best explanation of our valuation of science.

[11]*Metaphysica generalis* is first principles of all sciences. *Metaphysica specialis* is first principles of a specific science. Current fashion is to replace the word "metaphysics" with "meta-science", "intellectual framework", and under Einstein's influence even "research program". "Paradigm" is another option.

[12]When logicians mention that Aristotle's logic is a small part of Boolean algebra, they refer to the hard core of his logical writings, in utter oversight of all that is murky in his works. Some Wittgenstein followers see this as support for his anti-metaphysics and others hold the opposite view. Hence, analysis wants analysis.

metaphysics was allegedly the foundation of science; the new – Einsteinian – view of it is that it offers research programs (Agassi 1964).

As to the scholasticism that Bacon disdained, the most central concern it exhibited was with the dispute between realists and nominalists on meaning. Realists advocated the Platonic ascription of reality to abstract shapes: they occupy a region outside space and time, known as the Platonic Heaven. Nowadays, this view is somewhat different, as it ascribes reality no longer to shapes but to abstract sets or even to abstractions or to abstract entities – including symphonies, said Popper. As realism changed, its name changed too: it is Platonism. As this name is historically somewhat inaccurate, Popper has replaced it with a new name: essentialism. Most early-modern philosophers preferred not to discuss abstract entities. The exception was Leibniz, who did so boldly. Frege, Gödel and Quine followed him; Russell was a follower of Leibniz too, yet he opposed Platonism sternly as long as he could. The traditional alternative to realism, nominalism, is the view of all names as proper names.[13] Russell never held it, since he always named sets as wholes, no matter how reluctantly; yet he always considered them fictitious entities.

The medieval disagreement is obsolete. The main reason for its obsolescence is that it took place within the traditional referential theory of meaning that Frege has refuted (see below). Considering language as containing essentially only names, medieval philosophers referred to proper names as denoting concrete things and general names (class names?) as abstract. Realism is the view that general names denote the abstract entities that dwell outside space (in Plato's Heaven).[14] Nominalism is the view that general names may denote concrete things, possibly more than one. The modern word for realism is "essentialism" since in modern philosophy the word "realism" denotes the assertion that physical objects do exist.

As a label, "essentialism" is not too felicitous. Whereas Plato said, the characteristics of things dwell out of this world, Aristotle said, they are the essences of things that dwell in the very hearts of things (*in ipsissima res*). He considered science the search for the right descriptions of essences. These descriptions are essential definitions. Bacon opposed (not essential definitions but) the search for them: he said they emerge by themselves out of the vast collections of information – the databases – that researchers should seek. The essentialism of Aristotle, not of Bacon, is methodological essentialism. The definitions that are not essential are called nominal (even though they are often are not proper names but names of sets). Hence, there are four items here: ontological and methodological essentialism and nominalism.[15]

Traditional radicals did not know all this; it is a recent innovation. They simply ignored theories of meaning and thus also any theory of language. The only theory of

---

[13]Popper [1935] 1959 §14, Burge 1973 and Reimer 2002 show how difficult it is to decide which noun is a proper name proper.

[14]Descartes located souls too in Plato's Heaven. This raises the mind-body problem in a new way and renders it insoluble. Russell rightly said, Descartes had erroneously adopted the ancient theory of substance.

[15]All this began with Boole (Bar-Am 2008); for "methodological essentialism" see Popper 1945, Ch. 11.

language they recognized was the denotation theory of names, and they accepted associationism as an explanation of denotation as due to experience of constant conjunction, to use Hume's expression. Frege refuted the theory of language as names by sheer logic and Oswald Külpe refuted associationism empirically. Frege's theory of names and Russell's theory of definite description were pioneering in one way; Külpe's was pioneering in another. Popper – a grand-student of his – connected these two with Frege's view of meaning as truth-value.

The rise of modern logic looked radical. It was not. It began with Boole's contribution, and it is still unfinished. His move was revolutionary as was his recognition of the empty class or empty set that legitimized empty names. (Throughout this work, I treat the words "class" and "set" as synonyms.) Suppose we provide proper names to real things only, as Frege and Russell demanded,[16] while allowing for names for abstract entities like classes.[17] Russell viewed sets fictitious entities. Consider Samuel Pickwick, the fictitious person that the real Charles Dickens has imagined. Can we declare the word "Samuel Pickwick" not a name? Possibly names of concrete entities are equally problematic. For, how does the bond between a name and a thing stick together enduringly? There is still no theory even of naming relations.

Paul Feyerabend[18] introduced the word "essentialist" to the analytic literature, by suggesting that Wittgenstein's philosophy is anti-essentialist. Now already Bacon and Descartes were not as radical as they had hoped. In particular, critics pointed out repeatedly, they took the referential theory of meaning for granted. Russell and

[16]The legitimation of empty nouns (like "square circles" or "mermaids") invalidates the square of opposition that is traceable back to Aristotle.

[17]To avoid admission of the existence of abstract entities, Carnap [1934] 2014 viewed language as ink stains of given shapes. Now shapes are abstract entities, rendered more abstract by the isomorphism with other communication-tools (sound, semaphore, etc.). Carnap tried and failed to account for this in his [1928] 1967 *Die Logische Aufbau der Welt* by replacing shapes with experiences of shapes (Agassi 1993; Pincock 2005). (Quine admired it despite its obvious failure.) As in this context experience is in a sense ideal, the trouble reappears (Agassi 1988, 77–8). Friedman 1999, Pt. 2, opening of Ch. 5, praises Carnap's *Aufbau*, calling it "a contribution to radical empiricism". This is a context-less backhanded compliment.

[18]Feyerabend's 1955 posed Wittgenstein as anti-essentialist although essentialist expressions pervade his writings (Bradley 1987; Oderberg 2011). Feyerabend's paper gave rise to a literature. The most unfortunate contribution to it is perhaps the hilarious redefinition of the term (Hacker 2001, 341): "essentialism – that is, the view that there are objective, language-independent necessities". This is quite different from the traditional meaning (Modrak 2001, 156). It seems Hacker wished to offset the criticism of his view by citing Wittgenstein (*PI*, §§371 and 373) "*Essence* is expressed by grammar" and "Grammar tells us what kind of object anything is". Wittgenstein's definitions of the essence of language differs from that of Aristotle, who said it reflects the truth of things (Matusova 2015, 49; Arneson 2007, 23). Thus, Hacker exposes here Wittgenstein as engaged in a metaphysical dispute, Heaven forbid. Here is an anti-essentialist Wittgenstein observation (*PI*, §65): "instead of producing something common to all that we call language, I am saying that these phenomena have no one thing in common." To clarify Wittgenstein's position on essences it is essential to attend to Popper's distinction between methodological and ontological essentialism.

Wittgenstein realized that early in the day. Wittgenstein was a philosophically extreme radical; he tried to start philosophizing *de novo* with no language at all. His *Tractatus Logico-Philosophicus* came to illustrate language construction from nothing. (This renders his casting away the ladder he used for that construction central in his early philosophy.)[19] Russell's Introduction to it overwhelmed him. If you want to justify his assertion that Russell misunderstood him, you can find the justification here: Russell ignored this aspect of Wittgenstein's project.

2.3 The Poverty of the Analytic Scholarship

Traditional essentialism is one of the most powerful ideas in the history of western thought. Currently most philosophers are either language analysts or phenomenalists, and only the latter are essentialist. The former center on language since essentialism has lost its dominance (Wisdom 1947). This holds even for the few analytic philosophers who advocate new versions of essentialism, namely, new theories of natural kinds that fixes species and their characteristics. Its rejection heralded a few revolutions, including Darwinism: Darwin offered as a mere convention a definition of species to be groups that (as it happens) do not interbreed.[20] Opening essentialist discussions, leading language analysts Kripke and Putnam sowed untold perplexity within their school (Schroeder 2006, 248).[21]

In the early nineteenth century, philosopher William Hamilton claimed that science is superficial, at least as compared to metaphysics. Logic, he said, goes for the essence, for the innermost, and mathematics for the form, for the outermost. George Boole disproved this by rendering logic mathematical, by creating mathematical logic (1847).

One need not over-praise Hamilton's work. Boole could not do what he did were he not a student of algebra beforehand and were mathematics not evolved in his days to a new level of abstraction with the invention of group theory that is autonomous (namely, that ignores the identity of group members). The evidence for this is that Leibniz tried to mathematize logic and failed. Indeed, Boole's aim was not to refute Hamilton: he used his algebra in his discussion of probability, and he considered probability a useful instrument for the development of his theory of knowledge as probable, and this is due to the refutation of Newton's optics (1818) that unintentionally legitimized hypotheses. Still, the revival of the variance between

[19]The "New Wittgenstein" theory is the denial that Wittgenstein offered "*deeply significant nonsense*" (Popper 1945, Ch. 11 n. 51). Alas, this requires Wittgenstein's claim that self-reference is meaningless. It is false: the statement "this statement is meaningless" is demonstrably inconsistent (Popper 1945, Ch. 24, n. 4).

[20]Darwin 1989, 180. Rhees 1984, 160–1 reports that Wittgenstein had opposed Darwinism.

[21]Putnam and Kripke began the blurring of the distinction between the two leading schools of philosophy. They were both well aware of this change though they did not see themselves as its initiators. Probably Stuart Hampshire and Arthur Danto deserve this attribution.

the metaphysics of Plato and of Aristotle was a motive in the development of Boolean algebra. This is thought-provoking. It is no evidence that there is merit to either Aristotle's theory of definition or to anything that Hamilton said. Hence, even what in retrospect looks quite uninteresting may prompt better ideas. Hence, whatever is said here, it renders questionable the hostility to poor ideas, metaphysical or any other. Some hostility may have been useful, such as the hostility that Bacon showed towards scholasticism: it was reasonable and fruitful, as the scholastic metaphysics was as stagnant as he said it was, and thus as harmful to research at the time as he said it was. This need not always be the case, however.

Since Boole was not the only founder of Boolean algebra, it is obviously easy to exaggerate the value of Hamilton's contribution. Problems with logic troubled not only Boole, but also Augustus De Morgan, Ernst Schröder and later on also Charles Sanders Peirce. All of them were concerned with the basic ambiguity of traditional logic. What troubled them all was whether the copula, which fuses the subject and the predicate of a statement, belongs to metaphysics or to science or to mathematics.[22] All discussions of the elimination of metaphysics are confused about this, with the exception of Russell's efforts to dismiss Plato's Heaven. He insisted on the rejection of Plato's obviously metaphysical ascription of reality to some non-material objects. Unlike Russell, Wittgenstein (and Schlick) stressed that both the admission and the rejection of Plato's Heaven are equally meaningless, yet they differed about what this means. Schlick was explicitly a staunch realist[23]; what Wittgenstein's view was is under dispute. What blocks clearing the situation is his hostility to metaphysics: he perpetrated the confusion that he was so intent to combat.

The huge literature on all this is very poor, perhaps not surprisingly. Let me mention one essay of this ilk, by leading and highly respected analytic philosopher Bernard Williams. That essay, "Knowledge and Meaning in the Philosophy of Mind" of 1968 has the status of a classic and it engages a still growing literature. Williams' argument there rests on a classical distinction of Kant, and so discussing it requires the introduction of some background information.

One of Kant's most prevalent ideas is his distinction between two dichotomous divisions of knowledge. The first dichotomy is between analytic and synthetic, with analytic knowledge being true in accord with traditional logic. The second dichotomy is to *a priori* valid and *a posteriori* valid, with validity resting on thought and on experience respectively. These dichotomies are not of all statements but only of demonstrated ones. This is of paramount importance, since, obviously only a fraction of all statements are demonstrated. Kant took it for granted that all analytic statements are logically provable. In a general sense (of Gerhard Gentzen), Kant's

[22]Koopman 2010 ascribes to Hacker 2001 the exposure of the ambiguity of the status of the copula in the analytic tradition. Paper is tolerant.

[23]The literature on Wittgenstein on realism is huge (Stenius 1960, *passim*). Schlick 1959 asserted his assent to traditional realism while trying to abide by Wittgenstein's demand to avoid asserting it. Hilarious.

view still prevails; in Frege's sense of provability, some analytic statements (Gödel statements) are demonstrably not provable. Therefore, analysts have to specify whether they refer to provability in general (in natural deduction systems) or in accord with Frege's proof procedure. We are thus much more careful about proof than Kant was. We likewise agree that no synthetic statement is provable, contrary to Kant's central thesis that was that some synthetic systems (Euclidean geometry and Newtonian mechanics) are *a priori* valid though not logically valid. He called the rules of this kind of validity *transcendental logic*. He never specified them. What led him to transcendental logic was his simultaneous use of the two traditional dichotomies between proven (or at least provable) statements: the formal, between the analytic and synthetic, and epistemological, between the a priori valid and the a posteriori valid.

Proven statements:

Analytic statements a priori valid	Synthetic statements a priori valid
Analytic statements a posteriori valid	Synthetic statements a posteriori valid

The two dichotomies produce four possible kinds of knowledge. One of these we ignore: knowledge of analytic statements acquired *a posteriori*. These do happen, but we ignore them in the present discussion, as they are unproblematic. There remain three kind of knowledge, then. First, analytic statements are *a priori* valid. He defined analyticity as narrowly as possible (see below), thus rendering them unproblematic too. Second, synthetic statements that are *a posteriori* valid invite a separate discussion (see below). Third, synthetic statements *a priori* valid are provable by transcendental logic. (Their knowledge transcends experience.) A transcendental proof goes like that. Scientific theories are synthetic and (as Hume has proven, they are not derivable from observations, so that) they are not *a posteriori* valid. Hence, if they are valid then they are *a priori* valid. But – this is the transcendental move – scientific theories do exist; hence, their validity is *a priori*. Kant took for granted the validity of Newton's theory. The assertion of Solomon Maimon that it is in doubt he found most annoying: if it is doubtful then it is no knowledge and it has no validity and no proof. Most philosophers were on Kant's side; now they are all on the side of Maimon.[24] Only Boole stood out. He said, empirical research should help us decide whether Kant is right!

So much for the background to the towering paper of Williams. He proved that metaphysics is impossible thus. Were it possible, he argued, it would not be analytic, or else it would be a part of logic; and it would not be valid *a posteriori*, or else it would be a part of empirical science. Hence, if it exists, then it must be synthetic *a priori* knowledge. Unlike Kant, he declared synthetic *a priori* valid statements impossible. This is his whole proof. Wittgenstein had presented it earlier, albeit much more briefly and (rightly) not as a proof. As metaphysics is speculative, namely hypothetical, it is not knowledge. Hence it is not provable either *a priori*

[24]See Agassi 1975, Appendix to Ch. 19. See also Bar-Am 2008, 106–7.

or *a posteriori*. Hence, Williams' proof tacitly rests on the idea that all meaningful statements are provable; the "Vienna Circle" praised this idea with great fanfare. It is the *verification principle*. It makes metaphysics meaningless with no further ado. One way or another, then, Williams' proof is redundant. Moreover, as it is the verification principle that renders Williams' proof redundant; its negation renders that proof obviously invalid. It seems he was unaware of this fact nonetheless. The high esteem that his proof is accorded raises the question, what lowers the standard of the analytic literature so much? This question rests on the supposition that once the standard of publications in philosophy was significantly higher. Evidence for this is from the classics that have survived to date. This is an oversight of our ignoring most old philosophy papers. In addition, when discussing analytic philosophy, we rightly ignore many texts as sub-standard. It is nonetheless hard to relieve Wittgenstein of all responsibility for this decline of standards, since he was pretty much in charge of the philosophy scene in England. As to Williams' argument, published nearly two decades after Wittgenstein's demise, would Wittgenstein have approved of it? That depends on his attitude to the verification principle. The question whether he did or did not uphold it is still under dispute. In *Philosophical Investigations* he discussed in detail the learning of the use of words such as "pain" (*PI*, §245). The easy way for testing the understanding of names for pieces of furniture is not available for testing the understanding of names like "pain". This discussion of Wittgenstein reveals his view about learning to speak and even about learning in general. His views on these matters are amazingly old fashioned[25]: no situation assures us that pain is or is not involved in it: like all knowledge, the knowledge of the presence or the absence of pain is uncertain. Yet Wittgenstein was concerned with certainty all his life. It was a painful, hopeless struggle. His anguish is what was most credible of his many qualities.

It is not clear whether Wittgenstein's suffering was due to his clinging to some old-fashioned philosophy. (His politics too was old-fashioned conservative.) It is also hard to judge whether his innovations, whatever they were, depended on his peculiar worldview. Nevertheless, he is not to blame for the shallowness of the output of so many of his disciples; it is better to praise him for the value of the output of some of them. Nevertheless, let me confess my prejudice; anti-metaphysics is no aid for the flight of imagination that Faraday and Einstein deemed necessary for the possibility of intellectual progress.

[25]William Somerset Maugham, *The Partial View*, 1954, 155, considered it odd that philosophers speak of toothache when much worse pain abounds. This is amusing but unkind: philosophers were studying not pain but empirical knowledge: they claimed that I am certain that I suffer toothache. Clinical observations refuted this claim (Agassi 1975, 120–6). Wittgenstein described how a child learns to observe how others are in pain. As a language-acquisition theory, his *Philosophical Investigations* is inadequate; the theory developed vastly after he died.

2.4 The Rise of Current Anti-metaphysics

Hostility is an injudicious expression of dependence (Spinoza; Shaw).[26] Wittgenstein endorsed the traditional hostility to metaphysics, adding to it hostility to its practitioners. Hostility apart, he shared the traditional censure of metaphysics as empty; he added the idea that the concern with it is sick.[27] This is a variant of an intellectual tradition rooted in Bacon's radical denunciation of metaphysics. The traditional denunciation was of its speculative character. This allowed for non-speculative metaphysics, for metaphysics based on science. Kant agreed. He wrote a booklet on the metaphysics that will in future claim scientific status. Mach's attitude was humbler: he advised to avoid taking side in any metaphysical dispute, presumably out of dislike for unsettled disputes.[28] Bacon's interest was in recruiting people to research and his intent was the elimination of dogmatism since in his view dogmas destroy the ability to conduct fruitful research. Wittgenstein's interest was different. Scientific research and its management, he confessed, did not interest him. He spoke against the confusion that causes pain.[29] Sometimes he suggested that it is the inability to master, to control (*PI*, §§31, 150, 546), a painful sense of being lost (*PI*, §123), of "deep disquietude" (*PI*, §111). People who advocate metaphysics he denounced as slum-landlords. He saw his task, the task of a philosopher, as combatting metaphysics, and this includes prejudice (*PI*, §340) and superstition (*PI*, §§49, 110). It seems he never explained this and it never intrigued his commentators, though some of them noted with consternation his remark "Superstition is the belief in the causal nexus" (*TLP*, 5.1361) in the light of his distinction between error (for science to correct) and superstition (for philosophy to correct).[30]

Aristotle aptly called metaphysics first philosophy (*he prote philosophia)*; Bacon called it first philosophy (*philosophia prima*). It is a grand axiom system, the foundation of science. Kant viewed it as the presupposition that render science

[26]Thus, "Wittgenstein's farewell to metaphysics was not without sadness" (Gabriel 2009, 61).

[27]Naming this claim "the new Wittgenstein" (Crary and Read 2000; Read and Lavely 2011; Stokhof 2011) is outrageous. "The philosopher treats a question; like an illness" (*PI*, §255) is too obvious for that. The only worse move here is the efforts to ascribe to Wittgenstein ideas that have appeared long after his death that leading Wittgenstein enthusiast Juliet Floyd listed on a few occasions (treating hagiography as serious).

[28]Young Popper followed Mach's advice. He later changed his mind on that.

[29]Wittgenstein referred repeatedly to the pain that metaphysics causes; he never discussed it. He did suffer from it personally. The same holds for the acknowledgement that many a metaphysician from Maimonides to Russell found in it the very meaning of life, no less. That pain is no reason for dismissal already Kant admitted, as he compared metaphysics to a mistress with whom we quarrel but to whom we return. Scientific researchers suffered pain from frustration – not only volatile Pauli, but also placid Einstein – with no intent to cease. Hence, it is not the pain but the futility of metaphysical research that mattered to Wittgenstein.

[30]See Black 1971, 244. The view of causality as a meta-scientific principle (Hume and Kant) is unproblematic. Bohr, Kramers and Slater contrasted causality and statistics (1924). The empirical refutation of their idea has changed the situation – not in favor of Wittgenstein's philosophy, though.

possible: the principle of causality or of simplicity. Bacon required empirical justification for it; Kant offered intellectual one. This raises a general question: what is a justification of an idea? Why do we need it? Einstein said, the only justification a hypothesis can have is from its ability to explain facts. This justifies only science; it allows the justification of a metaphysics only when it is scientific. Williams, we remember, took it for granted that there is no such thing. Nevertheless, researchers do employ metaphysics: a scientific theory can conform or not conform to this or that metaphysical system (intellectual framework), and as such it may play a significant heuristic role in research (Whewell; Einstein). Bacon did not deny this option, but he denounced it as prejudicial. His dichotomy between prejudice and science is obviously false: science and metaphysics may be in conflict and this is no disaster. The searchers for the justification of science advise to ignore the metaphysics (Mach); those who view science as a critical debate take it as possible challenge (Agassi 1975, 231).

Even some leading anti-metaphysicians admit that metaphysics has contributed to science.[31] Mature Popper followed Einstein and asserted with no qualification that some metaphysics is heuristically valuable – systematically, though emphatically with no guarantee. When the heuristic that it suggest is unpromising, then it is in need for overhaul. All this, however, deviates from Wittgenstein sufficiently to count as irrelevant to the present study, except for one development: quite a few metaphysicians and some theologians are eager to count as members of the analytic tradition. Even among the leading disciples of Wittgenstein, some are sincerely religious. Let me ignore this here as much as possible.[32]

Apparently, then, analytic philosophy dismisses theology. Is it possible to be religious and reject all theology? Is this last question philosophical? If so, is this in agreement with Wittgenstein's philosophy? Many members of the analytic school are religious, including Anscombe who was a leading disciple and friend and a Roman Catholic convert.[33]

---

[31]Neurath 1973, 307–8, 360, Zilsel 2003, 9, 73, 75, 78, 89, 91, 114, 126, 131, 140, 145, 180.

[32]I do so reluctantly, since a philosophy indifferent to established religion is barren. Wittgenstein was ambiguous or ambivalent on this. Russell and Popper deemed this his greatest defect. He promised peace of mind not due to (Socratic) action but due to (Pyrrhus-style) inaction, namely, to the loss metaphysical curiosity. This idea is attractive, I do not know why. The first to oppose it was Spinoza, whom Russell followed. Wittgenstein followed Tolstoy, whose *magnum opus*, *War and Peace,* is a tribute to inaction.

[33]This we can admit: the anti-metaphysics, and more so the anti-theology, although seldom just, has served well as a pruning. Anthony Kenny (1993), a leading Roman Catholic Wittgensteinian, praises the unmasking of the confusion between metaphysics and grammar, viewing as grammatical the claim that minds have no location (akin to Linsky's observation of the national flag as having no size and contrary to Freud's (1923, I) true observation that whether I am my body or have my body is context-dependent). Kenny also claims that computers are mindless. Even science fiction does better, as it describes possible living computers, not to mention Leonhard Euler, who imagined a rhino with a human soul (as a criticism of the parallelism of Leibniz). Wittgenstein deemed meaningless the claim that we possess souls in the traditional theological sense, not in the sense of being alive. These two claims share a grammatical form and are equally universal.

Wittgenstein made significant contributions when he was a student of Russell. He then contributed significantly to Russell's development of his mature ideas and he then wrote a draft of his *Tractatus*, all prior to World War I. His early contributions count even though by now they are outdated. Historians count the *Tractatus* only on the supposition that it is not. Thus, they all praise his invention of truth tables[34] as they are fixtures. Not so his dismissal of the quantifiers ("all" and "some") that is generally rejected. Now truth tables replace the axioms of the calculus of statement composition. Every elementary course in modern logic begins with them. They look unproblematic. For example, the truth table for conjunction of two statements ("a and b") says that it is true iff both of its components are true and that the disjunction of two statements ("a or b") is true iff at least one of its components is true. (Note: "iff" is shorthand for "if and only if"; thus, "a iff b" stands for "if a then b and if b then a".) Yet the truth table for negation is not that obvious. It says, the negation of a statement (~a) is false iff it is true. What that means depends on the universe of discourse. To see what this means is to know the extension of universal statements: what do they denote? Frege said, all true/false statements have the same reference. This is far from satisfactory.

The other contribution of young Wittgenstein is extensionalism, the tenacious preference for extensional systems. Boole's principle of extensionality is the idea that sets are equal iff they have the same elements. Extensionalism is the preference of classes over properties and over any other characterizations of things (Quine 2008, 11). It is then opposed to essentialism (Bar-Am 2008, 123). Extension and intension are akin to reference and sense as well as to denotation and connotation. (Extensionalism is roughly treating extension in disregard for intension; the identity of the morning star and the evening star is extensional.) It renders logic metaphysically neutral (Bar-Am 2008, 125). Obviously, then, it has no full implementation (Bar-Am 2008, 137).[35]

All this belongs to logic. Wittgenstein contributed to it only in his early days. He later discussed metaphysics as a torment, and left unanswered the question, what was his attitude towards those who enjoyed metaphysics? As Russell was a metaphysician (Mumford 2003, 1), he did not think much of Wittgenstein's idea that the concern with metaphysics invites therapy. Did Wittgenstein think that Russell needed therapy? He must have thought about this question, since he reported

[34]Charles Sanders Peirce has priority for this idea. Truth tables played a great role in the early Wittgenstein philosophy; that put them on the map. Since the truth-table definitions of the connectives of the calculus of statement composition are formally equivalent to their axioms and to their introduction by their deductive force within the natural deduction system (Popper), the choice between these three options is possibly a matter of philosophical taste alone.

[35]Kripke 2008 notices that some thinking about sense must precede discussion of reference. Purely extensional systems have no known reference (Lakatos). Similarly, purely formal systems are utterly silent about their possible applications (Lakatos). This is the background to Wittgenstein's new idea of the ineffable. Russell's criticism of it is right: the ineffable in one language may be articulable in another, whereas the utterly inarticulate is not open to critical discussion and so it is not agenda. Nevertheless, the novelty of Wittgenstein's idea of the ineffable deserves attention.

"with pleasure" that Russell had said, "Logic is hell!" (Malcolm 2001, 57) Apparently, this refers to Russell at the time when he explored the option of logic as a replacement of metaphysics. Russell gave up this idea, considered metaphysics unavoidable, and agreed with Einstein's distaste for the fashionable fear of metaphysics (Schilpp 1944, 696). The only escape from this unpleasant situation is to take more seriously either Wittgenstein's therapeutic technique or his description of metaphysics as grammatically defective in a special way, as language idling, as tantalizing efforts to transcend language in a way that is inherently blocked (*PI*, §426). It behooves those who take Wittgenstein's characterization of metaphysics seriously to compare it with its traditional characterization, shared by Aristotle and by Schlick. To repeat, Popper mentioned ([1935] 1959, §85) ancient atomism as disproof of the idea that metaphysics is ungrammatical: it is a part of scientific atomism. Wittgenstein did not respond to this. Carnap did. He had no intent to put obstacles on the road of research, he said. Taken seriously, this means that only some metaphysics is such an obstacle, not all. This raises the question, which metaphysical doctrine is objectionable and why. The confused ones, of course. In the end of his life, Carnap expressed the hope that a metaphysics-free language should be soon available, forgetting all about atomism. Like others in the "Vienna Circle", Carnap found some metaphysics appalling and wished to contribute to the advancement of learning and to human welfare by pitching logic against it and against its obscurantism. Alas, this is impossible. Obscurantists will not listen to reasons against their views. It is possible to try to improve logic, though. Wittgenstein did that – when he worked with Russell, although he still was an undergraduate. He was understandably eager to strike on his own. Russell was eager to help (Landini 2010, 34). He did help, but Wittgenstein wanted his approval and missed it direly. Ray Monk, Wittgenstein's chief biographer, declared (1990, Ch. 1.4) that the relations of master-apprentice between Russell and Wittgenstein was reversed. Evidence goes the other way, despite Wittgenstein's fervent ambition to be one up on Russell (Diamond 2000, 263).

Diverse lines of thought intermingle here. I hope that the clarity of my exposition of Wittgenstein's work will help improve the discussion of what his new philosophy is and what it shares with the new logic: it is the clarity of the new logic (especially as opposed to the vagueness of traditional logic). Traditional logic suffered from paralogisms – from the failure of its formal apparatus to exclude some palpably invalid inferences. Plato reports that sophists played on this defect of classical logic (*Charmides*). The theory of judgment[36] was license to disregard the defect in logic (sophisms) that the sophists had discovered. By contradistinction, the new logic is formal: its rules of sentence formation and of inferences dispense with judgments. As Frege, Peano and Hilbert distrusted judgment, they all devised formal systems to

[36]The word "judgment" is obviously highly context-dependent. It often means a decision as to assent to any given proposition (or dissert from it). Here it means assent to an inference as valid. The new logic demanded treating formally every question as to the validity or invalidity of an inference, according to explicitly stated rules that fully determine each step of an argument. This demand seemed obvious until Gödel showed it excessive.

render it redundant. To ensure success, ignore meanings and look at a system utterly formally. Frege was the first to conceive of the system as a formal language, to come close to having achieved construction of a language adequate for mathematics, and to do that single-handedly, seemingly with no predecessor.

Chapter 3
The Waning of Essentialism

John O. Wisdom's *The Metamorphosis of Philosophy* (1947) says, [new-style] analytic philosophy has resulted from a radical change in our theory of definition, and the resulting devaluation of the place of essential definitions in science and philosophy.[1] Moreover, since most traditional and most current metaphysics are deeply involved with essential definitions, I would like to support the proposal of Stuart Hampshire, to salvage as much of metaphysics as possible.[2] This is the proposal to the followers of Wittgenstein to correct his erroneous demand to oust all metaphysics by limiting it to the demand to oust metaphysical systems that rest on essential definitions. This, I will argue, is tantamount to the suggestion to oust the pretense to know what scientific theories will win assent and keep it to the end of days. Even this is possible to restrict further: whenever possible, we should reinterpret essential definitions to read them as speculations or as conjectures. Perhaps

[1]Ben-Yami 2017, 408 cites Wittgenstein's explicit dissent from Plato's view of knowledge – the one that Popper has christened the theory of essential definitions. In the passages that Ben-Yami cites, Wittgenstein declared that he had no theory of knowledge, only instances of knowledge. This way Wittgenstein obviously took back his early doctrine of *showing*: what young Wittgenstein said only shows itself the mature Wittgenstein considered an unsayable universal: observation imposes truth-value; whatever mature Wittgenstein showed he could say, and it was particular. What should intrigue Wittgenstein biographers is that he had his mature philosophy articulated in 1932 at the latest. True to it, he spent his last two decades in diverse illustrations of his opinions on diverse philosophical questions while repeatedly rejecting all articulations of the opinions he wished them to illustrate. His reference to Plato's theory of knowledge is thus not to any doctrine but to his having a doctrine to begin with. This is my reading of Russell's view of Wittgenstein's mysticism as dominating his thinking in general.

[2]See the *Stanford Encyclopedia of Philosophy* Articles on Neurath, Carnap and Hampshire. Essential definitions restated as hypotheses are often false. Whereas the debate rages about the question, are we really good/bad, the hypotheses that we are good/bad are obviously false. Nevertheless, we can ask, does goodness prevail in crucial moments as optimists say or are the pessimists right? This once-essentialist question is interesting.

J. Agassi, *Ludwig Wittgenstein's Philosophical Investigations*, Synthese Library 401,
https://doi.org/10.1007/978-3-030-00117-9_3

surprisingly this proposal of mine tallies with comments made by even the most ardent and significant members of the analytic school, such as Neurath and Carnap, not to mention Hintikka: insofar as metaphysics was ancillary to science, they all agreed, they did not object to it. This concession fully reverses their radicalism. It was a great loss that they ignored Wisdom.

3.1 The Background to Anti-Metaphysics

Wittgenstein regularly presented his contribution as also a contribution to social welfare (Pichler and Säätelä 2005, 95). Decades after his death, Richard Rorty won the status of a leader within the analytic school due to his claim that philosophy has not ended, that it will remain a topic of conversation (Rorty 1979, 394). He increased his fame by moving from elitism to democracy. How democratic was Wittgenstein's philosophy? Is his anti-theology democratic? Since churches are often anti-democratic, perhaps yes; yet since in some versions popular theology encourages tolerance, perhaps not. Democracy must allow all of us the right to make mistakes. For, the aim in introducing it was to minimize the damage due to error, not to prevent all error in advance. Is rationalist philosophy (science included) opposed to religion? This is a difficult question. Traditional theology is the effort to answer it in a way friendly to both religion and reason. The introduction of the hostility to metaphysics, however, as the doctrine of prejudice, sharpened the question, and it emerged more forcefully than ever before. The scientific community then declared semi-officially a kind of truce between science and religion, a truce that Darwin's disciples broke semi-officially. Where religious tolerance dominated, this did not matter much. Elsewhere it did. Yet the situation had philosophical aspects that required attention. One of them was logic, no less; the theory of meaning, to be precise. Yet mature Wittgenstein said theology could not contradict science since it is meaningless; even two seemingly conflicting theologies, he said, cannot contradict each other (1938, 1) – being meaningless.

Philosophy began as rationalistic or as pro-science. Logic, Aristotle taught, is the soul of rationality. His logic is the logic of nouns, and his science is the search for the perfect taxonomy, a classificatory system, a system of nouns, and thus of things. Hence, logic naturally is the logic of science, the exposition of our knowledge: he took biology to be the paradigm case of scientific theory, seeing it as a taxonomy, as a theory of classification. Biologists before Darwin viewed the biological characteristics behind any proper hierarchical classification as resting on essential definitions.

Classical logic was out-of-date already when it appeared. It survived with the aid of the theory of essential definitions that was an integral part of its taxonomy. By definition, abstract entities dwell in Plato's Heaven. This raises a tough problem, as the term "essence" is Aristotelian and "form" is Platonic (they appear as interchangeable already in Aristotle's texts and still in Bacon's); Aristotle denied the existence of Plato's Heaven.[3] His ascription of essentialism to Socrates or to Plato concerned method: he acknowledges that he had inherited his doctrine of the method of seeking definitions from the Pythagoreans via Socrates. Definitions, he said, are statements concerning the nature of things. Platonic forms and Aristotelian essences compete as views of the nature of things, of definitions. The fusion of the views of Plato and of Aristotle on the nature of things is traditional. Boolean algebra absorbed traditional logic while removing the discussion of the nature of essences from logic to science. He did not disallow such discussion; he merely stressed that it was external to logic, that logicians, qua logicians, have no use for it. He replaced them with classes. His very first proposal was to allow the construction of any class we wish and thus also any subclass or superclass for it as we wish. Moreover, he declared that the identity of a class might be determined with no reference to the meanings of the terms involved, merely by pointing at its members. We may identify a class as one that satisfies some conditions, but we may determine it arbitrarily; and we declare identical two classes defined differently by reference to its members alone, thus:

Every two sets are identical iff all members of one are members of the other.

In symbols,

$$(\forall a)\,(\forall b)\,\big((a = b \text{ iff } (\forall x)\,(x \in a \equiv x \in b)\big).$$

(Note: the expression "iff" is in the meta-language; "≡" is in the object language; ordinarily, both translate to "if and only if". Addicted to ordinary language, Wittgenstein rejected the very idea of the meta-language.)

The initial objection of Antisthenes to the definition of Man as a talking animal was that Man is equally a featherless biped. This objection that demolishes classical essentialism by sending it to the impossible task of distinguishing between essential and other (accidental) definitions, Boole accepted wholeheartedly. Accordingly, in Boolean algebra, in direct opposition to essentialism, Boole's *Principle of Extensionality* holds: a class is identifiable by its members, and all classes are equally "legitimate".

This principle allows logicians to ignore class-members: it is possible to speak of classes and of sub-classes instead, thus:

Two classes are identical iff they have the same subsets.

To put it more explicitly,

Every two sets are identical iff they share all subsets.

---

[3]Plato's term is "*eidos*", meaning shape; Aristotle's term is "*ousia*", meaning substance or essence. To avoid ambiguity, Quine has suggested the quasi-medieval term "quidity" or "whatness" for idea-or-essence.

In symbols,

$$(\forall a)\,(\forall b)\,\big((a = b \text{ iff } (\forall c)\,(c \subseteq a \equiv c \subseteq b)\big).$$

The move from classical logic to the principle of extensionality is very radical and hides many radical transformations of logic (Bar-Am 2008, 123). The chief radical transformation is from classical logic (of terms) to modern logic (of statements).[4] The elimination of metaphysics moved from analysis old-style to analysis new-style, from adverse terms (with no reference) to adverse statements (pseudo-statements). Both ideas fail. First, as long as we permit with no condition the concoction of compound terms we cannot exclude terms that do not denote ("goat-stag"). Second, the same goes for the composition of metaphysical statements: whatever metaphysical statements are, it is impossible to guarantee that the rules of statement composition will forbid composing or deducing them from proper statements.[5]

The modern idea that truth-value is a property of a statement (or a proposition) is so well entrenched that it conceals the historical fact that Plato discusses truth as a property of names (*Cratylus*, *Phaedrus* 265e). Ordinary parlance is full of instances of this, such as the expressions "true friend" and "true friendship" (where truth is more or less genuineness). Yet the discussion in question is of names rather than of nouns, and then it is a theory of reference right away: Tom's true name is "Thomas", and that name is true of him, not of Dick or of Harry. Plato's idea that a true name, the truth of a name, is language-invariant, introduces the concept of the ideal language at once and without deliberation. This is a serious defect, tolerated because the discussion concerns not truth by convention but truth by nature, not the received notion but the real one. Plato presented as central to all philosophy that dichotomy between truth by nature and truth by convention and the search for the natural in disregard for the conventional. He restated this dichotomy with regard to language, despite the obviously conventional character of names.[6] This way Plato imposed equating the view of the truth of a statement with the truth of a name: nouns are the essence of language, and nouns are essentially proper names. Now a name can be true or false, depending on its reference. To know if it is true, we need a pointer to the named object. The pointer may be an act of pointing, known as "ostensive definition" (though it is no definition, of course), and it may be the word "this". The knowledge of the designation of a name is what Russell christened knowledge by acquaintance (despite his view that statements, not names, are objects of knowledge). Now Plato said, true names name ideas; thus, when we say, "Socrates is wise"

[4]The distinction between old-style and new-style analyses is neither exclusive nor exhaustive. In particular, both traditions include discussions of whether the verb "exist" is a predicate.

[5]Young Wittgenstein claimed to have done so by constructing a system with a very poor language plus some observation terms like names of colors or sounds, and nothing else: not even nouns or proper names.

[6]Unlike definite descriptions, Linsky observed, proper names are *ad hoc*. Hence, Kripke 1972, 127 was in error when he recommended conflating them.

we really mean "Socrates is wisdom", where "Socrates" is a pointer and "wisdom" (or any equivalent word) is a (true) name. This is too unsatisfactory. Yet it proves exact the terminology of Aristotle's famous dictum (*Metaphysica Gamma*, 7):

> It is false to say of that which it is that it is not or of that which it is not that it is, and it is true to say of that which it is that it is or of that which it is not that it is not.

(The modern version of this replaces some uses of the word "that" here with "the proposition".) This hardly differs from Plato's assignments of names (*Cratylus* 431b):

> The right assignment of them we may call truth, and the wrong assignment of them falsehood.

Plato stresses there (439b) that "the knowledge of things is not to be derived from names": knowledge of ideas (essences) comes before names. Thus, knowledge is of ideas, not of the words are arbitrarily used to denote them.

In Aristotle's view, essential definitions of concepts serve as both explanations and knowledge. He also offered a theory of probable knowledge that is not necessarily of essences and thus not of concepts as such. This looks clearer, but is not clear at all. Plato's treatment of names (*Theaetetus*, e.g. 202 and 207) is less clear: do proper names differ from what we call nouns or class names? How? According to medieval (and Renaissance) logic, terms designate the objects they name. It distinguishes between three kinds of categorical (subject-predicate) statements. Singular statements use proper names for subjects; universal and particular statement use only nouns. Particular statements have unspecified subjects: the classical medieval discussions on the matter are confused. They centered on a different matter: the dispute between the realists, who followed Plato and the nominalists who followed Aristotle: both agreed that terms (should) name directly ideas/essences and only indirectly things too. The question was, do terms name only things or also ideas/essences? If ideas/essences should have names, do they need separate names? This was the major problem in the medieval theory of meaning and the major dispute in logic.

Happily, Boolean algebra leaves all this behind. It includes two separate lists of names: of objects and of classes. It thus calls for no discussion of reference or of designation. It is thus no theory of meaning. As classical logic is profoundly a theory of meanings, with a theory of essential definitions as its culmination, and as Boolean algebra ignores meanings altogether, the transition between them was stunning. The new was too powerful to ignore and it left a lacuna.

The new logic is forceful in many ways. For one thing, it is clear and unambiguous. Unlike classical logic, it distinguishes once and for good between objects and classes, and between class membership (Socrates is mortal), class inclusion (all Greeks are mortal), class equivalence (All and only Spartans are brave; no taxation without representation) and diverse kinds of identity (water is H_2O; Humans are rational animals; Greeks are Helens; Marcus Tullius is Cicero). Although these distinctions are central to the past progress of logic, the ordinary use of copulas (the words "is"/"are"/"is not"/"are not") is usually unobjectionable.[7]

Traditional logic was stagnant, partly because of pervasive confusions. Clearly, however, the terms that Boolean algebra discusses are hierarchical: it allows only (whatever we recognize as) objects as class members; a class can be a subclass of another class, not its members. (Abstract set theory allows classes to be members of classes.) In biology, taxonomy comprises an Aristotelian subdivision (*definitio fiat per genus proximum et differentiam specificam*) of the universal class of living things neatly into subclasses.

The need to differentiate class membership from class inclusion clearly demands a new terminology: "$\in$" (the Greek letter epsilon) to denote class membership and "$\subseteq$" to denote class inclusion (and also "$\subset$" (the horseshoe) for the proper subclass relation). The term Aristotle introduced to denote a high superclass such as "living things" is "category". (Literally the word means predicate; roughly, categories are predicates that have very wide extensions.) Unfortunately, the doctrine of "the" categories is very vague; complaints about its vagueness were voiced already in Antiquity, although not clearly. Modern logic makes the shortcomings of that doctrine trivial, and this complicates careful and empathic historical reconstructions. It may well be the most confused part of the Aristotelian corpus. However, we have no record of a deviation from it before Kant and Hegel, who claimed to have offered alternatives to it. Boolean algebra renders any theory of the categories superfluous; it therefore waned and dropped out of the logic textbook within a relatively very short span, regrettably with very few comments about its shortcomings.[8] Many who are unschooled in modern logic are tempted to return to it or at least to discuss it; this is still the chief obstacle that prevents dialogue between the familiar with modern logic and others, including many leaders among the followers of the Kantian and Hegelian traditions. Rorty and his friends attempted to bridge the gap between analytic philosophy and these traditions. This initiative sounds possible as it comes to remedy the quick dismissal of traditional logic. It is insufficient. A clear discussion of the problems is necessary, and these will involve determination and scope of terms (see below). Thus far, neither party is interested. Hintikka's call to study scope failed, to his great disappointment.[9]

Anti-essentialism grants some reasonableness to the idea that metaphysical/theological terms denote nothing and so are meaningless, so that their elimination from language will incur no loss and may be beneficial; but only with the recognition that the idea is false. To repeat, the idea belongs to the anti-metaphysics tradition of the seventeenth and the eighteenth centuries. Kant terminated it. Bolzano and his

---

[7]Wittgenstein said, he loved clarity, meaning, he wanted all vagueness cleared, even if it is harmless (Sass 2001, 110, 115), yet he never complained about the vagueness of the ordinary copula.

[8]The most remarkable exceptions being Boole's and De Morgan's notes on it.

[9]"No Scope for Scope?" Perhaps Hintikka's failure to raise interest in scope is due to his view that the only way to handle scope properly was to reform Frege's theory of the quantifiers: others found it prudent to wait for less radical a solution.

followers took Kant's idea seriously, and Boole allowed for names that do not denote even before his epoch-making introduction of the empty class (the class that has no members).[10] This is the demise of essentialism. Students of classical logic still find it hard to notice that though the empty class denotes nothing, it is not nothing. This raises the question, what is it? What is a class? This question has led some modern logician (Quine 1947, 1990, 27–8) to Platonism-without-essential-definitions of sorts.

3.2 Anti-essentialism

The concept of a class ("lions") seems paradoxically more intuitive – less abstract – than that of a common name ("a lion"). This is deceptive: a common name denotes both a class and any individual member of it. Boolean algebra has rendered this obvious. It led to the development of modern logic far beyond the dreams of its pioneers, simply because essentialism was pervasive and so its elimination invited a radical overhaul. It thus raised anew the problem of meaning: what does a class name denote? To let a class-name denote a class is to side with essentialism against nominalism. Therefore, it seems Boolean algebra brings back the problem of universals. Not so, as Boolean classes differ from Aristotelian allegedly natural taxa.

For the versed in traditional logic, Boolean algebra was disturbing. It rendered the three basic laws of logic obvious (a = a; a and non-a = 0; a or non-a = 1) and a part of a rich set of such statements (the most striking of which was de Morgan's rules: not (a and b) = not a or not b; not (a or b) = not a and not b; etc.). And yet the loss is obvious: the whole doctrine of the syllogism has disappeared. (They turned into statements; thus, Barbara, all men are mortal; all Greeks are men; therefore, all Greeks area mortals, became the statement {(a ⊆ b and b ⊆ c) /(a ⊆ c)} replaces the following inference: the statements {(a and not b = 0) and (b and not c = 0) together entail (a and not c = 0)}).

Two technical terms enter here. First, the conditional statement, if a then b, in symbols, a ⊃ b; this is a name of a sentence of a certain form. It is constructed out of two sentences, a and b, by writing the word "if" and then the words of the sentence a and then the word "then" and then the words of b. For example, when a is the name of the sentence "today is Tuesday" and b is the name of the sentence "I am a king" then if-a-then-b is the name of "if today is Tuesday then I am a king". When a is "in the beginning there was the word" and b is "the word was God", then if a then b is "if in the beginning there was the word then the word was God". The reason we name this kind of sentence this way is to remind us how we have constructed it. At times, it may be better to use another name, such as "c", and then "c" is the same sentence as "if a then b".

The other technical term is "entails":

---

[10] See the first three lines of the first chapter of Boole's *The Mathematical Analysis of Logic,* p. 15: "...every conceivable class of objects weather actually existing or not..."

"a entails b",[11] "a, therefore b" and "b follows from a" are all synonyms.

In symbols,

$$a/b.$$

Another relation appears already in Aristotelian logic:

a and b are mutually deducible

In symbols,

$$a//b.$$

Thus, "all humans are mortal" and "no human is immortal" are mutually deducible:

"all humans are mortal" // "no human is immortal".

Notice that the same idea may appear thus: "all humans are mortal iff no human is immortal", except that instead of speaking of two sentences now we have one. It looks tempting to say, (a // b) iff (a iff b). This is ungrammatical because the first pair of brackets contain a sentence and the second pair of brackets contains a name. To correct this, we may write instead of the name "a iff b" the sentence {"a iff b" is true}: {a // b} iff {(a iff b) is true}. This is grammatical but it is false in case a is "it rains" and b is "the pavements are wet": a iff b is true, but a and b do not follow from each other.

Back to Boolean algebra. Its pioneers were not anti-essentialists, nor were they revolutionaries. The inner logic to Boole's principle of election – all classes are equally legitimate – led to a revolution (Bar-Am 2008, §16). This was a break away from obligatory essentialism; it was far-reaching. The central example for this is the logic of relations. A part of the essentialist view of language was the hope that statements of relations are at bottom nothing but categorical. This hope is the source of the failure of the great Leibniz, who began the study of the logic of relations. By contrast, Augustus De Morgan, the co-founder of Boolean algebra, said clearly the opposite, and from the very start. Boolean algebra could not serve as a logic of relations; Frege improved matters a few decades later, as he replaced the predicates of classical logic with n-place predicates with n being any number. Obligatory essentialism impeded also the progress of the doctrine of inference, and the essence of inference was troublesome ever since Antiquity. Unfortunately, Boolean algebra did not clear the situation; perhaps it is still problematic. Boolean algebra did not include a theory of inference: it replaced traditional rules with intuitive ones: tautologies are true and they take over the role of traditional inferences via a simple version of the deduction theorem intuitively taken for granted. That theorem says, an inference can appear as a tautological conditional:

[11]The word "entail" differs in logical and in legal discourses. Legally one may speak of actions entailing obligations, for example. Not so in logic, where only statements entail statements. Of course, it is easy to see similarity here, but let us ignore it.

a entails b iff if a then b is logically true.

In symbols,

$$a/b \text{ iff } \vdash a \supset b.$$

The theory of meaning as reference makes the theory of language a theory of terms: nouns comprise the basic part language. Aristotle's logic, resting on this theory, is a logic of categorical (subject-predicate) statements: its rules inference are syllogisms, all variations on "'all S are M', 'all M are P', therefore, 'all S are P'" (known traditionally as Barbara). Already medieval scholars realized that some valid inferences are not syllogisms. The most famous of these is the *modus ponens*[12]:

a, if a then b; therefore b;

In symbols,

$$a, a \supset b/b.$$

It became central in modern logic when Boole did away with the syllogisms. This makes inferences in Boolean algebra intuitive like in geometry and the Aristotelian theory of inference disappears. Frege compensated for the loss by the use of the *modus ponens* as the only rule of inference.

The results were greatly liberating, but puzzling too: its great asset was that its inferences were purely formal, yet it included no theory of inference. Frege mitigated this trouble somewhat by describing inference as truth preserving: it was no news that the truth of the premises in an inference whose form is valid guarantees the truth of its conclusions. Frege reversed the idea: any inference form that does that with no exception is valid. The syllogism became then no longer "*the*" proper valid form of inference. Frege used substitution and the *modus ponens* as the only rules for his system, and only for deriving truths from known truths; this is by far not a theory of inference. He used it as a mere interim stage, as he saw that it sufficed for his purpose – of proving that arithmetic is analytic. The theory of inference was then left fallow for a few decades, and logic centred on logically true sentences instead. Russell was trying desperately to explain what these are and here Wittgenstein, then a student of Russell, made his first contribution. It would be impossible without the elimination of obligatory essentialism.

Officially, obligatory essentialism enters logic through the doctrine of definitions. They describe essences and new terms: this way terms denote essences. Does every legitimate term name an essence? This is not clear. Aristotle claimed to have allowed names to denote either essences or accidents (properties that are not essential). He also had a theory of primary and secondary essences, which perhaps messes things

[12]Arguably, already Aristotle realized that; and already the stoics developed an early version of the logic of propositions; yet it remained marginal; a kind of curiosity. All that remained of it that signified somewhat was the *modus ponens*.

up more. Leibniz rectified this by declaring all proper terms names of essences and putting all essences on a par. Already the scholastics offered a few conditions for essential definitions. These are overruled as essentialism ceased to be obligatory.

Before that, when essentialism was problematic but still daily fare, it was challenging. The idea of modern logic is the idea of the utterly formal, of the computable. It turned to bring about new problems of its own, but these were strictly different from the utterly formal as represented by the computer. And the first step towards this change was Boole's removal of essences from logic to metaphysics and the next was the elimination of the doctrine of judgment from logic, where judgment is the use of healthy sense to dismiss glitches in logic, such as the liar's paradox (sentence a: "sentence a is false").

The elimination of essentialism and judgment from logic imposed attention to the degenerative power of the so-called "existential import". Traditional logic allows inferring "some x are y" from "all x are y". Now "some" means, "there exists at least one" so as to render "all a are b" and "some a are not b" contradictory, the way "no a is b" and "some a are b" are likewise contradictory. That the inference of "some x are y" from "all x are y" is invalid is thus all too obvious: "all mermaids are maids" is true but not "some mermaids are maids". This was taken up rather lately, and it turned out to be a most significant opening of new horizons. For, what makes existential import seem a valid inference is its validity for all cases except for empty terms, and we often (but not always) tend to avoid using empty terms. Yet we do so only when we know that they are empty. Traditional logic ignores the empty class (very much the way traditional mathematicians ignored the number zero). This was the source of the dissatisfaction with the traditional referential theory of meaning and thus for the rise of analytic philosophy. Already Aristotle was aware of names that have no reference, of names that do not designate: he spoke of the square-circle and of goat-stags. The claim that logic discusses essences allowed the persistence of the traditional theory of meaning, and the pretense to know what there is before logical inquiry begins. For, when the question arose, what does an empty term denote? It led to the scholastic dispute over the question, is the essence of the non-existent object possible? The traditional wording of this question was, does essence precede existence?[13] Any inconsistency and any other inadequacy of logic was set aside with reference to Aristotle's distinction[14] between syllogism and enthymeme (= incomplete syllogism) that is the theory of judgment (Sigwart 1895, 178–80).

[13]Existentialists revived this terminology in talks on the possibility to alter character (=essence). This is a new use of an old name; it is responsible for much confusion.

[14]Since this distinction is a part of Aristotle's rhetoric, it was often ignored; the transition from syllogism to enthymeme was then furtive; not surprisingly, then, its rejection was tacit. Kant's theory of judgment went together with his austere view of logic that forced logicians to increase precision gradually and thus dismiss Aristotle incidentally. The most striking instance of change follows the comparison between Mach's dismissal of paradoxes and the view of them a major catastrophe that ensuing logic took for granted.

Classical logic, an inadequate and erroneous conflation of brilliant formal logical insights with metaphysical hubris, was endorsed for millennia as the most general,[15] all-embracing and far-reaching foundation that science can ever have. This grand theory gave way to a new logic. It is clearly better but also with the obviously great limitations that began with Boole's barring of essences from logic. It is impossible to comprehend the work of Russell and Wittgenstein without noticing that the development of Boolean algebra denuded logic of its metaphysical component and that Russell was trying to return to logic its status as overall philosophy, as he indicated in his 1914 *Our Knowledge of the External World as a Field for Scientific Method in Philosophy*. The title of its Chapter 2 is, "Logic as the Essence of Philosophy". The successful part of the program was the irreducibility of two-place predicates as already Frege had practiced it.

Wittgenstein found deeply impressive the contrast between the bombast of the old logic and the prosaic character of the new. (This enabled him to dismiss the old logic with scarcely any knowledge of it.) Russell and Wittgenstein worked together on the idea that logic will engulf metaphysics and thus replace it. Russell soon gave it up.[16] Wittgenstein never did (*PI*, §371–3: "Essence is expressed by grammar. . . . Grammar tells what kind of object anything is."). He never gave up Russell's program as expressed in 1914. It was his finest hour: though an undergraduate student, he took the part of Russell's program that was quite impressive: embed the demand for verification in the ideal language of logic (*TLP*, § 5.5563; *PI*, §97).[17]

3.3 Existential Import

Existential import is the most problematic of presuppositions of classical logic. The standard excuse for it is that discussing non-existent entities is pointless and may even lead to absurdities, such as a sentence and its negation both being false. Modern logic strengthens this excuse, since it allows attributing any property to

[15]Kant declared logic, arithmetic and geometry complete and final. He included Newtonian mechanics within geometry (*Prolegomena* §38). He wished chemistry to be scientific (i.e., worded mathematically).

[16]This contrast connects with another: Russell gave up the hope to establish certainty; Wittgenstein did not.

[17]"In pure logic, which, however, will be very briefly discussed in these lectures, I have had the benefit of vitally important discoveries, not yet published, by my friend Mr. Ludwig Wittgenstein" (Russell 1914, end of Preface);

"In the above remarks I am making use of unpublished work by my friend Ludwig Wittgenstein" (Note 55, appended to the remark that the logical constants are not names).

the non-existent.[18] Nevertheless, discussing non-existent entities is far from pointless: we want to discuss the advisability of creating a new kind of machine; we want to discuss the impossibilities of certain constructs (e.g. the greatest prime number or the energy-generating machine), that is to say, we may want to know of any object that we can conceive of, whether it may exist. As we do not know what the furniture of the world is, and as we expect the use of logic to help us improve our views of the world, logic should not prevent the discussion of the possibly non-existent. If it does, we have committed the essentialist fallacy: allowing our empirical conjectures and our metaphysical preconceptions to function as presuppositions to logic. This leads to a hopeless conflation of our methods and our goals. The wish to know whether atomism is true, for example, leads to efforts to generate refutable versions of it – by ascribing to atoms certain qualities and by assuming that they exist. (Remember: the non-existent has any property.) It would be therefore impossible to say that atoms do not exist had logic forbidden empty nouns.

Universal statements do not say what exists. They only say of certain things that they do not exist; existential statements assert that some things do exist. In the study of the empirical world, a refutation of a universal theory is possible only because we consider assertions about observed things as implying their existence. Hence, science is inclined to realism. Notoriously, however, observations may be erroneous: observations of things in dreams, mirages and in other sense illusions are sufficiently familiar. Whether an observation report is true is a question of fact open to further study. What is interesting here is that things are different in logic and in mathematics. Mathematics admits the existence of everything that is consistently conceivable, that is to say it excludes from its ontology only those items whose existence leads to contradiction. (Axiom systems usually include axioms asserting the existence of the objects that they discuss. Thus, Boolean algebra asserts that its universe of discourse is not empty: $V \neq \Lambda$.) This assertion is very different from the Aristotelian existential import: what Frege and Russell called existence is not what we mean when we discuss the question whether Homer existed or not. A still different sense of existence is Aristotle's notion of existence that implied that we must have some insight into the furniture of the real world before our logical quest can begin. Hence, we may dispense with the concept of existence of, say, numbers, or leave it and postulate for sticks and stones an additional, different kind of existence (Lejewski 1954). In any case, the existence of abstract entities like numbers and institutions, the love of truth and symphonies, may recommend a diversity of existence.

The case of problematic names has gained more attention even than the case of fictitious names. Talking about Samuel Pickwick, the hero of *The Pickwick Papers*,

[18]Proof: the assertions "x possess the property p" and "x does not possess the property p" contradict each other unless "x exists" is false. This sounds unintuitive but examples for it are abundant: mermaids are both human and non-human; the impossible event will happen at the end of the days.

the famous novel by Charles Dickens,[19] as if he exists is no more problematic than *reductio ad absurdum*, the method of proof that opens with assuming the negation of the theorem to prove. Empty names are essential even in ordinary discourse: historians have the right to discuss the possibility that Homer is fictitious. Allowing for this may cause confusion. Thus, when the herald in Lewis Carroll's *Alice in Wonderland* declares that nobody is coming, and the king is impatient, as he is waiting for Nobody in vain, the king is confused, presumably because he does not allow for a noun with no reference. Clearly, the word "nobody" as used here[20] is not a proper noun. Proper nouns exist that name nobody, such as the name of Samuel Pickwick. What are they? If we do not allow ourselves to name anyone and anything unless we know that they exist, then we thereby greatly limit our language, indeed we may be turning it into a formal deductive system (Bar-Am and Agassi 2014). We do not want logic to prejudge such matters. At least one leading logician, Quine, has found names so irksome that he suggested (2008, 499) to do away with all of them (by converting, say, the unique Pegasus to the only item that pegasizes, that has the proberty of being a pegasus. We may express this idea thus: $(Ex)p(x) \bullet (y)(p(y) \supset y = x))$.

Existential import has troubled quite a few philosophers, perhaps because they found the criticism of Aristotelian logic excessive. Discussions of Aristotle's philosophy strike me as excessively emotional. All this sound and fury is quite unnecessary, since Aristotelian logic allows for *ad hoc* rectification of its defects. This is the already mentioned theory of judgment. Others seem to have found existential import troublesome as if it shows some defect in ordinary language. This is not quite true, and more-or-less for the same reason. What is troublesome is not our everyday language but rather the hope to have an everyday-language that grammatically excludes noun phrases that have no denotation. Usually, everyday discourse is too context-dependent to yield simple rules with no exceptions. How Wittgenstein's disciples handle this fact is a different matter. There are quite a few alternatives to the formal logic taught in introductory logic courses. We need not discuss these. Rather, we should glance at traditional logic in its best wording. This may help us better comprehend Wittgenstein's *Philosophical Investigations*.

In Boolean algebra, the question of individual names does not arise: in its simplest version it handles classes only, and names in it are then class names, not individual names. Individuals are introduced there, if need be, by unit classes, so that in it "Socrates" denotes the class which has one and only one member, namely, Socrates: it is a non-empty class all of whose members are identical with each other.

---

[19]*The Pickwick Papers* by Dickens (1837) introduces the term "in a Pickwickian sense" that means not really.

[20]Daniel Cohen observes this. The great Wavy Gravy – once a poet named Hugh Romney, and always called Mr. Gravy by the *New York Times* – ran a campaign "Nobody for President". Because it was almost a truism that "Nobody cares about you" "Nobody will solve income inequality" "Nobody gives a damn" and so on, so that by the time these truisms add up, one almost had a picture of NOBODY (Ms. Nobody let us say) that captured your vote. "Nobody's perfect". In the Bay Area Wavy Gravy is still a historic (hippie) and compassionate activist icon.

The Boolean wording of categorical (subject-predicate) statements is so much simpler and clearer so that it demanded the correction of the classical error. Within Boolean algebra, classical statements appear as assertions of non-existence or of existence of members of different classes, as a class must be empty or not:

$$\mathrm{x} = 0, \text{namely, x has no members.}$$

Alternatively,

$$\sim (\mathrm{x} = 0), \text{namely, } \mathrm{x} \neq 0.$$

For compound classes there are eight options. Thus,

$$\begin{aligned}&\mathrm{x} \circ \mathrm{y} = 0 \,\text{or}\, \mathrm{x} \circ \mathrm{y} \neq 0; \qquad \sim \mathrm{x} \circ \mathrm{y} = 0 \,\text{or}\, \mathrm{x} \circ \mathrm{y} \neq 0;\\ &\mathrm{x} \circ \sim \mathrm{y} = 0 \,\text{or}\, \mathrm{x} \circ \sim \mathrm{y} \neq 0; \sim \mathrm{x} \circ \sim \mathrm{y} = 0 \,\text{or} \sim\\ &\mathrm{x} \circ \sim \mathrm{y} \neq 0.\end{aligned}$$

The signs "$= 0$" and "$\neq 0$" describe universals and particulars respectively, non-existence and existence respectively. Thus, "all a are b" is a $\circ{\sim}\mathrm{b} = 0$ and its negation "some a are b" is a $\circ{\sim}\mathrm{b} \neq 0$.

To complete the picture one needs to designate complements of compound classes. The most famous example is De Morgan's theorems, for example:

$$(\sim \mathrm{a} \circ \sim \mathrm{b}) \equiv (\sim (\mathrm{a} \cup \mathrm{b})); (\sim \mathrm{a} \cup \sim \mathrm{b})(\sim (\mathrm{a} \circ \mathrm{b})); \text{and so on.}$$

These theorems go well beyond classical logic; they face no problem of existence; the problem reappears in a wider context.

Perhaps it is best to speak of things in the subjunctive mood, as if they exist, without asserting that they do.[21] This is apparently a reasonable proposal. Strangely, traditional logic forbids it. Nor is it easy: Russell began the new philosophical analysis when he decided to agree with classical logic about names, except that he limited his discourse to proper names and to descriptions of unique entities. In retrospect, it is therefore difficult to see why classical logic excluded nouns that designate nothing, as it seems so unreasonable. Yet the fact is that the correction of this error started an avalanche of disputes that are still going strong. We may thus sympathize with the traditional error.

Consider then a name that possibly denotes nothing. It provides an example of the kind of exercise known as [philosophical or logical] analysis. It is a process different from what Newton called analysis, which is the process of empirical induction or of mathematical proof: it is not supposed to lead to discovery. Still, Russell performed a few philosophical or logical analyses that were very interesting and very useful to the development of mathematics. All his life he regretted that he had never made a discovery within mathematics proper. (It turns out, incidentally, that he did. This,

[21]Leading philosopher Husserl called "bracketing" his act of suspending judgment about existence.

however, is nitpicking.) He did not consider philosophical or logical analysis part of mathematics proper. He took the kind of analysis that he had inaugurated to be specifically philosophical or logical. Yet it was through his analysis that he came to develop a formal language, as distinct from the idea of formal systems.[22] The difference between the two is obvious and clear-cut: a language proper should include as legitimate, that is, as meaningful grammatical constructs, false sentences, including contradictions; whereas a mathematical deductive system should be consistent. It was hard to consider a contradiction legitimate and Russell did it as he distinguished between a contradiction that is refutable yet legitimate and a contradiction that is provable – a paradox – that is unmanageable and that somehow slipped into our deductive system with disastrous results.

Boolean algebra became a more or less formal calculus of classes with Boole's introduction of two complementary classes, the universal class (the class that contains everything) and its complement (the class that contains nothing). They occupied a prominent place at once. The question raised in the literature repeatedly was, whether the empty class was a proper class? It is a discussion parallel to the discussion of the question, is zero a proper number, that is to say a discussion that mathematicians and accountants are oblivious to, while philosophers may fervently partake in? Moreover, is zero positive or negative? These questions are essentialist, or else they rest on the obligation to stick to ordinary use (that is impossible to comply with prior to the specification of the ordinary; Watkins 1957b). It is also objectionable since it conflicts with the liberty to use terms freely,[23] and it is most inconvenient to refuse to mention an empty class or the number zero. Of course, there is a price to pay for including the empty class in the family of respectable classes – just as there is a price to pay for including zero in the family of respectable numbers. This last point is part of ordinary algebra: when dividing one number by another we should take care to ensure that the divisor is not zero. This burden is due to our decision to consider zero a number. It is a price that mathematicians are ready to pay. Their usage has entered ordinary discourse, even though not quite systematically. In some ordinary contexts, we still deem speaking of zero as a number as illegitimate: the statement "I have written a number of books", means "the number of books that I have written is equal to or greater than 2". The same holds for the place of the empty class in ordinary parlance. Consider "Smoking outside the designated areas is forbidden". The exclusion of the empty class implies that designated areas exist. Do they? How do people understand this? Empirical evidence suggests that the

---

[22]Hilbert had developed a formal presentation of geometry in 1899. Whitehead and Russell developed a formal language in 1910. The contribution of Frege was most significant for both. He was the first to state the distinction between a formal system and a formal language and even came close to having developed the first formal language proper in his 1879 *Concept-Script: A Formal Language for Pure Thought Modeled on that of Arithmetic*.

[23]The liberty to use terms freely should be obvious but still is not. Lewis Carroll has voiced this:

> 'When I use a word,' Humpty Dumpty said in rather a scornful tone, 'it means just what I choose it to mean – neither more nor less.' 'The question is,' said Alice, 'whether you can make words mean so many different things.' 'The question is,' said Humpty Dumpty, 'which is to be master – that's all.'

educated diverge from the common here, justifying Russell's (1953, 304) contention that ordinary language depends on the level of education. This holds particularly in logic, where parlance is often stilted.

3.4 On Stilted Language

Modern logic begins (1847) with Boolean algebra as a calculus of classes, a part of an artificial language of sorts: before the categorical statement "all humans are mortal", say, enters the calculus it undergoes translation. It becomes the statement, "the class Humans is a subclass of the class Mortals", or, alternatively, "the class of immortal humans is empty". This is a bit awkward and it is a stilted manner of speaking; it may bother those who do not see the advantage of the use of stilted speech. (They often hate legalese no less.) The bother may then lessen with the realization that traditional logic too was conceived as an effort towards a stilted scientific parlance; at first sight, "all humans are mortal" may not seem stilted, but let us replace words by their complements. A complement of a human is something that is not human, and so when we wish to speak of the complement of "humans" we have to speak of things that are not human, whom we call at times "inhuman", "non-human" and "not-human". The choice may depend on the place of a word in a statement: we tend to speak of non-humans when we use "non-humans" as a subject, but when we want to use the word as a predicate we prefer the word "inhuman" to "non-humans" or "not human". It really does not matter if we center on information, and if, to that end we agree to ignore niceties. There is no trouble here, but both logicians and poets may find upsetting this logically irrelevant variance. Logicians insist that for a syllogism to be valid a word in it has to appear more than once; at times poets want to avoid this repetition by replacing a repeated word with a synonym. Whether a given expression is awkward or elegant is not always a local phenomenon: some experience with writing brings about familiarity with this: a long discourse may render an initially awkward expression elegant. The rewording of an expression in a text to avoid stilted language may easily cause the language in another part of the text become excessively stilted. This raises a question: where is it preferable to introduce an element of artificiality in the text for the sake of a (seemingly) "natural" flow? This question has no general answer. There is no general rule that dispels the awkward sense that formal logic often conjures up. The question whether to change direct speech to indirect speech in one place of a text or to change indirect speech to direct in another frequently occurs to writers who face the task of introducing some parallelism into their texts. Readers not in the habit of writing who read only well-written texts may be unaware of all this. The new student of logic or philosophy of language may bump into what seems a needlessly irritating artificiality. The artificiality comes to make the application of rules rigid, which is fine for students of rules. This fact troubled Wittgenstein very much, as is clear from many parts of his *Philosophical Investigations.*

Already Plato found all this troubling. He demanded that a name should name an idea (essence) but he had to admit in displeasure that the choice of a word to represent it is arbitrary. Students of logic lose their sensitivity to the stilted language used in logic. In introductory modern logic, instructors expect students to translate familiar statements into the stilted language that logicians prefer. Usually, this happens without discussion or explanation. This way, instructors convey the impression that there is no trouble concerning the matter of translation back and forth between smooth and stilted statements, much less debate about it. This is false. Quine has produced lengthy discussions of the problems it involves, claiming that no translation is smooth: "*traduttore traditore*" says the Italian proverb: all translation is treason. Any teacher of logic can tell you that many students fail formalization exercises, at least for a while, until they get certain awkward translation standards internalized, not suspecting that these involve translation conventions that are still under dispute among experts.

In mathematical physics, the translation of a problem in physics to the language of complex numbers may help solve it in disregard for the physics it represents; to recover the lost meaning of the equation one then converts the solution back from the language of complex numbers to the language in which meanings are again obvious. This is very useful for calculations and is unproblematic for physics, so that physicists are happy with it and they do not complain. Should grammarians also be pleased with the logicians who convert familiar statements to stilted parlance and back? At least Quine objected. He said, any translation might be problematic, and this indicates that formal and natural languages are very different entities.

The reform of language goes after clarity and precision; this is not always called for (*PI,* §§69–71); at times it clashes with flexibility that is useful too (*PI,* §§81–1). The funniest case is the disjunction of a class with itself, the class a or a. It hardly ever comes up in ordinary situations,[24] but certain proofs may require it, and naturally a discussion of any case of a or b that does not explicitly exclude the option b = a allows a or b to become on occasion a or a. When the disjunction is exclusive, then either a or a but not both is the empty class but when the disjunction is inclusive then a or a is the same as a. Likewise, a and a is a, or $a^2 = a$. This formula is so important that it has a name: *idempotence,* (Boole called it "the index law", see Boole 1847, p. 18).[25] If we want numbers to represent Boolean algebra, then it seems easy to restrict the numbers to 0 or 1, since these are the solutions to the equation $x^2 = x$ that expresses idempotence. As it happens, two numbers, 0 and 1 will do nicely for no and yes. (When discussing support, indifference and opposition, say for a hypothesis from empirical evidence, then the use of three numbers (−1, 0 and 1) may very well be more useful.)

[24]Were Shakespeare's neologisms natural or stilted? Some he meant as stilted yet they became common. So is the vulgar "am I right or am I right?"

[25]TLP used a funnier formula: neither a nor a, which is the same as non-a. He did that since he wanted to use one undefined symbol, Sheffer's dagger (for neither-nor).

The general picture of the rise of Boolean algebra begins with clarifying the copula and the introduction of a few simplifying symbols. The result was exhilarating, since it included the liberation of logic from obligatory essentialism, and it raised new doubts, cast even on the strongest intuitions. The present study is not the place to air them. My discussion of them here comes to help comprehend the move of Wittgenstein from his early naturalism to his later pluralist view of language.

Essentialists defended the awkward expressions that logic employs by reference to some (metaphysical?) intuitions that allegedly justify them. (It is amazingly easy for philosophers to consider a metaphysical idea non-metaphysical simply because they happen to endorse it. But let us skip that here.) Modern logic has introduced new awkwardness, such as the sentence "there is a thing that is a man and has grey trousers", to quote a tidbit that displeased Wittgenstein (1975, 268). Learning the stilted language of modern logic was thus disturbing; logic teachers impose it in the hope to grant insight into the inner workings of all human languages. Later on, some logicians found this naturalism untenable. Wittgenstein drew the farthest consequences from this and declared the possibility of a multitude of logics, that "we use the words 'all' and 'any' in a mass of different ways of which Russell takes no account" (*ibid.* 269). It was probably much more unusual then to say, "There must not be a Russellian and a non-Russellian logic, in the way in which there is a Euclidean and a non-Euclidean geometry" (*ibid.* 172). This renders metaphorical Wittgenstein's 1969, §56 "everything descriptive of a language-game is part of logic". This situation raises perplexing questions: what (if anything) is the strictly literal meaning of this observation and what is its role within the teaching of Wittgenstein. The answers to this question are controversial, and all parties to the controversy admit that it is significant, but they disagree on why.

There is some advantage to stilted speech (remember the proverbial monotonous "it does not compute"), perhaps for some stilted conduct in general, for putting people and things in some ready-made pigeonholes, which is stereotyping. This causes strong reactions, at times very hostile, but with very little argument. Thought, after all is virtually impossible without pigeonholes; in a sense, this holds for all meaningful words. Nevertheless, the objection is understandable: things are complicated and sometimes objections are hard to articulate, much less to discuss. Yet, to repeat: it is impossible to avoid stereotyping; it is only possible – and greatly advisable – to increase awareness of the shortcoming of our commonsense modes of thinking and perhaps from time to time try to be critical of this or that way of stereotyping, in attempts to improve matters somewhat. This holds particularly for prejudices – including those against people and against cultures, including languages.[26] Nor is this all: psychologically, it is common to hear not the exact wording of a message but its corrected version: without noticing, listeners delete grunts and repetitions and correct what seem grammatical errors – at times erroneously. Wittgenstein praised analysis as the method of revealing the formal structures that we

[26]Are dialects, jargons, pidgin, creole, koiné, Swahili, are these stilted or natural? Is PMia a kind of pidgin?

follow in daily discourse, and he first deemed them the ones used in mathematics too. What he thought later in life is under dispute, but it is agreed that he chased grammatical errors that masqueraded as deep thoughts, and in hostility to pretense he demanded that we should seek the structure of daily discourse as such a pursuit will help avoid such errors and thus become better people. This is a far-reaching hypothesis. In addition, it is much more optimistic than the views that Wittgenstein voiced in other contexts.

3.5 Reinterpretation in Logic and in Science

Surprisingly, it is possible to view Russell's study of meaning as more-or-less the effort to restore the reference theory of meaning (Quine 1967, 663, 665). Frege's refutation of this theory was from examples of identity and of the empty class. The case of identity Russell covered by his description of it as the identification of the x that is the morning star with the x that is the evening star. He added to formal logic the expression "the x such that", namely, his "principle of abstraction", namely, the idea of selection from a set all items that possess a certain characteristic the single object that has some addditional characteristic. This ,enabled him to separate the morning star from the evening star and identify them with no inconsistency. Thus, cogently, the planet seen in the morning is identical with the planet seen in the evening. Thus, likewise, prime numbers are numbers that are … The empty class and the empty name are harder to dispense with. Obviously, the square circle and the greatest natural number share non-existence yet they differ in sense. This seems to conflict with the principle of extensionalism. This is serious, since we hardly allow for doubt as to whether Achilles existed or not and we do allow for doubt as to whether Homer existed. As we want to allow for a rational debate between the holders of the competing views on Homer's existence, we want to hold the usual distinction between Homer and Achilles.

Whereas Boole had no problem identifying the square circle and the greatest natural number, Frege had; as he wanted to keep the distinction between them, he introduced sense to perform this task. Russell objected to Frege's sense. He considered the difference between the morning star and the evening star not in sense but in description: some entities described differently are identical. He considered false both the sentence "the present king of France is bald" and "the present king of France is not bald". This seems paradoxical as the second sentence looks like the negation of the first. He resolved this paradox by his theory of definite description. The sentence means, he said, there exists one and only one object such that … It is possible to continue the sentence in different ways. The point of the analysis is this: as the sentence in question includes the claim for existence, it is no longer paradoxical that both it and its seeming negation are false, since its negation is usually misconstrued.

This is not the first systematic reinterpretation of a text that allegedly solves a difficult problem. Perhaps the most famous case in the scientific tradition is conventionalism-instrumentalism of Duhem and Poincaré, the view that scientific theories do not mean what they say; that what they say is a mere *façon de parler,* merely a conventional mode of speech. Some cases fit this theory with no controversy: the theory of elasticity assumes that the medium is continuous although we all agree that its structure is atomic. The assertion that the medium is continuous, we all agree, is a mere *façon de parler.* The conventionalist-instrumentalist idea is that this holds for all science: science then is but applied mathematics. This theory is strange. The rationale behind it is that since rationality is certitude, and since truths by nature are unknown and only truths by convention are certain (since they are artefacts), and since science is certain, it is truths by convention. Young Wittgenstein considered certain only two kinds of statements: tautologies and observation-reports. Hence, scientific theories are either observation-reports or tautologies. This is his version of conventionalism-instrumentalism. His doctrine of perspicuity (McCutcheon 2001, Ch. 1) enabled him to add a point (*TLP*, §6.36): "There are laws of nature. But of course one cannot say that: it shows itself."[27] In other words, scientific theories are not utterly arbitrary (*TLP*, §6.342): "what *does* characterize the picture is that it can be described *completely* by a particular net with a *particular* size of mesh".

This is new, interesting and valuable: the logical status of Newtonian mechanics insures it against refutation; the demand for its ability to explain all the phenomena in its possible domain of application takes this insurance away. This is of great heuristic value, but as we do not know this domain of application, we may ascribe scientific status to the theory, said Duhem (1905, §8), on the understanding that every attempt to extend the known domain of application of a theory is a personal risk of the researcher. This is a remarkable achievement and it justifies Wittgenstein's view of the importance of the preliminary discussion prior to theorizing that runs throughout his *Tractatus Logico-Philosophicus*. Now, this idea is obsolete, of course, since by now Einstein's replacement of Newton dispenses with the demand for certainty and Popper's demarcation of scientific theory as refutable replaces certainty as the mark of rationality with openness to (empirical) criticism (Munz 2004, 3, 11, 29–30, 59, 81–2, 133).

Still, the advantage of Wittgenstein's observations is there. It dismantles Galileo's criticism of conventionalism-instrumentalism that still is the strongest: it views science as stagnant in oversight of the goal of scientific research that is the search for the laws of nature. Those who wish to see science grow, Galileo said, see the adherents to instrumentalism as people who play the game heads I win, tails I do not lose. Consider a theory as a picture (say, a document granting the right to enter science, or a passport of sorts), and you see that adherents to conventionalism-instrumentalism say, you can consider a picture genuine; otherwise, rather than say that it is not genuine, say that it is not a picture. This is clearly unfair. Wittgenstein abolished the unfair aspect of that theory. He said, not all frameworks are as

[27]Kant declared the law of causality synthetic *a priori* knowledge. Russell tended to agree. Young Wittgenstein agreed but forbade saying so. This raises the question, did he allow saying this in ordinary language? His *Philosophical Investigations* assiduously ignores this question. What does his *Tractatus Logico-Philosophicus* say on this? Advocates of "the new Wittgensteinian approach" discuss this matter (Crary and Read 2000, 225–6), especially Putnam. I wish I could report what he says on it.

successful as the received scientific theories, and their success renders them less arbitrary, even though one can only see this, not say it.

Galileo's criticism is applicable to Russell's view of empty names: when we realize that a given proper name does not denote anything, we declare that it is (not a name but) a (false) definite description. This is unpleasant. Fortunately, this criticism does not hold for mathematics, since in mathematics existence is merely consistency. In mathematics, the major objects are not individuals but sets. (In particular, in the Frege-Russell system, numbers are not objects but sets of sets of objects; the set theory known as the von Neumann-Bernays-Gödel theory does away with objects altogether.) In a broader context, any blurring[28] of the difference between a proper name and a definite description diffuses this criticism.

This leads to one central question: what are sets? Where do they reside? Clearly, they do not reside in space-time. This is the same as to say, they reside in Plato's Heaven. To say that they are fictitious is cold comfort. Indeed, it had Russell upset. As Landini has stressed, Russell found more upsetting his reduction axiom, the one that declares the mathematics of sets of sets the isomorphic to the mathematics of sets of sets of sets, of sets of sets of sets of sets, and so on. He viewed it as fictitious, although more so than sets. Quine, on the contrary, declared sets real – as real as pieces of furniture, he said. Neither view is comfortable. Both invite a deeper look at what services we expect our ontologies to provide, and at what we mean by the term "real".[29]

This leaves open the question, from the modern viewpoint, is Galileo's criticism of the conventionalist-instrumentalist view of science just? Galileo and young Russell sought certitude, and this search fuels the drive for conventionalism-instrumentalism. Russell then adopted Einstein's view of science as series of approximations to the truth. This still allows for the conventionalist-instrumentalist view of science, as Arthur Stanley Eddington 1928 tried to show. Yet it renders the doctrine pointless acrobatics. Indeed, already in 1911, as Poincaré recognized Einstein's special theory of relativity as superseding Newton's mechanics, he recognized that this rendered pointless his own conventionalist-instrumentalist defense of Newton's mechanics. Did Wittgenstein know this when he published his *Tractatus Logico-Philosophicus*? He could, as Russell was always *au courant* and as he went over the manuscript text of it line by line during their meeting in The Hague in 1919, prior to its publication.

The picture of the situation at the time is highly problematic. This problematic character had driven Russell to the study of logic that made him famous even before the publication of the *Principia Mathematica* that was the peak of his achievement. This point is seminal: Wittgenstein's interest in the foundations of mathematics had led him to Frege and Russell and he therefore joined Russell in his researches

---

[28]Identifying all names by descriptions is impossible (Popper [1935] 1959, §14; Agassi 1995b, 265).

[29]For more detail see Leonard Linsky 1967, Ch. 4. Pelletier and Bernard Linsky, 2009, 60 find "much in common between empty proper names and improper descriptions, from an intuitive point of view. . . . And whatever account is given for non-denoting proper names should also be given for improper descriptions . . . In any case, we should care about the present King of France".

(18 October 1911), as Russell had hoped that the new logic would offer a scientific (i. e., certain) philosophy, including a firm view of science. It was the ability of Wittgenstein to partake in Russell's investigations that rightly impressed Russell so. Wittgenstein left in 1914 to enlist in the Austrian army; when he met Russell again (December 13–20, 1919) at The Hague they discussed – line-by-line – the manuscript of his book that gained him his fame and his academic post. This book offered the vision that Russell gave up, of science as prescribed by logic. The vision was a failure, and one that was pivotal in the development of the philosophy of science. Yet this was not clear enough at that time. In 1928, Eddington still advocated the instrumentalist-conventionalist view of science that both Einstein and Russell had already rightly rejected. This indicates the great difficulty that the philosophy of science still posed, the complex situation in which modern logic offered a new challenge to the theory of knowledge. We should not underestimate the challenge and the difficulty. Wittgenstein's endorsed the instrumentalist-conventionalist view of science in 1922 that Eddington advocated in 1928 and that many philosophers still advocate, despite the claim of Galileo and Einstein that this philosophy is too conservative to encourage research.

We have to go over all this slowly and see how modern logic came to solve difficulties in mathematics, how it then challenged thinkers to develop a new view of science, and how the view of science as certitude repeatedly pushed philosophers to the instrumentalist-conventionalist view of science, no matter how reluctantly. The demand for certitude was hardest to let go.

3.6 Methodology, Epistemology and Ontology

Beginning students of any field of study have hardly any idea about what to expect from their teachers. They usually have some idea about the field of their chosen study, either from high-school courses or from the popular literature on it. The case of philosophy is scarcely different in this respect. Nevertheless, the sense of surprise that students experience in the introductory courses may be different. What is so very unusual is the sense that accompanies the very idea of universal doubt. The story is that Wittgenstein experienced it as an adolescent, that it hit him hard, and that he tried hard to overcome it later in life. This was not unique. Russell had a similar experience. He began his logical investigations in effort to establish certitude. He found no solace in them as he learned – perhaps from Wittgenstein – that only tautologies are certain, since they are utterly uninformative.[30] Hence, no theory of the world – no ontology (the theory of the *Greek "on"*, the Latin "ens", the English "entity") is disallowed. Russell concluded that there is no disproof of solipsism, of

[30]The Polish school of logic, incidentally, denied that: they found somewhat informative even the most trivial of tautologies, all humans are human. In this study this is overlooked and the Russellian or Wittgensteinian idea is here taken for granted.

the idea that only I exist, and that nonetheless we should dismiss it as contrary to commonsense. (Let us remember that Russell took science very seriously on the ground of common sense. Now we scarcely need that; for upholding realism suffice it to admit my inability to have imagined all the artworks that I was fortunate enough to enjoy.)

In line with the observation that solipsism is irrefutable, the demarcation of science as certitude thus was meant to allow for solipsism, even though, naturally, no one ever took seriously the solipsist ontology or even the idealist ontology, since we all take it for granted that some eons ago dinosaurs roamed the earth. The name for the theory that allows for both solipsism and commonsense realism is neutral monism. It differs from universal skepticism minimally: it allows for the certainty of my experiences, but not of the existence of the entities of that experience. The reason for this is empiricism, or more precisely, empiricism Locke-style. I find this rather silly, but better philosophers than I am are of a different opinion (Quine 1988).

Moreover, the reading of idealism or of neutral monism as ontologies has some adherents: Lenin 1909, Pepper 1942, and Carter 1989. Lenin ascribed neutral monism to Ernst Mach, who responded, saying, Lenin had totally misunderstood him. This stymied commentators, although it is obvious: Lenin ascribed to Mach not so much neutral monism, which is incontestable, but as an ontology rather than as a methodology.

We learn one profound if simple insight from Wittgenstein: any alternative to any metaphysics is a metaphysics. (This follows from the idea that metaphysics is meaningless in Russell's sense: it has no meaning. The difference between falsehood and meaninglessness is just that: the negation of the meaningless is equally meaningless.) Hence, Wittgenstein rejected realism, idealism and solipsism with equal resolution. His neutral monism was as methodological as that of Mach. This dispels of most of the commentaries on Wittgenstein's first book as sheer confusion, though as understandable and so as scarcely culpable. Nevertheless, we should leave it behind. This was the idea that Schlick stated when he ascribed to Wittgenstein realism and this is what Hintikka had in mind too as he denied that Wittgenstein ever sponsored solipsism: neutral monism is not an ontology but a methodology.

Although all this is of great historical significance, incidentally, it is behind us since we have rejected the demarcation of science as certitude (1917 or thereabout). This we had to do anyway, or else Newton's physics would have ceased to be scientific, which is intolerable (Popper [1935] 1959, §9). It is a strange fact: what is specific to the culture of science is the admission of alternative options as significant possible alternatives, and as ones that do not lose their significance after they become outdated. The paradigm case is the contrast between the views of matter as continuous or as atomic. (The atom is allegedly *atomos*, namely uncuttable or *individuum*, namely indivisible. The negation of atomism is the claim for continuity.) We have versions of both continuity theory and of atomic theory of both matter and light, and all of them have versions that belong to applied mathematics, to metaphysics and to science. One may overlook these facts, of course. This happens both to the ignorant and to the obscurantist, not to mention the one who does not care about what one says as long as one has an audience. The present study would be utterly impossible were it possible to suspect Wittgenstein of these characteristics. The honorable attitude is to suggest that he was *bona fide* and in error.

Chapter 4
Logic and Mathematics

It is Frege, not Boole, who is the father of modern logic. What exactly is modern about modern logic? Why did Frege develop it? The answers given here are these. Modern logic is both comprehensive and fully formal. The comprehensiveness in question is the sufficiency for the purposes of mathematics. (Other kinds of comprehensiveness are possible.) Fully formal systems are such that computers can use them. The idea of a fully formal language is Frege's; he offered fully formal inferences. Russell was the first to offer a formal language proper. These days, when computers are parts of everyday life, developing a formal system is hardly a challenge. At the time, Russell noted, only two formal systems were available, his and that of Wittgenstein. Most significantly, it was Frege who rendered logic unmistakably the logic of statements; analysis of concepts thus turned into (post-Wittgenstein) analysis of statements. (Regrettably, on this too many analytic philosophers are not *au courant*.)

Frege's aim was to disprove Kant's view of arithmetic by presenting it as a part of logic proper. To this end, he had to deviate from Kant's view of logic. His achievement renders standard modern logic – of Frege or a variant of it – far stronger than classical logic. This is not surprising, since demonstrably the valid part of classical logic is a minor part of Boolean algebra. Still, all this raises anew the question, what is logic? It was in an effort to dodge the question that the terms "symbolic logic" and "mathematical logic" entered the general use. It did not work: it raised the question, where is the dividing line between logic and mathematics? Does abstract set theory belong to logic or to mathematics? Quine has raised forcefully this important question; Hao Wang has discussed it at some length. Since this study concerns Wittgenstein's *Philosophical Investigations* that scarcely touches on this, let me try to avoid discussing it here. What matters here is Frege's theory of meaning.

J. Agassi, *Ludwig Wittgenstein's Philosophical Investigations*, Synthese Library 401,
https://doi.org/10.1007/978-3-030-00117-9_4

4.1 Words, Statements and Inferences

Frege aimed at basing arithmetic on logic.[1] To that end, he fashioned modern logic as primarily the calculus of statement formation and composition plus the calculus of functions; he ignored the fact that classical logic is a theory of nouns and of inferences; rather, in contrast with classical logic, instead of subject and predicate, his system is the analysis of propositions into functions and arguments. This is not quite true, as his system had to include the theory of quantification: standard "all humans are mortal" became in his system "(x) (Px ⊃ Qx)", namely, "for every object, if it is human then it is mortal". (Note: here terminology follows tradition and abbreviates "(∀x)" to "(x)" to read "for all x".) Statements are made of words the way brick-houses are made of bricks; inferences are made of statements the way neighborhoods are made of houses. Tradition endorsed the claim that the most general assertions in science are definitions, that definitions introduce names of essences, that language is essentially a vocabulary, a dictionary consisting essentially of nouns. This helped gloss over the distinction between nouns and statements, or between words and statements. The confusion between statements and rules of inference is permissible, since the central theorem in the theory of inference, the deduction theorem, offers a bridge between the two distinct items that logicians often treat as one. It is this: an inference is valid if and only if the conditional statement with the conjunction of its premises as antecedent and its conclusion as consequence is a tautology or a theorem. *Tautologies are true by virtue of their structure, namely, they are true in all substitutions.* The most basic theorem in this context is Gentzen's completeness theorem for any logical system:

Inferences with tautologies as their conclusions are valid regardless of their premises.

In symbols,

$$\vdash a \leftrightarrow \vDash a.$$

Here the (meta-linguistic) symbols read thus: a is the name of any statement; "↔" is "if and only if"; "⊢" is, "is a theorem" and "⊨" is, "is a tautology". Similarly,

(a entails b) iff (if a then b is a tautology) iff (if a then b is a theorem).

In symbols,

[1]This is the initial story; not the whole story. Anachronistically, all math rests (allegedly or in truth) on abstract set theory and the Frege-Russell system is an attempt to base abstract set theory on logic. This raises a plethora of problems. In what sense is the Frege-Russell set theory the same as ZF? Did Peano get the natural numbers right? These questions grew with the growth of model theory and abstract algebra, especially the Skolem-Lövenheim theorem and non-standard arithmetic.

$$a/b \leftrightarrow \vdash a \supset b \leftrightarrow \vDash a \supset b,$$

where the slanting line "/" stands for the deducibility relation between premise a and conclusion b; "⊃" is called "horseshoe"[2] and "a ⊃ b" reads, "if a then b" and is the name of the one conditional statement composed of two statements.[3]

Modern logic differs from classical logic in that it concerns statements, not rules of inference, and in its concern with mathematics, not with taxonomy. Traditionally, mathematics centered on proofs that employed intuitive inferences with obviously true axioms as premises. For millennia attempts to develop a logic of statement composition and a theory of inference were abortive[4] – despite impressive successes, particularly in analyzing hypothetical inferences, such as the *modus ponens*,

a and if a then b entail b.

In symbols,

$$a, a \supset b/b.$$

No less important is the ancient rule, the *reductio ad absurdum:* a statement that follows from its negation is a theorem.

In symbols,

$$\sim a/a \rightarrow \vdash a$$

or even

$$\sim a/a \leftrightarrow \vdash a.$$

And, of course, the most central rule ever, the exclusion of contradictions.
The formal way of saying this exclusion sounds odd to begin with. It is this:
Any statement follows from a statement and its negation;
Any statement follows from a contradiction:
Ex contradictione (sequitur) quodlibet:
From a contradiction whatever you want (follow).
In symbols,

[2]The resemblance of the symbol for the conditional with the symbol for subclass is annoying. It has its historical roots, but we should ignore this here.

[3]The deduction theorem is intuitively quite unproblematic. It allowed the view of inferences as conditionals of sorts, and Russell and his followers used the word "implication" for both. Quine succeeded in decreeing a taboo on the use of "implication" in precise texts.

[4]Kneale and Kneale (1962) present this way the history of logic between Aristotle and Boole.

$$a, \sim a/b.$$

It resembles saying that a chess game is over when the king is lost by saying that afterwards all moves are equally permissible and equally impermissible. Thus, the exclusion of contradictions is the cornerstone of logic: affirming a contradiction is affirming every statement: endorsing a contradiction is departing from logic.

Frege had only rudiments of a theory of inference, and in his masterpieces he limited his rules of inference to substitution and the *modus ponens*. Substitutions make for an inference-scheme. They are present in logic since its creation[5]: logicians used words like "mortal" or "Socrates" as variables, to be substituted at will; this process is common to all logic or to all languages. In addition to substitution, Frege had only one rule: the *modus ponens*. He justified it in the traditional way: it is truth-preserving: it never leads from true premises to false conclusions. We view Frege as the inventor of the calculus of statement composition. This is not quite accurate, as many logicians used parts of it much earlier. It is accurate if we mean that he presented it as a calculus proper.[6] Even this far-reaching move did not undermine the authority of Aristotle. His having joined it with the functional calculus (see below) and his use of them as the basis for arithmetic have stabilized modern logic. This began the erosion of Aristotelian logic that is still in process.

Modern logic belongs chiefly to Boole and Frege, but also to Peano, Hilbert and Russell. (This verdict is of Gödel 1944, 125). Frege wanted to derive the basic properties of arithmetic from logic. Peano developed an elegant symbolism and managed to express precisely a big collection of complex mathematical theorems with it. *Principia Mathematica* of Whitehead and Russell proved that $2 + 2 = 4$. Earlier, Leibniz had offered an informal proof for that. The task of proving arithmetical theorems within a fully formal system is formidable. Nevertheless, Gödel observed, Russell's achievement was incomplete. Hilbert achieved that, though this assessment comes from different places, and it has to do with his sharp formulation of the very idea of a formal system that imposes the distinction between object language and meta-language, plus his effective attempts to employ these ideas in order to seek proofs for the consistency and completeness of mathematics. This search challenged Gödel to prove it futile. It is amazing, in hindsight, what a profound and rapid change logic underwent in this process within less than a century at most – beginning with Boole 1847, through 1913 PM of Whitehead and Russell, and culminating with Gödel's proof of 1931. Yet the first revolutionary step was Frege's 1879: he was the first to formulate the calculus of statement composition as an axiom system. Russell and Hilbert offered variants of it. Others joined in and offered a few variants (Hazewinkel 2013). He was also the first to introduce into logic predicates with more than one subject, a move that Russell always deemed

[5]Arguing for the validity of any specific inference, Plato illustrated it with a few obvious substitutions.

[6]The very concept of a calculus was an innovation related to closure under some operation that evolved out of group theory and Boolean algebra.

basic, and for diverse reasons, logical and metaphysical.[7] Hilbert was the first to create a fully formal system: Euclidean geometry. Russell created the first fully formal language (PMia). Alan Turing has entrenched it all with the theory of computers (of Turing machines).

What is formal and what is informal (MacFarlane 2002, 26 *ff.*)? Making up a language is creating items – words, sentences, and rules of inference. Not all rules are formal. *To render as clear as possible what move is permissible and what not, a formal system is necessary*, say akin to that of checkers (draughts) but not of poker. It is not easy to analyze this last sentence, but this is of little concern here, where it is significant that the system of Wittgenstein's *Tractatus Logico-Philosophicus* is formal and that of his *Philosophical Investigations* is not. Computers can judge this: they operate only in formal systems: a computer is a glorified abacus: its formality appears in its stilted style ("This does not compute"). In a formal system the sign for negation, for instance, always appears with the same words. Customarily these are "it is not the case that" or "~"; the negation of "p" then is "~p", the negation of "q" is "~q", *etc.* Alternatively, when a sentence is named a, then the name of its negation is non-a, often written as ~a. The sign "~" in the object language and in the meta-language reads differently. If you want to prevent confusion, you can write "¬a" and "~p" where the first item is a name and the second item is shorthand; the sign "¬" is a part of a name whereas the sign "~" is shorthand for the expression "it is not the case that".

The development here mentioned – the establishment of mathematical logic – changed current views on logic and on language radically, although they did not lead to unanimity. Since the disputes about it are metaphysical, they should leave Wittgenstein's deliberations behind. They do not.

Conclusions that follow (validly) from tautological or analytic premises are likewise tautological or analytic. This is a most basic statement in logic. It permits the development of a language limited to tautologies or analytic statements (with the ability to undergo extensions by adding to its vocabulary descriptive terms). Tautologies are supposed to convey no information. This truth may easily mislead. "There is no largest number" is doubtless informative, yet, being a theorem it is judged uninformative. Supposedly, the full explication of the situation comprises the observation that the information in "there is no largest number" is about numbers and not about the world that science studies, not the objective reality, not the contents of our space-time. If we agree that "there is no largest number" is not about anything within space-time, and that we use the word "informative" as shorthand for the expression "not about any object within space-time", then the situation is supposedly clear. It is not.

---

[7]Russell's conversion from Hegelianism to realism with the help of Moore had a lasting influence on him; Hegel made much of the limitation of logic to categorical propositions and spoke of internal relations as relations between essences; what he said of them is very confused. Russell found in the irreducibility of two-place predicates to one-place ones means to overcome his Hegelian bent. This served Wittgenstein as a model in his effort to explain the attractiveness of meaningless metaphysics.

The heart of the trouble is metaphysics. Whether it studies the world of thought or of science is not clear. Is a symphony in space-time or not? Opinions differ. Carnap said, this question is unclear. This absence of clarity, he added, characterizes metaphysics and thus condemns it. The atomist metaphysics of Democritus shows Carnap obviously in error here. Even though both numbers and symphonies are not in space-time, assertions about the former are tautologies and about the latter metaphysical. Why? Possibly, Carnap's view has a significant kernel of truth. Yet the situation is baffling.

A historical glance at things might help. The distinction between the informative and the uninformative follows the ancient Greek distinction[8]between truth-by-convention and truth-by-nature: what is *a priori* valid is valid for no better reason than that it does not oblige anyone but those who agree to it and on the force of nothing but their agreement (convention). Since mathematics rests not on observations but on proof, it was tempting to suppose it to be truths by convention, until Kant declared its truths to be synthetic, namely informative. He did that after he admitted analytic truths to comprise nothing more than the laws of identity and non-contradiction. Under the law of identity he allowed not only a = a, but also ab = a. Boolean algebra goes much further but perhaps Kant would have allowed it. Frege went further in order to disprove Kant's view that arithmetic is synthetic. Russell claimed then that the axiom of infinity is not included in logic even in its extended Frege-style. Hilbert tried to show that mathematics could do with the axiom of infinity even while declaring the very concept of infinity utterly meaningless. Later Wittgenstein adopted something like this idea as he insisted that theologians could not contradict each other; yet he did not condemn them for this as he condemned metaphysicians: what he wished to condemned is pretention. Therefore, the verification principle is less significant than the claim that a metaphysical assertion is verified when it is not. As it is possible to choose to superpose humility or hubris on any expression, Wittgenstein's idea of the limit of language fades out.

The search for ways to say what should remain unsaid is not new. Thus, when Newton did not wish to state a hypothesis, he tucked it in a mathematical assertion: if two celestial bodies interact according to his law of gravity, he said, then one moves in an ellipse as the other rests in one of its focuses. A simpler way to say it is, if the gravity between two bodies varies according to the inverse square of the distance between them, then it varies according to the inverse square of the distance between them. This expression has the form "if p then p" or "$p \supset p$". A less boring tautology is "$p \supset (p \vee p)$" (for example, if it rains then it rains or it rains); a still less boring form is "$(p \vee p) \supset p$" (if it rains or it rains then it rains) or "$p \supset (p \vee q)$" (for example, if it

[8]Popper found it outrageous that the dismissal of metaphysics generates a new verbal category: significant nonsense. In a letter to Engelmann Wittgenstein considered it exciting: "(And therefore any proposition that does not have that form must be nonsense, but yet nonsense that is not meaningless)." This is scarcely serious. Hilbert's meaningless is serious as a technique; his attitude to the infinite is oddly different. His disciple Robinson is the only one who has noted this, regrettably without discussion.

rains, then, it rains or today is Tuesday). Still, Newton had to do it as he did since unlike my alternatives, his was informative – in a sense different from the one defined above.

My examples from the theory of statement composition have similar examples in Boolean algebra, every class is a sub-class of itself ((a) a $\subseteq$ a) or simply (a $\subseteq$ a), and in the theory of inference: every statement follows from itself (a/a). These are idle examples. (Wittgenstein declared them meaningless since they are obviously correct! He likewise declared "2 is a number" meaningless, although he knew that "1 is a number" is the first of Peano's path-breaking axioms for arithmetic.)

Why is it so useful to allow for such trite claims as a $\subseteq$ a? Now this question is not limited to mathematical logic. We have learned similar things in arithmetic classes already: $0 + 0 = 0$, $1 + 0 = 1$, etc. These formulas puzzle because of the oversight of their aims. The aim of logic is not to recommend making inferences; it is to state rules as to what deductions are permissible and what assertions are true under any substitution. The road to any progress, including the development of a truly formal logic, is teeming with obstacles. Their removal involved explicit permissions for pointless options. Some of these pointless options were under dispute. The mathematician Brouwer, whose lecture in Vienna brought Wittgenstein back to Cambridge,[9] made his magnificent mathematical career by refusing to permit the view of the double negation of any statement as logically equivalent to it (a // ~ ~a). Most mathematicians consider this trivially true.

Logic and mathematics texts are mostly informal. Utterly formal texts are very hard to write and harder to read. (Computer programmers do that regularly.) Logic is indifferent to the way one may complement a term to negate a statement. In formal systems, whenever more than one way to do so exists, a definition has to render them equivalent. Not so in informal systems. This fact troubles hardly anyone. A less obvious case is that the complement of "free". It is possibly "not free" and possibly "unfree". Yet it is hard to choose between kinds of complements to "free", such as "slave", "imprisoned", "bound", "working", "busy", or "engaged" – especially since the last term may mean both "busy" and "not available", not to mention "not in the marriage market", for example. There are two sorts of varieties of complement here. One is a variety of prefixes that designate complementation ("not-" "non-" "un" and at times even "in" as in "infrequent" or "ir" as in "irresponsible"). This kind of variety raises no problem; it is very easy to instruct a computer to treat them as interchangeable, and even some rules as to which of them to prefer under what conditions. The other kind is a complementation only in the sense that it names the complement differently by allusion to the choice of different classes to serve as the universal class relative to which complementation is to be fashioned, the class that is

[9] Rhees (1984, 16), "... in the summer of 1931..., under the impact of Wittgenstein, young men went about saying: 'It's absurd to say that 2 is a number—what else could it be?'" (Cp. *TLP*, §4.1272.) Wittgenstein likewise eliminated infinity, since it shows itself. (See *TLP*, §5.5351.) Black (1971, 297) observed that Wittgenstein viewed infinity too as obvious and unproblematic. Some people love problems and then they tend to see problems everywhere; others do not.

the disjunction of each class and its complement. The rule of complementation is rigid: it holds under the conditions:

$$a = \sim b \leftrightarrow b = \sim a \leftrightarrow \{a \cup b = 1 \text{ and } a \cap b = 0\};$$

(to avoid confusion, some texts replace 1 and 0 with $\bigvee$ and $\wedge$; "$\wedge$" is pronounced "lambda" as it looks like the Greek letter by that name;) also, $\{a \bigcup b = \bigvee$ and $a \bigcap b = \wedge\}$ makes the two complements.

Negation then is the wrong word for complementation and it is still in common use since classical logic allows it due to its fusion of the calculus of classes with the calculus of categorical propositions. Thus, like double negation of a statement is that statement and the complementation of a class is the initial class. So we can read

$$a // \sim\sim a$$

as a statement in the calculus of statement composition and as a statement of Boolean algebra, depending on whether we read a as the name of statement or of a class (and then "~" designates negation or complementation respectively).

To return to the discussion of formal class complementation, by definition,

$$b \subseteq \sim a \equiv \sim (b \subseteq a):$$

the complement of any class is relative to the universal class. Clearly, then, the choice of "free" as the complement of "slave" or of "imprisoned" is not different from the choice of any other homonym. Logic requires excluding them – often by adding to them different qualifying words or syllables or other signs. Alternatively, we may send the words that are homonyms to different universes of discourse.

Before leaving all this a few words about the similarity between Boolean algebra and the calculus of statement composition. The formula we have just discussed, $b \subseteq \sim a \equiv \sim(b \subseteq a)$, where a and b are names of classes, is a part of the calculus of classes. To make it a part of the calculus of statement composition we have to view a and b as names of statements and also as names of the contents of the statements a and b, where the content of a statement is the class of all of its (logically valid) consequences (Tarski). This explains the similarity between Boolean algebra and the calculus of statement composition. We need not discuss this here.

4.2 The Universe of Discourse

The universe of discourse is an aspect of Boolean algebra that has contributed to philosophy, thereby refuting the claim that they are irrelevant to each other. In Boolean algebra, for every class a there exists a complement class ~a. The universal class is called ⋃, ⋁ or 1, namely a-or-not-a (a or ~a; a v ~a) and its complement, the empty class, is called ⋂, EC; (the Greek letter Lambda), 0, or Ø. (The name 0 is confusing as it may be read as zero, so it was converted to Ø.) It is easy to prove that both are unique:

For every class, every item is a member of it or of its complement
and
no item is a member of both a class and its complement.

In symbols,

$$(a)\ a \text{ or} \sim a = \vee; (a)\ a \text{ and} \sim a = \wedge.$$

$$(a)\ a \vee \sim a = \vee; (a)\ a \bullet \sim a = \wedge.$$

The universal class ⋁, to which everything belongs, is no less problematic than its complement. The very concept is awesome. What does it include? Consider, for example, the complement of the class of all clowns. Is the class of those who are not clowns the class of all members of the circus or of all employed people or of all humans or all vertebrae or of all living things? Of the whole universe? Does the universe include the letter c? *To dodge this question, logicians limit the universe freely to the universe of discourse* ⋁. This might solve the technical part of the problem, but render the metaphysical problem behind it worse. Wittgenstein refused to allow it in his formal language. Russell considered this a great limitation: it prevented Wittgenstein from admitting even that at least three things exist: this innocuously seeming sentence tacitly refers to the universe of discourse.

The function of the universe of discourse ⋁ is eliminative: logicians leave open the question, which class is the universal class; at most, they are interested in its size: is it finite or infinite? How is this to be decided? On what grounds? These questions they ignore. Would mathematicians care more about it? Hardly. The universe of discourse comes in handy, say, in the study of non-Euclidean geometry: the question, which geometry is true, does not concern mathematics as such, as mathematicians relegate it to physicists. They are concerned with the question, what does a Euclidean world look like? What does this or that alternative to it look like, and how do they compare? They do not ask, what does all this exactly mean? They dismiss this question, or leave it to logicians and philosophers of logic.

We assume that the universe of discourse is not empty:

$$\mathrm{V} \neq \Lambda.$$

This assumption is unproblematic: it can be overruled, but to no good purpose: the empty universe of discourse is uninteresting. Why then do logicians make this assumption? Why do they not take it for granted tacitly?

This is a central question. For millennia, geometricians took for granted an assertion that Pasch discovered (in 1882) that it is an axiom, namely, that it does not follow from the other axioms of geometry. The study of axiom systems raises at the time questions of completeness and of decidability. In the 1930s, Kurt Gödel put this study in the center of the foundations of mathematics. Earlier, positivists like Poincaré dismissed it. Even Wittgenstein, who died in 1951, found Gödel's undecidability theorem unacceptable, perhaps because the verification principle says, every properly worded statement is testable (and hence decidable). By now axiomatics and decidability comprise legitimate parts of mathematics and computer-science.

Things looked very different in the nineteenth century, when many mathematicians looked down at the foundations of mathematics. As Boolean algebra was from the start a version of logic, and not even an algebra proper (Hailperin 1981) it did not interest mathematicians. When Frege tried to apply logic to mathematics in the late nineteenth century, most peers ignored him. Russell developed similar ideas in ignorance of his work. With the success of Russell, it became a bit harder for mathematicians to ignore all the applications of logic to mathematics. This development owes much Schroeder's and Huntington's presentations of Boolean algebra as a formal system. This gave it the kind of legitimacy that mathematicians are after, especially after its integration within general algebra. The story of abstract set theory is more complex, as it has many versions offering many ways to overcome its paradoxes. Its first axiom system that appealed to mathematicians was ZF (or ZFC) of the early twentieth century. As it was essential for Frege and for Russell, it is not hard to appreciate the pioneering character of their output. During its evolution, most mathematicians were not interested in it, and philosophers were less interested in it, as most of them did not comprehend the deviations from traditional logic that it enticed nor did they care for mathematics and its foundations proper: the pioneers of logic were an isolated lot. Just when Boolean algebra began to interest mathematicians, the calculus of classes or sets developed well beyond it into the full-blown abstract set theory that often serves as the foundation of mathematics, whereas Boolean algebra is only a small part of algebra. As a highly developed part of mathematics, Boolean algebra nowadays goes far beyond the concern of logic. It is, significantly, the first stage of any study of computer theory. Mathematicians find no problem in the choice between Boolean algebra and abstract set theory: it is a matter of convenience, decided by the context.

The claim that the rules of a formal language come to suit given epistemic aims has met with vigorous objections. Most leading thinkers, including mature Russell and Noam Chomsky, insist that all formal languages are artificial, and it holds only for artificial languages. This view is traditional and naturalist (since the chief traditional aims of logic were epistemic) and goes back to Aristotle at least. It conflicts with another, equally traditional view, of logic as utterly conventional, a game of sorts, whose meaning does not and cannot lie outside its rules. The dispute between naturalists and conventionalists is pervasive in western philosophy. It is irresolvable because it rests on a misleading and yet indispensable dichotomy. (A dichotomy that splits a to b and c is the view of b and c as exhaustive and exclusive relatively to a: $b \wedge c = 0$ and $b \bigvee c = a$.) The ancient, misleading, and yet indispensable dichotomy that the dispute between naturalists and conventionalists rests on is the division of all truths into truths by nature and truths by convention. It is the most basic dichotomy in the history of metaphysics from antiquity to the end of the nineteenth century. The ancient efforts and failures to apply it are central to philosophy. It is still very popular. Advocates of the modern dispute about the meaning of formal rules overlook two significant facts. First, most of ordinary discourse is not within a formal language. Second, most of our opinions are at best intelligent speculations, putative truths, hopefully scientifically improvable. Improvements undermine the dichotomy: it is impossible to improve truths by nature and there is no need to improve truths by convention. Since the disjunction applies to knowledge, to fully demonstrated statements, it is thus usually much too simplistic.

Thus, all significant human creation, including art and science and even much of mathematics, are neither truths by nature nor truths by convention, but rather putative truths. I will regretfully leave this point without elaboration, though it is very relevant to Wittgenstein's later philosophy, as he then combatted the naturalism of his youth as too narrow and perhaps as false (he seldom admitted error outright, and the process of trial and error had no role in his philosophy). Did he give up the dichotomy between truth by nature and truth by convention? It is hard to judge, since he never spoke hypothetically (of putative truths). Moreover, ordinary-language philosophy ignores the status of putative truths. Ordinary-language philosophy does not attend sufficiently to ordinary language. Consider an utterly uncontroversial case: nineteenth-century English views on sex and gender. Since we recognize them as untrue, we can hardly discuss them sympathetically, except as putative truths within a limited universe of discourse.

That it is better to replace naturalism with putative naturalism is obvious. Not so regarding conventionalism. Its advocates take it for granted that the introduction of formal terminology into any discourse is a mere matter of convenience, as tradition has it. This is not true. Often the significant aspect of the formal terminology upsets informal logicians and novices alike. The prime example for this is Frege's terminology that led to a dispute between Frege and Peano. The latter's terminology too has turned out to have been crucial in the history of both logic and mathematics.[10] Terminology often imposes the recognition of the vagueness of ordinary expressions as compared with their formal substitutes. And since ordinary expressions are vague in different ways, the choice of the way to sharpen them matters.

Postulating a universe of discourse is the simplest way to postulate putative naturalism or putative conventionalism as you wish. This does away with ever so many self-doubts that make Wittgenstein look so Kafkaesque. Some language analysts prefer ordinary terms to formal ones and they argue for their preferences.

[10]Landini 2007, Appendices, observes that Wittgenstein had hoped that his new notation would usher a revolution in mathematics: it would have the structure of the name of a mathematical theorem indicate its method of proof, like a map (Russell 1922) or the way some names of organic compounds indicate their spatial molecular structures. This was the return of John Wilkins' idea of a perfect language, in a modern guise. Landini clarifies Wittgenstein's puzzling remark "The seed I'm most likely to sow is a certain jargon", said in self-deprecation (Wittgenstein 1975, the very end of the book; also Malcolm 1967, 53). Mathematicians rightly ignore it (Agassi 2009, 522).

This is often a misdirected effort: language is a tool, a considerably large toolbox, and the evaluation of tools is (not absolute but) relative to the tasks that they are supposed to serve. This is the idea that Feyerabend has ascribed to Wittgenstein but this is not so easy. Liberal as Wittgenstein's idea of language-games is, it still is not easy to apply, since a tool may serve diverse aims, and thus language serves diverse aims, even though its chief aim is communication. Consider lies. They are parts of our lives. Worse, they are more powerful when people insinuate them rather than assert them openly. This often confuses. Strangely, this trouble does not rise within formal systems that ignore insinuations. Insinuation is at least a two-tiered activity. To the extent that cross-examination procedures are formal, a major use of them is to make insinuations explicit. Now an explicit insinuation is an oxymoron. The rubberstamp example for this is Collingwood's question: "Have you stopped beating your wife?" He observed that both the affirmative and the negative answers to it sound like admission of guilt. This is manageable with ease, as it is the task of the attorney for the defense to correct false impressions by inviting the accused to explain: "Why did you not stop?" The reply "Because I never began" does this very well.

An example nearer home is the metaphysics implicit in a text; at times even its author is not aware of it. Historians of ideas are familiar with this, as one of the tasks that they undertake is to formulate and discuss implicit metaphysical ideas and even metaphysical frameworks of great thinkers. As the dichotomy between truth by nature and truth by convention is metaphysical, Wittgenstein could not object to the suggestion to jettison it. Moreover, in his youth he was a naturalist about observations and conventionalist about theory. He then spoke of the diversity of functions of language and of the limitation of descriptive formal logic, ascribing to each function a language game (Lewis 1979, 339). . What his opinion was on conventions in general is under dispute (Jarvie 1983, 118). The dichotomy of truths to those by nature and those by convention has to go anyway, with or without his approval.

Consider then logic as the theory of the formal aspect of language and consider its development, beginning with the classical (medieval) and the Boolean systems. Some categorical statements are translatable into statements of the classical form "SaP" where the letters "S" and "P" stand for their subject and predicate. Translated into Boolean, "SaP" becomes "$b \subseteq a$" where a and b are classes. Now "$a \subseteq b$" is the same as "$a \sim b = 0$" that reads "the class a and non-b is empty". Are the SaP and $a \subseteq b$ the same? Most logicians say, definitely yes. Quine says, definitely not. Are they more or less the same? This depends on context. The moment we see that "$a \subseteq b$" is "$a \sim b = 0$" and that classically SaP (all S are P) entails SiP (some S are P), we see the force of Quine's objection. Most logicians dismiss as erroneous the classical claim that SaP (all S are P) entails SiP (some S are P): SaP/SiP is invalid. They translate the classical Aristotelian categorical statements to Boolean ones about the existence or non-existence of members to some classes (we write "$\sim a = 0$" as "$a \neq 0$"):

$$\text{SaP} \leftrightarrow \text{a} \cdot \sim \text{b} = 0 \qquad \text{SeP} \leftrightarrow \text{a} \cdot \text{b} = 0$$
$$\text{SiP} \leftrightarrow \text{a} \cdot \text{b} \neq 0 \qquad \text{SoP} \leftrightarrow \text{a} \cdot \sim \text{b} \neq 0$$

What is left of the square of opposition then is obvious, since a ~b = 0 obviously contradicts a ~b ≠ 0 and a · b = 0 obviously contradicts a · b ≠ 0 (since a ≠ 0 is ~ (a = 0) by definition).

A given class, say, humans, contains all and every human; generally, the class C contains all and every individual with the predicate or property P:

$$(x)\, x \in C \equiv P(x).$$

Obviously, P is in a different domain than C: moving from P to C deletes the difference between any two classes that happen to have the same memberships (no matter why). This is tacit already in classical logic, as is manifest in the classical debate that raged then as to whether essence precedes existence or not. It is sheer confusion, since the problem holds for inessential qualities no less than for essential ones. Hence, the retreat to essentialism here is obviously a mere excuse. What then is the problem? There is no need to worry about contradictions here: a decision to distinguish between class identity and class equality suffices for the acquisition of different empty classes. This is indeed how we know mermaids from square-circles. This raises the possibility that Wittgenstein investigated, that possibly intensions hide in rules,[11] and rules determine extensions (class membership). Wittgenstein was right: the more extensional our language is, the more the intensions behind the scenes signify (Bar-Am 2008, 41–7).

The formula "(x) x ϵ C ≡ P(x)" may disturb the connoisseur. For it should be a translation rule rather than a fusion of two calculi (Boolean algebra and Frege's functional calculus). They will observe that by Boole's principle of extensionality, it is quite possible that also "(x) x ϵ C ≡ Q(x)", since it may happen that "(x) P(x) ≡ Q (x)", namely, that two predicates hold for exactly the same individuals (at times analytically so, as in the case of the properties "triangular" and "trilateral"). Predicates from the classical logic of categorical propositions appears here, once in translation into the class calculus and once in translation into Frege's functional calculus. It is thus better to view it as a rule of translation between three calculi. All this is unproblematic unless one wishes to have one overriding system, like that of young Russell or of Wittgenstein. Only the latter dispensed with classes altogether (*TLP*, §6.031: "The theory of classes is completely superfluous in mathematics"). This may do for young Wittgenstein, but not for Russell who tried to capture all extant mathematics and not for mature Wittgenstein who endorsed ordinary language. For, in elementary school classes reign supreme. They do so also at the horse races, where classes of horses appear in stereotypes with no qualms. For, a stereo-

---

[11]Rijke (1997, xii): "There is still no *general* proof theory of modal or intensional logic." The reason for this is that modal and intensional intuitions diverge.

type is a type, and a type is a class. Though "horsiness" is stilted, "fast" is not, and the conjunction of horses with fast yields fast horses – as a class.

What then is a class? The Platonic idea of horsiness, or horsiness as a property, is the class "horse", the class that includes all and only horses. (Does "horse" include or exclude "mare" or "foal"? The true answer depends on context, not on a rule.) The names "Universal", "form" and "essence" serve in the present context as synonyms. The present question concerns the empty universal (mermaid). Classical logic made no allowance for it as it allowed deducing from the premise "all S are P" the conclusion "some S are P". Since "all P are P" is always true in the classical system, it follows that in this system "some P are P" ("PiP"), to mean that a P exists, also obtains. In a less stilted language, whatever P is, there exists at least one thing that is a P (Bar-Am 2008, 140): there is no empty class, or, rather, there is no word for it. This means that whatever class happens to be empty, there is an agreement not to mention it; if a term does not refer, it has no use. Some modern experts in both modern logic and Aristotle's texts read his original texts to make this suggestion. It is easy to observe, however, that whether he made it or not, it is impossible to follow, since at times one has to discuss a quality at some length before one can make up one's mind as to whether there exists anything that is in possession of that quality. Thus, Aristotle discusses atoms at length to prove their non-existence.

What barred the empty class from classical logic still makes it hard for novices to swallow it. The classical theory of meaning that belonged to classical logic was confused, and logic was the last thing to consider confused. The admission of the empty class makes nominalism impossible, as it was the very denial of the existence of classes. Nominalists declared class names collective names of their members; this does not work for the empty class. How then did they understand SeP (no S is P)? Classical nominalism leaves no room for naming nothing, and logicians overlooked this or else they limited their discussion to essences (assuming that existence precedes essence). We cannot imagine computers with no Boolean algebra, and we cannot use classical logic instead as it is very limited and it is very limited because it is confused. To clear the confusion, it was necessary to give up classical nominalism even as an option! Realism was relatively clear: it says, classes exist. So the debate was now on the question, is realism true? Russell said, no, and, ascribing to Frege the affirmative answer, he developed his theory of definite description that he was always proud of. In 1940, over a decade after Frege's demise, Russell conceded: he could not rid logic of realism. As it happens, nothing much changed. Quine disagreed with Russell about a few matters, all of them interesting. His realism stood out. It isolated him, his tremendous prestige notwithstanding, since he took offense at the refusal to do him the courtesy of recognizing his admission of classes as real things – "like pieces of furniture", he said. The main reason why this does not matter here is that in the present context it is easy to keep it out of the universe of discourse – to ignore it.

The discussion here shows the most significant advantage of Boolean algebra over classical logic. This advantage became significant when it served the study of the foundations of mathematics. Why was there a need for these foundations? The answer is, there was a need to get mathematics out of two major troubles that it endured in modern times. One was the need to present new theories consistently, a need felt strongly ever since Bishop Berkeley spotted (allegedly or in truth) an inconsistency in Newton's infinitesimal calculus (Wisdom 1939). The other was the loss of the foundations of geometry when alternatives to classical geometry gained legitimacy,[12] thereby deeply shaking the sense of assuredness about the axioms of geometry. The matter of precision, though merely instrumental, was a very strong factor in the growth of mathematics. The oversight of it is due to the defensiveness of historians of mathematics prior to John O. Wisdom's 1939 and 1953 discussions of the tremendous impact that Berkeley's criticism had as a motive for its defenders to develop it. The matter of precision comes closest to our story in the case of Peano, whose interest was in precision, not in the foundations of mathematics (Segre 1994). The most significant of these neglected aspects of the story are responses to Kant's theory of knowledge (Russell 1897, §51: "Kant's views on this subject, true or false, have so dominated subsequent thought, that whether they were accepted or rejected, they seemed equally potent in forming the opinions, and the manner of exposition, of almost all later writers." This explains Russell's reluctance to take seriously anti-metaphysics unadorned.)

4.3 The Place of Kant in the Story[13]

Trying to comprehend Wittgenstein, we may try to comprehend Frege and Russell, the two who have influenced his researches most. Frege had a terrific ability to concentrate on one challenge. His study of the foundation of arithmetic invited a theory of logic and a theory of knowledge as its framework. He had neither. He refused to develop any idea in those directions prior to developing first the very details of his use of them. That took him all that he had. He developed a firm grasp of what he needed in order to develop what he wanted. First, he needed a theory of inference. He had none but decided, rightly as it happens, that he could do with substitutions and the *modus ponens*. He criticized Boole's approach to logic in two papers that he failed to publish. Second, he needed a theory of knowledge. Again, he had none. For a substitute, he criticized the two extant views – while avoiding the search for an alternative to them. These were the views of John Stuart Mill and of Kant. He dismissed Mill's proposal, as he had offered psychological foundations instead of intellectual ones. He found these irrelevant (Wettersten 1995). He found it

[12]The legitimation of the diversity of geometries was the embedding one in the other or of all of them in a wider system.

[13]Whether Wittgenstein was a Kantian is too fuzzy a question to be under a helpful dispute since obviously one way or another Kant has influenced us all. Leinfellner (1982), conclusion, declares the output of Wittgenstein but a "linguistic variant of Kant's conceptual transcendentalism". This is trivially true or too farfetched, as you like it.

less interesting how we come to hold that 1 + 1 = 2 than that we take it for granted that for every number n the number n + 1 = n' (where n' is the immediate follower of n). He wanted proofs for arithmetic; Kant had proven that there is no analytic proof for it, as its truths are not analytic. He never proved its truth. Frege scarcely explained his response to Kant beyond noticing that he (Frege) followed Leibniz in considering the truths of arithmetic analytic. He said he was following the search of Leibniz for a *characteristica universalis* (a universal language) without the conflation of Leibniz of analytic and synthetic knowledge. He made it his life task to show this.[14]

Kant aroused forceful responses, positive and negative. His sweeping general statement concerning human knowledge is simple: it is demonstrable but not by means of traditional logic and not by means of traditional induction; the means for that demonstration is what he called transcendental logic. The gulf between the pros and the cons of his transcendental logic was unbridgeable. He defended transcendental logic and its contents by reference to transcendental arguments. Their form is this: *were* a *false, then* b *would be impossible; but* b *obtains; hence, b is possible; hence,* a *is true*. This form of argument is valid. There is no need for transcendental logic in order to apply it as it is a (modal variant of) *modus tollens*. As Solomon Maimon has noted clearly, the trouble is with its application is this: the claim that b obtains is a hypothesis.[15] Maimon said, this is begging the question, a *petitio principii*, asserting what is to prove, smuggling in that which requires legitimation. Worse: Kant assumed that b is uncontestable, that scientific certitudes obtain prior to the application of transcendental logic. This renders transcendental logic utterly circular and thus redundant.

It is hard to comprehend the folly of this idea and the tremendous acclaim that it has won. This is not to say that Kant's introduction of the very idea of synthetic *a priori* knowledge was a waste of time. The very rejection of it, of his view that such knowledge exists, sharpened the understanding of Hume's proof of the impossibility of empirical theoretical knowledge.

Born almost half-a-century after Kant's demise, Frege tried to prove arithmetic truths analytic; he did not discuss geometry, much less empirical science. This, again, was his outstanding sense of proportion. Already then he had to alter logic and with tremendous and most profitable general results, yet not fully satisfactorily. He tried to develop the idea of a fully formal deductive framework. Russell achieved this idea in 1905. He also refuted the hope of Frege – and of young Wittgenstein – to establish arithmetic as a part of logic: as it turns out, only finite arithmetic is purely logical. This led young Wittgenstein to finitism, to the grand metaphysical denial of the existence of infinity. This denial he kept for life.

---

[14]Kitcher (1979, 252) suggests this. "Frege regarded Kant as having settled all of these issues. His primary disagreement with Kant concerns the suggestion that arithmetic is synthetic *a priori*."

[15]Solomon Maimon, *Autobiography*, Ch. 28: "while I hold the Kantian philosophy to be irrefutable from the side of the Dogmatist ... I believe that it is exposed to all attacks from the side of the Skepticism of Hume."

The history of the effort to found geometry on arithmetic altered radically with efforts to refute Kant on a different matter: contrary to his view, there can be alternative geometrical systems. That there can be alternatives to Euclidean geometry follows from Kant's assertion that it is synthetic. That the alternatives must stay useless follows from Kant's claim that Euclidean geometry is *a priori* valid. He considered the option of non-Euclidean geometry, and he rejected it – as possible but useless. So did Poincaré around 1900. For, non-Euclidean geometry will destroy the unanimity of science.[16] The appearance of non-Euclidean geometries in itself is not contrary to Kant's view. What is contrary to it is the idea, initiated by Carl Friedrich Gauss, that a crucial experiment between different geometries is possible. Writers stress that the crucial experiment that Gauss envisaged was not performable. This is a misunderstanding: the mere indication that the applicability of a new geometry is conceivable suffices to demolish Kant's view; the crucial experiment that Einstein devised in 1917 and that Eddington performed in 1919 makes this incontestable. Yet already decades earlier, Russell had suggested that geometry need not be Euclidean. Conventionalists argued against this. They developed a new idea that extended the domain of logic very far: it was the idea of implicit definition: *the set of objects that a given axiom-system describes is any set for which the axioms are true*. This was a radical change in mathematics, in logic, and in the theory of meaning. In particular, it excludes Euclid's geometry, as it is a mere approximation (see next paragraph).

Frege,[17] Russell[18] and Wittgenstein[19] ignored all this; Quine (1964, 71) declared it a tool for introducing as much arbitrariness as we wish. Wittgenstein's idea that understanding concepts develops by playing language games that employ them, and such games are parts of logic, goes much further than the idea of implicit definition.[20] It does not work for science. Whether it works for commonsense is much harder to judge. It does if instead of the system being true we speak of it being

[16]This is intriguing. Traditionally the aim of the search for certitude was to close all disagreement. It did not, since claims for certitude are disputable. The demand to exclude metaphysics from science rested on the observation that metaphysics is the deepest source of disputes. This raises problems that Mach cut short by proscribing (not metaphysics as such but) metaphysical disputes. The problem of demarcation between science and metaphysics turned then into the demarcation between scientific and metaphysical disputes: the scientific ones are resolvable. Perhaps they are marginal in science. Einstein and Bohr rendered this doubtful. This renders the problem of demarcation of scientific theories into the problem of demarcation of reasonable/rational/scientific disputes. All this leaves Wittgenstein way behind; it requires but a minor modification of Popper's view, perhaps only of emphasis.

[17]Dummett (1981, 654; 1991a, 123).

[18]Quine (1960, 71), Quine (1964), Gandon (2012, 82).

[19]See the emptiness of the exegetic and apologetic Baker and Hacker (2009), Ch. 14. Conventionalism.

[20]Wittgenstein (1969, §56), Harvey (1961), Wang (1962, 235).

sufficiently close to the truth for the purpose at hand. The same theory comprises then different language-games: there are simple ways to present the way a teacher, an engineer, a mathematician or a researcher uses it.

Duhem, Poincaré and Hilbert needed the ploy of implicit definitions for the support of the idea of unanimity within science. That idea was (and still is) very strong. Thus, when Einstein openly defied the unanimity of science and went against Newton, the whole of the edifice of modern theory of knowledge collapsed and the search for a new one was on its way.[21] Hence, it seems, Kant's importance was in his replacement of objectivity with inter-subjectivity, and of traditional essentialism as well as of traditional theory of knowledge with his claim that the unanimity of science rests on the status of all mathematical and scientific theories about the world as *a priori* knowledge that does not touch the world in itself. Einstein rejected this. He thus destroyed the obligation to have unanimity in science: "Newton forgive me!" is how he expressed his idea of the new freedom.[22]

Logicism, the view of mathematics as a part of logic, is an old idea in a new context – of modern logic. The novelty of the idea was as Kant sharpened it. Boole and Frege advanced it, and sharpened it more. By then Kant had refuted the idea of essential definition, thereby reopening more precisely all old epistemic questions. He had refuted the doctrine of essential definitions by logical analysis akin to the one that Hume had offered. Yet, whereas Hume was a nominalist, Kant was not, and (unlike Hume) he refuted (not the essential definition of cause, but) the very method of essential definitions in general. Hume went so far as to deny that the axioms of geometry are certain. He took for granted the claim that only logically true statements are certain, and without discussing the status of arithmetic. When Kant repeated Hume's criticism, he did so by distinguishing the status of certitude from that of logical (or analytic or tautological) truth and came up with the idea that arithmetic is certain despite its not being analytic. His analysis being unsatisfactory makes many historians of logic reluctant to praise it; this is narrow-minded.

Gauss derided Kant's view of analyticity: it must be false or trite, he said. This is indeed so, but the derision is unbecoming: Kant's analysis is trite – as all valid analysis must be. This is "the" paradox of analysis. It rests on the erroneous claim that everything proven is trivial, and that repeating a (trivial) statement is useless.[23] The same holds for any repetition of any item of information. It may serve a different end, such as to convey urgency (*Genesis*, 41:32). In mathematics, repetition serves other ends. Edmund Husserl said that Frege's arithmetic is trite as the logic on which it rests is extensional. Poincaré derided Russell's arithmetic proofs as trite: we do not need Russell, he observed, to teach us elementary arithmetic. The paradox of

[21]The Einsteinian revolution rendered Bayesian subjectivism or inductive probability popular among philosophers of science. This resettled the question of unanimity. In my reading of the view of Bruno de Finetti, he says, no matter how diverse the initial subjective probability may be, information soon equates the diverse a posteriori probabilities of any hypothesis. As there is no discussion of how fast this happens, its advocates do not take the idea seriously.

[22]Einstein (1949, 31). See Agassi (1975, 404 *ff*).

[23]To repeat, this was one of Wittgenstein's biggest mistake. See Pascal (1973, 25).

analysis appears in different fields, including mathematics and physics. Poincaré did not handle it there: he admitted that physical theory repeats known information. Even the foundation of mathematics, the theory of implicit definitions, is in his view quite trivial. (Quine 1964 complained that it is too arbitrary.) Except for information, Poincaré viewed (1905, Ch. 5) all truth as *a priori* given and so as trite! It is trite as information, not trite at all otherwise.

4.4 Logic, Mathematics and Meaning

The fragmentation of instruction confuses. Courses on the foundations of mathematics belong to philosophy, not to mathematics. Elementary logic courses ignore meaning. Elementary analytic philosophy courses begin with the discussion of Frege concerning the meaning of the words "morning star" and "evening star". These courses begin with no background, with no mention of a problem, much less its significance. That there is any connection between the calculus of statement composition and the theory of meaning students may be ignorant of, as the presentations of these courses are not historical. Otherwise, they may find it intriguing that Frege is the individual behind both ventures.

Popular opinion often considers Wittgenstein responsible for the current popular philosophical interest in meaning. This makes the fact a matter of fashion. It is not merely that. The whole study of analysis will be easier and much less puzzling for the novice if there was agreement to discuss meaning and meaninglessness separately. For, they address different issues altogether, even though they share the initial problem that gave rise to them and that emerges out of Frege's refutation of the traditional, ancient theory of meaning. (The biblical myth of Adam giving names to animals is translatable to the theory of language as a system of nouns and of meaning as reference!) The responsibility for the interest in the meaninglessness is almost entirely due to Wittgenstein, although the start of the trend is due to Russell, who introduced meaninglessness formally in order to overcome his own paradox. Russell also discussed meaning in order to improve upon Frege's theory of meaning that came to replace the classical theory of meaning that he had refuted. Russell discussed meaninglessness formally and meaning analytically. By contradistinction, Wittgenstein discussed both. This raises the question, what did mature Wittgenstein see in the meaninglessness? He said, metaphysics is tempting, but deceivingly: it is meaningless. This is why he exposed it as sham. Wittgenstein condemned engagement in metaphysics. Russell condemned as sham excessive vagueness and excessive pretentiousness, but he said little about these and used only commonsense to distinguish the sham form the valuable.

Russell thus viewed Wittgenstein's doctrine of ineffability as central and dismissed it as a version of mysticism. Since mature Wittgenstein claimed that rather than respond to a request for a definition, say, of knowledge or of virtue, he could only offer examples for them, perhaps Russell's judgment is a bit too severe. Yet, clearly, Wittgenstein's claim here is insufficient; his view of his philosophy as therapy goes much further than his doctrine of the ineffable. One offers therapy because it hopefully works. The theory that explains its success may be known or not-yet known; Wittgenstein claimed that his therapy works with no hope of finding that theory – since it is ineffable. His therapeutic theory then becomes too elusively subtle to face criticism. Was this Wittgenstein's intention?

4.5 Logic and Mathematics

When non-Euclidean geometry acquired legitimacy equal to that of Euclidean geometry, the old image of mathematics as self-evident was gone. The project that engaged many mathematicians and philosophers was to find what mathematics was all about. A simple mathematical project of the search for the limit of the applicability of the theory of Fourier series forced Cantor to develop abstract set theory that altered mathematics beyond recognition. He also discovered a paradox: the set of all sets is demonstrably bigger than it is. Unlike your yacht, that I only thought that it was bigger than it was, this set is harder to explain away. Frege's logical research would be inconceivable without abstract set theory. He found Russell's paradox shattering, but not Cantor's paradox that he considered resolved. He debated with Cantor the question of the consistency of his system (Ebert and Rossberg 2009), but not with Russell. These days, when the idea of formal systems is clear enough, checking whether a paradox is real or apparent is less problematic; the question, is a system paradox-free is insoluble, as Gödel has proved for any system that is rich enough to be of mathematical interest. This did not trouble Wittgenstein. Consequently, Hintikka (1996, 82, 102, n17, 174) spoke of his "so-called philosophy of mathematics"!

His first concern with it was the same as Russell's: to dispose of Platonism in mathematics, of metaphysics. This is inconceivable *naïveté*, the one that goes with nominalism, sensationalism and a vague Realism that together are less metaphysical than Platonism but not less difficult to maintain. Unsurprisingly, Russell and Wittgenstein failed to dispose of Platonism. Russell admitted this openly (1940, concluding chapter). Not Wittgenstein. The reason for Russell's failure was his theory of definite description that is but a rudiment of a theory of meaning. (Wittgenstein rejected it with no discussion.). Meanings are obviously Platonic objects of sorts. Young Wittgenstein had a theory of meaning that he gave up (as too narrow) and mature Wittgenstein declared repeatedly he had no theory of meaning. As to the philosophy of mathematics, as he did not admit the distinction between the object language and the meta-language, as he ignored the paradoxes, he left the field. Clearly, his early theory was finitist, and one that Russell viewed as obviously inadequate for mathematics proper. He never tried to improve upon it. Russell said he had lost interest. Now such a loss is not objectionable; declaring that mathematics is free of metaphysics nonetheless is objectionable, if not downright ridiculous. Wittgenstein invented a system of naming of logical theorems that should show their proof. He clung to it to the end (1976, last sentence). It is a dud. What he stressed that deserves notice is that even were logicism true, the concerns of mathematical logic and of mathematics proper are very different. On this, Russell agreed. What the difference is exactly is an open question that deserves study (Rodych 2018, conclusion).

Chapter 5
Logic and Language

The great change at the turn of the twentieth century, in philosophy in general and in logic in particular, was the transition from the view of logic as the logic of science – of proven informative truths – to the view of logic as the logic of formal languages – of correct speech, of following grammar. In artificial systems, the rules of grammar are worded in advance; proper formulas – strings of words – are well-formed (wff), and then they are true or false (Frege), within the language to which they belong (Tarski). For natural languages this holds only partially; they are given and their grammars are only partially given; and so is synonymy within them (Tarski; Quine). This was progress, as it led from the old concern with concepts to the new concern with statements and contexts. Yet the progress caused some regress: the traditional center of logic with inferences shifted to a concern with tautologies. This led analytic philosophy away from formal logic. It returned to logic, only partly under the influence of Wittgenstein, since his later philosophy was concerned with fragments of language and neither with language as such nor with its global problems, whereas the concern with logic is global by definition. Wittgenstein expressed concern for this – repeatedly and on diverse occasions.

5.1 Proper Form Versus Truth-Value

The rise of language analysis as a central philosophical activity in the twentieth century enabled Russell, Wittgenstein, Carnap, and others, to forge an up-to-date version of traditional anti-metaphysics, and the new Wittgenstein thesis that philosophy is but language analysis – *sans* metaphysics: analysis should erase all metaphysics as ungrammatical. This obviously is over-simplified, simply since the subject matter of metaphysics was determined historically (and fixed in Aristotle's *Metaphysics*). It is impossible to prove successful avoidance of metaphysics with no clear-cut criterion for metaphysics. Still, logically, the question of proper speech is prior to the question of truth: we have to speak correctly and then ask whether what

J. Agassi, *Ludwig Wittgenstein's Philosophical Investigations*, Synthese Library 401,
https://doi.org/10.1007/978-3-030-00117-9_5

we say is true. Though the aim of science is to seek truth, the matter of truth is dangerously metaphysical, as the truth of an assertion somehow relates to the fitting of the assertion to the real world, and the real world as such is historically the subject matter of metaphysics. To show this is very easy: take the earliest metaphysics reported: Thales said, everything is water. Now one may agree or disagree, but one can hardly feign incomprehension. One can protest that one's beer is not all water; this expression of disagreement displays a level of comprehension that suffices for defying Wittgenstein.

Now and then questions of meaning and of truth arise in common situations. Encounter with a statement in a distorted or doubtful form, or suspicion that a statement is distorted – say, because a wise person pronounces it yet it seems obviously false – normally leads to a request for a clarification or to a conjecture about its content. Such conjectures often come up unnoticed. Russell offered a lovely example: "I thought your yacht was larger than it is". Literally, this sentence is absurd, he observed (1905 final paragraph) yet it makes good sense after it undergoes interpretation.[1]

The concept of inference that Gentzen employed (the calculus of natural deduction) was surprisingly forceful. Frege had introduced rules of inference formally: the rule of substitution and the rule of detachment (the *modus ponens*). The first rule says, whatever is valid for one set of terms holds also for any of its proper substitutes; variables may replace any of their permissible values and *vice versa*. In other words, if an inference is valid with one set of terms, it also holds for the other. This is not very satisfactory: on the one hand, it is less obvious than it sounds. We speak of inference schemas rather than of inferences, so that substitution is included in the very basis of the discussion and invites no further specification. The rule of detachment (the *modus ponens*), then, is the only genuine rule of inference that Frege allowed. Inferences are entirely formal, and so they win assent without debate. It looks as if Frege left out a lot, such as the rules of classical logic¸ including Barbara:

$$\mathrm{SaM, MaP, /SaP.}$$

Worded in accord with the rule of detachment, the rule is this:

[1]Whereas semantic ambiguities are obvious, syntactic ones are not always obvious. Chomsky showed this by presenting some. (In normal reading of the sentence "I love her cooking", he noted, the object of the verb "love" in it is "cooking" with "her" as its qualifier. Grammar allows reading "her" as the object of "love", and "cooking" as its qualifier. This reading is cannibal; it would be normal were we cannibals.) Groucho Marx exploited syntactic ambiguity to conjure funny misunderstandings. (His famous quip, "I shot the elephant in my pajamas", where the adjective may qualify the noun and the pronoun; only the semantics of the quip makes it clear that the adjective goes to the pronoun not to the noun as he intentionally misunderstood the sentence.) Russell offered his amusing case in order to illustrate his view that we always employ language analysis (even if unnoticed). Language analysts often viewed metaphysics as confusion due to ambiguity (Williamson 2002, Ch. 3. Rehabilitation of Vagueness).

$$\text{SaM, MaP/SaM} \cdot \text{MaP; SaM} \cdot \text{MaP, (SaM} \cdot \text{MaP)} \supset \text{SaP/SaP.}$$

This invites a justification, and is easy to find: [(SaM · MaP) ⊃ SaP] is a tautology, and so one can safely add it, and the single premise (SaM · MaP) is equivalent to the two premises "SaM" and "MaP". This is all obvious, and the crux of it is the justification. Hence, rather than write the rule as presented here, it is easier to write the tautology instead of the *modus ponens*:

Barbara as a tautology: ⊢ (a ⊃ b·b ⊃ c) ⊃ (a ⊃ c)
Barbara as a *modus ponens*: SaM · MaP, {(SaM · MaP) ⊃ SaP}/ SaP

The equivalence of these two versions is a case of the deduction theorem.

All this indicates that there is a theory behind Frege's considerations. It offers proof that Frege's view that the *modus ponens* suffices. Tarski proved it. The theories developed in the wake of Gentzen's theory do so directly, more forcefully, and more generally. They are known as "natural deduction", since they rest on the view of logic as a theory of inference, in contradistinction to the systems of Boole and of Frege and Russell that comprise systems of tautologies as was just illustrated. To show this, let me notice that the discussion here smuggles a rule of inference that Frege and others used without notice: a, b / a·b. The truth table for "and" justifies it, of course. It is the unnoticed use of this rule of inference that is the point here. In natural deduction its name is "I·"; that of its inverse, a·b / a, is "E·".

Let me try to present a skeleton of Popper's version of the Gentzen-type natural deduction theory. Its point is that it justifies all and only valid inferences. The justification of all of them is one: they preserve truth. The concept of inference-scheme comes to explain this. An *inference-scheme* has no descriptive word, only logical constants and variables. Any inference displays its scheme upon the replacement of its descriptive words with variables. A sentence treated in that way becomes a *matrix.* Let us then go back to the starting point.

A language comprises a list of formal words, lists of replaceable variables and lists of replaceable descriptive words that may replace the variables. An inference-scheme is a non-empty set of matrices with one of them designated conclusion. An inference-scheme yields an inference by binding variables by quantifiers (the words "all" and "some") or by replacing variables in it by descriptive words. (Thus, the scheme "p(x)" becomes one of the following sentence: "(Ax)(px)" or "(Ex)(px)" or "p(a)".) An inference whose premises are true and its consequence is false is a counter-example. All inferences are invalid that follow any inference-scheme for which there exists a counter-example. An inference-scheme with no possible counter-example is valid: all inferences conforming to it are valid. Thus, by definition, a valid inference transmits truth or else it retransmits falsity. (A valid inference with a false conclusion has at least one false premise.)

Notice this: a variable is free when it appears with no quantifier, as in an inference scheme. When in an inference-scheme a quantifier is bound, there is no need to replace it to make the inference-scheme an inference proper. This is the difficult part of logic, and it can become even more difficult, but we will leave this for experts.

Still, this much is obvious. Putting words for the variables in a matrix to make it a sentence is natural: it is like replacing the word "somebody" with a proper name. "All people", then, may also qualify as a replacement of "somebody".

This is it. The trouble with it – with the theory of natural deduction – is that it is not natural. It holds only for formal systems, not for natural languages, since in natural languages the division between formal and descriptive words is not a partition. This is the point made by Tarski and by Quine: they said, the dichotomy between the analytic and the synthetic does not hold for ordinary discourse. The example above (of transforming a classical syllogism to a *modus ponens*) suggests that more than that is involved: adding a tautology to the premises of an inference. It does not change its status as valid or not. Tautologies are, as they say, necessary truths – given the rules of sentence-formation, the rules of inference generate tautologies as theorems. The stress here is this: valid rules of logic validate tautologies and *vice versa* (the deduction theorem). Yet the division between analytic and synthetic statements is not sharp (Quine 1988).

The deduction theorem explains the move back and forth between logic as a theory of inference and logic as a set of tautologies. The bigger innovation here that makes the matter at hand interesting, is the move made at the turn of the twentieth century from logic as the study of thought to logic as the study of language. What this amounts to is not clear at first glance, since speech expresses thought. Nevertheless, the difference is enormous. It is, quite decisively, the ability to abstract the thinking process from the ideas thought. The English word "thought" at times means thinking and at times ideas. We may want to distinguish between them. This is not as easy as it sounds, and for the strange reason that thinking is concrete and thoughts are abstract.

The transition from the theory of thinking to the theory of statements or propositions is the revolution that is modern logic and modern philosophical analysis. It was very hard to make. When Russell published his *A Critical Exposition of the Philosophy of Leibniz* in 1900, its reviewers found it hard to comprehend – presumably since the transition from thought to propositions invites metaphysics. The very claim that ideas or propositions exist – as objects, that is – sounds too Platonic, as ideas are not to be found in space and time. It is still hard to come to terms with.[2] It was not Russell, however, that began this move; it was Frege. Michael Dummett has put it in his article on Frege correctly but too briefly.[3] He said, "Frege's distinction between objects and concepts cuts clean across the traditional method of posing the problem of universals". The problem of universals is the outcome of the traditional theory of meaning: does a class-name name a class or every one of its members? The reason for putting this so is the traditional view of logic as a system of classification. Hence, its meanings are systems of names, and names denote. This theory – the

[2]When Popper spoke of World 3 in the 1960s, he puzzled peers greatly. The expression is due to Charles Sanders Peirce and to Frege; Russell endorsed a limited version of it. Note: even were language not a World 3 object, ideal language surely has to be.

[3]Edwards 1967, 230; Dummett 1978, 99.

theory that names denote – is traditional. It is historically an idea that many thinkers across time have found most attractive (*Genesis*, 2:20), even though it does not do what it is supposed to do, not even under the best conditions: it does not tell us where ideas reside. It gives the impression that if all there is to language is names, and if names mean their bearers, then somehow thinking is secondary to recognizing objects.

Admittedly, logic flourished when it replaced thoughts with assertions. Efforts to get rid of thoughts altogether, however, are futile. Many sorts of abstract entities occur (including matrices and inference-schemes). People deem computers proofs of the possibility of doing away with thoughts. Now computers perform elementary operations like moving an item from one list to another. The computer moves not abstract items but concrete ones: the numerals 0 and 1, they say. This is false: numerals are abstract. A computer can delete and add an item; it cannot move it around. We consider the process to amount to moving an item. Thus, the computer performs a concrete act and humans read it as abstract. (Computer language proper, assembly language, is quite different from the usual computer languages.)

This is where analysis old-style comes in. It is the alternative of John Locke to both Platonism and nominalism. It is his conceptualism: ideas sit in the mind. *The endorsement of Locke's view on the problem of universals has generated conceptual analysis old-style.* This makes the problem of universals belong to psychology. Frege called it "psychologism" and criticized it. His rejection of the theory of language as names, to quote Dummett's stilted summary again, has "cut clean across the traditional method of posing the problem of universals". Frege was thus the first to ask, Dummett adds, what is the meaning of a statement? This, in his view, is how Frege inaugurated language analysis: where do meanings reside?

Dummett moves too fast. Frege said, assertions convey their own meanings, and these reside in propositions. (As synonymous sentences share meaning, they share a proposition.) He then offered a way of doing away with propositions: formalism. Some formalists, including young Carnap, took linguistic entities to be physical entities such as ink-marks on paper (i.e., printed words) or uttered sounds (i.e., speech). This is perhaps the most unintelligent assertion that one of the most intelligent philosophers of the century has produced. For, the ink marks had to be of given shapes and they had to comprise a system of alphabet. (With no knowledge of the alphabet of a message, it is impossible to copy it, except by producing a precise replica of it as only few artists can.) The ink marks are signs, viewed as meaningless, as distinct from symbols: these are meaningful signs. Strings of signs placed in a proper sequence are well-formed formulas (wffs). Proper sentences are well-formed formulas and propositions are their meanings. In the tradition of modern logic, the following formulas describe meaningful sentences:

$$\text{symbol} = \text{sign} + \text{concept},$$

and, accordingly,

$$\text{statement} = \text{sentence} + \text{proposition}.$$

This is cheating. In truth the formula should read,[4]

$$\text{sign} = \text{symbol} - \text{concept};$$

and

$$\text{sentence} = \text{statement} - \text{proposition}.$$

Matters are not symmetrical here. The idea of denuding statements of their meanings, is a strong tool of logic. It is the invention of Hilbert, who used it for looking at mathematics in a simplified way in order to check it for consistency, as that is so basic for all formal systems. It is by now a standard technique, since computers never handle symbols: they handle signs. Already Maxwell indicated the idea, as he noted that a researcher might take a recognized scientific formula and change its meaning (Harman 2001, Ch. 4). The philosophy of mathematics known as formalism goes further than anything that Maxwell or that Hilbert had in mind, as it is considering signs [logically] preceding symbols. Despite opposition from Wittgenstein (1958, 2), this idea was very popular once, and for obvious reasons: it is often very useful to forget what a statement means. Yet for getting rid of meaning altogether this will not do. The distinction of signs[5] as signs is the outcome of viewing them first as linguistic entities, namely, as symbols. This is an established empirical fact. Hence, if we want a theory as to what constitutes a language, we need a language within which to assert that theory. This is the story of the use of an instrument to forge it; it is obviously frustrating. This, perhaps, is the major difference between the works of Wittgenstein when he was young and developed a detailed theory of language, and mature Wittgenstein, who took language for granted, upon trust, no less, and who only illustrated the diverse ways in which language functions.

[4]*TLP*, §3.32: "A sign is what is perceivable as a [virtual] symbol" (my translation). This is quite different from Carnap's idea of any possible symbol as initially a sign.

[5]Although Carnap was a pupil of Frege, he viewed language as initially a system of signs. The set of signs grew from spoken and written to whistles, gesticulations and semaphore. The addition of speech to Carnap's ink spots is Putnam's (as a silent correction); the addition of semaphore is my response to Putnam. The addition whistling is Ramsey's; the addition of gesticulating is Sraffa's, (Both responses to Wittgenstein preceded Carnap's work.) Dr. Samuel Johnson has topped us all when he said, his sending an empty tobacco box to his tobacconist comprised a clear message. Concrete though the box was, when it served as a message it became highly abstract in a context-dependent manner (as it pertains to a purchase); its grammar is a mystery.

5.2 Laws of Thought Versus Laws of Language

The distinction between the views of logic as the laws of thought and of language, to repeat, is the breakthrough that belongs to Boole; yet he himself labeled his major book, of 1854 *An Investigation of the Laws of Thought.* He introduced there the probability of a scientific theory as the surrogate for its truth. Twentieth century books that emulate him are labeled (not *the laws of thought* but) *inductive logic*. The difference is indicative: Boole referred to proper thought yet his title refers to thought. Unbelievably, Wittgenstein left this error untouched when he repeated the idea that logic is the laws of thoughts and of judgments.[6] He seldom fell into this error, however, as he was regularly concerned with articulations [of thoughts], not with thought.

The revolution in logic was its transition from the logic of science to the logic of language. It thus left science in the lurch.[7] This is not to say that inductive logic now is as backward as its predecessors were. It is much more backward, but due to influences beyond control. Initially, the logic of science was the logic of certitude (or, at a pinch, of probability), as traditionally science was identified as certitude (or, at a pinch, as high probability). The initial, etymological meaning of the word "science" is certitude.[8] Tradition declared science perfect knowledge and scientific method the procedure assured of success. The denial of the possibility of certitude (like Maimon's denial) sounded as the denial of the very possibility of science. This allowed for Newton's theory, for example, as this was called philosophy or, fully, experimental philosophy. Today we call experimental philosophy science with no reference to certitude (the way we call the chemical elements "atoms", namely, indivisible, despite their being divisible). This raises the problem of demarcation of science. What was young Wittgenstein's view of it is clear. What was the view of mature Wittgenstein of it? Clearly, in his *Philosophical Investigations* he refers to natural science with great respect. Yet, he does discuss this.

Russell's view of the matter is clear. He started in search of certitude. His interest in logic began, he reports, as he had hoped that logic would show the way to certitude. He concluded in disappointment: achieving certitude is impossible.[9] He thereby rendered final Frege's divorce of logic from epistemology (Wettersten

---

[6]Wittgenstein 1975, *Lectures on the Foundations of Mathematics*, 272 to the end of the book.

[7]Popper stayed within the logic of science and refused to move to the logic of language. As he was so very daring in other respects, his Viennese peers ascribed to him this transition despite his clear and unwavering defiance. On this, their conduct was both improper and unintelligent. By ignoring Russell on meaning while writing on logic proper, Popper followed Russell more closely than Wittgenstein did.

[8]Frege used "science", "scientific" and "scientific work" to mean utterly certain, utter certitude, and his own research into logic. Both "science" and "certitude" won a special meaning in philosophy and left their ordinary meaning behind. The demon of Descartes destroys the utter certitude of the philosophical sense, not certitude in the ordinary sense – not even in primitive languages!

[9]Since then a few authors, e.g., Hintikka, devoted books that repeat this lesson: utter certitude is impossible.

1995); logic was a theory of science no longer. This was the first step leading to the separation of the study of scientific method – methodology – from logic proper.

The problem of induction was central to modern philosophy at least since Kant acknowledged the justice of Hume's analysis. Hume's own solution to the problem of induction was psychological, as was the solution that Kant has offered – no matter how differently they viewed both psychology and knowledge. (The main difference is in Kant's assurance that all people must share the same solution, so that it cannot be empirical. This is his transcendental logic that he never described.) Frege's separation of logic from psychology raised the problem of induction afresh and in a new light. This was an obvious achievement.[10] Russell's *Problems of Philosophy* of 1912, written from the viewpoint of the new logic, became standard at once. He spoke there (Ch. 13) of probable opinion. Wittgenstein rightly disagreed: he separated induction from probability (*TLP*, §§6.31, 6.363; 4.464, 5.15–5.156, 6.34 *ff.*), saying, probability is a formal system that cannot possibly account for learning, much less for learning the laws of nature. He triumphantly stated the divorce between logic and psychology, with psychology taken as just one science among others (*TLP*, §§4.1121, 6.3631). On this Popper fully agreed with Wittgenstein. Once we give up the Kantian idea that our mind bestows certainty upon experience, the laws of thought become the laws of all thinking, right and wrong, not necessarily the laws of proper thinking, nuch less the laws of *a priori* knowledge: then they become no longer guarantees. This is no small matter, considering the inability of the members of the "Vienna Circle" to assimilate this contention into their official philosophy despite their acknowledgement of debt to Wittgenstein. The exception was the proposal of Hans Reichenbach to bring inductive logic closer to logic proper by viewing probabilities as degrees of truth. Popper refuted him ([1935] 1959, §§8, 80) independently of views about inductive logic. The divorce of logic from psychology allowed Popper to argue ([1935] 1959, §8) that science has no use for induction. Most analytic philosophers still swear by inductive logic without taking the trouble to try to refute Popper. This way they forego the benefits of his criticism.

The divorce between deductive and inductive logic that Frege and Russell effected was a break with tradition: Aristotle instituted inductive and deductive logic as isomorphic, as sharing the same formal characteristics. The inductive logic that Aristotle spoke of is that of Socratic dialogues, of cleansing the mind and thus setting it on its way to the acquisition of certain knowledge.[11] Deductive logic uses this knowledge. Inductive logic was thus traditionally a part of traditional logic – from its earliest days to the days of Bishop Richard Whately, whose text (*Elements of Logic* of 1826) Boole had used. Since influential Bacon criticized Aristotle severely, logic suffered neglect. Hume's critique of induction and Kant's austere view of tautologies pushed traditional logic further aside. The crisis in

---

[10]Although already Faraday suggested that priority in the discovery of problems wants recognition, nothing has happened on this matter.

[11]Aristotle offered an alternative, *pro-tem* method. Its popularity sent the original to oblivion.

physics of the turn of the twentieth century brought methodology back to center-stage and Russell to examine logic as a part of this process.[12]

The two chief advocates of inductive logic, Aristotle and Bacon, both introduced stopgap methods that eclipsed the originals; until the middle of the eighteenth century, all this escaped criticism, perhaps because Newton viewed induction as the only empirical method. The writings of Hume were very unpopular, yet they began a change. Induction slowly changed to become the laws of (mathematical) probability or the laws of abstraction, or preferably, both in close cooperation and mutual support. Abstraction was central item in nominalism, as at least seemingly it conflicts with Platonism. Abstraction is the process by which, when we see a class of things sharing a characteristic, we allegedly develop the concept of that characteristic.[13] Whately's discussion of inductive logic presents abstraction this way. The nominalist bias of his view is obvious: it takes the abstract entities as class characteristics, class characteristics as classes, and class names as the names shared by all their members, so that classes need not exist: they are fictitious *interim* entities, mere nomenclatures (Quine 1967).

Russell's *Problems of Philosophy* (1912) set the trend. He admirably reworded Hume's critique of induction neatly and described its import broadly. One of the most striking of this is the stress on the surprising contrast between empiricism and this nominalist bias. Russell said, the abstract quality "red" is more concrete than red things, since, according to [Locke's] empiricism, objects are constructs of sensed *qualia*.[14] That redness is an abstract, namely, a universal, is a clear fact that analytic philosophers constantly ignore – because they are in conflict here about things being primary. Their nominalist bias forces them to prefer to see a single red rose as primary and so as prior to its redness.[15] Their empiricist bias, however, goes the other way, as they present the rose as a complex with redness as one of its components (and its shape as another). This still is the source of much confusion. It is one of the great benefits from the separation between philosophy and psychology that it renders discussion of observation-reports methodological and thus free of all perception theory (Popper [1935] 1959, §§8, 27). What made Popper so hard to

---

[12]Monk 1996 and 2000 life of Russell does not do justice to his most significant characteristic: from an early age, Russell was politically deeply involved, feeling the weight of the world on his shoulders (Monk 2001). Russell retained this attitude as a matter of course when he entered philosophy: he considered the problem of rationality politically most important and the problem of induction he saw as a cardinal a part of it.

[13]Russell's principle of abstraction that is a part of his theory of definite description is not methodological, since it begins with the given abstract property and posits it as a selection rule.

[14]This recognition of the vagueness of what is inductively primary appears also in Wittgenstein's only published paper. I suggest that it forced him to stop discussing science altogether.

[15]Carter 1989, 4.6: "Russell's Bundle Theory". Bishop Whately wrote as a *naïve* empiricist. Russell wrote as a methodologist. Carter wrote as a metaphysician. He follows his intuition when he disagrees with Russell. The value of his work is in his view of sense-data theory as ontological, though it began as methodological. (Pepper 1942 did this earlier, cast his net much more broadly, and had more interesting results.)

comprehend is that he founded a new kind of empiricism: he offered no discussion of perception since his empiricism was without foundations (Agassi 1963).

Frege's anti-psychologism came to fruition only when Russell and Wittgenstein forced perception theory to yield to language as the focus of the new logic (Wettersten 1995). Now language theory has its referential theory of meaning as central, and there the basic dispute is between nominalism and realism/essentialism. As Frege has refuted the referential theory, we may sympathize with the investment that he and his followers, Russell and Wittgenstein included, of efforts to replace it. The starting point here is ontology. We all wish to assent to the view that whatever exists dwells within space-time (Bunge 1979a, b, 278). Our language should reflect this.[16] Frege seems to have violated this when he postulated his *drittes Reich* (third domain) – as did Peirce with his Threeness and Popper with his World 3. It is possible to go further and ascribe this violation to the very naming of abstract entities, including classes or sets.[17] The hardest problem to face when discussing the background to Wittgenstein's philosophy is Russell's criticism of Frege, a milestone in the philosophical development of Russell and then of Wittgenstein. What was the disagreement? Russell said he rejected Frege's metaphysics, namely his recognition of Plato's Heaven. Where, however do Russell's sets dwell? This question does not obtain, because he deemed sets fictitious. By the same token, he could endorse Frege's third realm as similarly fictitious. He ridiculed logicians who assumed that such things as logical connectives ("and", "or", etc.) exist somewhere in the sky. Yet this existence differs from the existence of pieces of furniture. Quine took deep offense, let me testify, from the refusal of commentators to do him the courtesy of recognizing that in his view sets exist just like pieces of furniture. Likewise, Bunge said that nations exist just as much as individuals do. Now, evidently, no one denies that sets differ from pieces of furniture in degree of abstraction. Why then did he insist that all existence is the same? We do not have to go this way with him (Lejewski 1954) as we have no adequate theory of existence anyway. Russell spoke rightly of his robust or vivid sense of reality that prevented him from assuming the existence of sets. Could he not postulate different kinds of existence without impinging on his robust sense of reality?[18] Bunge objected sternly to Popper's World 3; would he have changed his mind had he read this Threeness as fictitious the way he himself reads all mathematics and even space-time? For, we all share the view that there is but one world. Would Russell admit Frege's *Drittes Reich* or Popper's World 3 on these terms? Popper did identify his own World 3 – that

[16]This does not pertain to poetry. Hence, the wish is to have a language for mathematics and for science separate from the language of poetry. This is impossible for obvious reason: we cannot predict the use of language – scientific or poetic.

[17]Mathematicians admitted sets casually; Cantor did this with a theory that had religious overtones. Frege followed him with his conception of mathematical existence. Dedekind and Hilbert legitimized abstract set theory; Zermelo and Fraenkel made it a fixture in one way and Whitehead and Russell in another (Tait 1997a, b, 35–6). Set theories then proliferated.

[18]Landini rightly wonders why the axiom of reduction has worried Russell this way more than sets.

includes a conglomerate of social institutions – with Frege's *Drittes Reich.*[19] Is it possible to overlook the objections of Russell and of Bunge as over-sensitive?[20] Russell admitted that this is possible when he admitted that his robust sense of reality did not allow him to admit the ghostly Platonic Heaven. He was proud of his theory of definite descriptions most as it repeats Frege's achievement without dependence on Plato's Heaven, even though not completely. Nevertheless, we should remember the admission of Russell (1940, Conclusion) that he could not destroy it fully. Since Wittgenstein acquired his anti-metaphysics from Russell, perhaps they share this very failure. Wittgenstein is thus to praise for his thoroughness that made this failure obvious.

Traditional empiricist philosophy took sense perception seriously as the foundation of knowledge. Whether the observation of the rose precedes that of its redness or not, we do not know. Hence, the foundation for science that empiricism can offer is shaky at best. We may then give up this function of empiricism and leave the question of what we perceive primarily to perception theory to decide. Though the analysis of perceptions was the task of methodology, with the loss of certitude this task has left it anyway. Scientific researchers do not have to speak of redness, unless it turns up in their scientific theories and in their observation reports that they use as means for their tests. Russell's unease about the redness of the rose was logical: is it a universal or a particular?[21] The problem has survived, even though rather shamefacedly.[22] Wittgenstein too struggled with names of colors (*PI*, p. 235). Their meanings are apparent in their use, as he repeatedly insisted, since they are very much language-specific,[23] and yet his discussions show that this adage, the meaning of words is apparent in its use, is a problem, not a solution.

The classical empiricist theory raises the odd question, what makes me identify this rose with this drop of blood rather than with that drop of milk? Carnap, the most distinguished analytic philosopher Vienna-style, answered this question ([1928] 1967, 92). His aim was to present a system in which "*all physical objects are reducible to Psychological ones*". (This seems to be the opposite of the materialist program of reducing all mental objects to material ones; if one is sufficiently confused, one can identify the two with ease.) Carnap's answer (§31) then seems to be the obvious one: this drop of blood is more "color-akin" to this [red] rose than that drop of milk is. This, Carnap stresses, is not quite a universal truth. Why then is this red rose color-akin to a drop of blood more closely than a drop of milk is? Carnap approached the problem formally. He considered basic (namely, as not open

[19]Popper admitted he could not use Frege's label because of its repulsive political association.

[20]"Frege himself never explicitly draws or suggests" (Kluge 2013, Appendix) a comparison between Plato's view and his own, yet Russell and Bunge took for granted that he was a Platonist.

[21]Russell 1956a, 103, "On the Relations of Universals and Particulars".

[22]See, for a conspicuous example, Hempel 1965, footnote on p. 269. The problem there concerns not sets but law-like statements or, more generally, universal statements. By contrast, Popper [1935] 1959, §14 noted that the difficulty is general: it applies not only to sensations.

[23]Brakel 1994 is an extensive survey of the situation. It is amazingly complex.

to further analysis) the concept perceptual kinship (§70) and created an axiom system around it. (He added a single non-phenomenological term of kinship, "x remembers partial similarity between y and z", as a primitive term, axiomatically defined; §119.) Why take this term (remembered similarity or anything else) as central? Did that suffice to adjudicate rightly between the normal and the colorblind?[24]

Carnap's taking the concept of perceptual kinship as basic is an expression of his inductivism. To ask why we associate sensation x with sensation y is to seek a cause for their association. The inductive canon is the suggestion that rather than to seek causes, rather than to try to answer questions about causes, the right procedure is to gather many facts and then abstract theories from these facts. These abstract theories will then constitute causal explanations. The claim that both the rose and the drop of blood and other things remind me of each other is the claim that I create for them a new common name: "red". Redness, then, explains the resemblance: the resemblance is in redness. As a description of the learning process, this is easily refutable by the psychology of infants. By contrast, as a description of the theory of knowledge that Carnap's study ends with, this is easily demonstrable: the answer to the question whether a statement is true lies in the extant list of all names of each of the things to which it refers. For example, "all swans are white" is true iff every object with the name "swan" also has the name "white". This, Popper observed, renders all empirical general statements necessarily true or necessarily false. Carnap's fans now concede this; regrettably, they do this awkwardly and reluctantly (Agassi 1981a, b). In time, they will learn to be more courteous.

5.3 Between Logic and Psychology

The transition at hand was from the concern of logic with proper reasoning to its concern with rules for forming and transforming sentences. Frege viewed the former as psychologism, deemed it a failure, and demanded a divorce. Russell's *Problems of Philosophy* (1912) agreed. This put an end to a great chunk of logic, such as the traditional theory of judgment.[25]

The fault of old-style analysis always goes back to the absence of the empty name (or the empty class). This absence prevented the identification of a universal statement with the assertion that some class is empty. It is thus less surprising than

[24]Note that whereas colorblindness is a deficiency, most perception failures are due to (magical) excess.

[25]It is hard to identify the theory of judgment. Despite its having occupied many pages of important ancient and mediaeval texts, it is absent from the histories of Kneale and Kneale and of Russell. My claim – that it comprises efforts to smooth rough edges of logic – is conjectural. Kant used this term to name the theory of the process by which one decides that a proposition is true. Russell and Wittgenstein together found it hard to characterize such judgments (Hager 2012, 163). Hintikka 1993, 80 took Wittgenstein's view of it to be his phenomenological language.

it sounds at first that the idea of universal laws as prohibitions is twentieth-century: Popper stated it in the thirties ([1935] 1959, §§6, 15, 31), Sir Edmund Whittaker did so in the forties (1949, §25). If so, to what do the laws of nature refer? If meanings are references, then this question endangers the very meaning of scientific statements. This is significant, even to young Wittgenstein, although he banned universal statements from his language. He did not notice that reducing universals to conjunctions he lost their characteristic as prohibitions. This bothered Schlick, the speaker for modern analysis: echoing John Stuart Mill, he said, strictly speaking, universal statements are meaningless.[26]

Modern discussion is of extension and intension, of reference and sense, of denotation and connotation. These sets of pairs differ from each other, but we may ignore this for now. The replacement of the traditional one factor with the modern two is indispensable, as meaning links language with the world. One is our understanding – of a word's or a sentence's contents, its sense, connotation, or intension. The other is the world that we describe – the reference, denotation or extension. Consider a true statement, and suppose that we see how we unite its sense and reference; this way, whatever it is, will not work on its falsity, for want of reference. Here not truth but falsity is the key, and the centering on logic. Logic as viewed traditionally centered on science, on knowledge, on certainty; this led attention away from falsehood. Quite generally, the difference between old-style and new-style analysis is often this. The aim of analysis is to get things right, and in the old style right means true whereas in the new style it means proper, true-or-false. Indeed, Frege's concept of truth-value, of being true or false, is a great innovation. And so, if we want a piece of analysis to be correct by the new standards, then we should see to it that it has a truth-value, rather than that it has this or that truth-value. Of course, we want to know the truth, but the suggestion is that before examining questions of truth we should examine questions of truth-value; if a sentence does have one, then we should try to find out what it is. This sounds not problematic at all; it took Russell to see here a difficulty (1912, Ch. 12):

> Our theory of truth must be such as to admit of its opposite, falsehood.

The *naïve* reference theory of meaning renders falsehood more problematic than truth. Before discussing factual falsehood, however, we may discuss analytic falsehood, namely, contradiction.

The view of contradictions as meaningless rests on the traditional referential theory of meaning that Frege banished. Nevertheless, he had trouble with it. Russell overcame this trouble by resolving a bigger problem, that of paradoxes, of demonstrably true contradictions, of statements that are demonstrably true-and-false simultaneously.

---

[26]Schlick received an appreciative letter from Einstein for his critique of Cassirer's conventionalist reading of the theory of relativity. He then joined Wittgenstein and this forced him to reversed judgment. He thus exchanged following Einstein with following Wittgenstein (Agassi 2005, 321).

Russell's paradox is quite general and independent of any specific system. Russell found it by examining the disturbing famous ambiguity between class inclusion and class membership. Consider the simplest categorical proposition, SaP. If S is class, then the copula "a" in it is "is a subclass of": SaP reads, the class S is a subclass of the class P. If S is an individual then the copula "a" in it is "is an element of": SaP reads, the item S is an element of the class P. Of course, readers trained in Boolean algebra will have no trouble speaking of that individual item as the unit class, as a class that has only one member; that will render the copula of SaP mean systematically sub-class. The artificiality of this should not matter. What matters then is the ambiguity: we may refer to an individual and we may refer to its unit class. When we say that the individual is the sole member of a unit class, we have to note the difference between class membership and class inclusion. Even earlier than Russell, the convention developed of writing "$\subset$" or "$\subseteq$" for class inclusion and "$\in$" for class membership. ("$\subset$" denotes proper subset and "$\subseteq$" denotes subset-or-equal-set.) What Russell was still troubled about was the ambiguity in ordinary parlance: he wished to free logic of it. Now there is a theorem, easily provable though it sounds strange: every class is a subclass of itself: ($\forall$a (a $\subseteq$ a)). (Proof: otherwise, there would be a class that is not a subclass of itself, contrary to the principle of extensionality.) This led Russell to the following idea. Suppose no class is a member of itself ($\forall$a (a $\notin$ a)); since all classes are subclasses of themselves, this will sharpen the contrast.

Russell tried to show, then, that no class is an element of itself. Call such classes normal. Normally, that is, classes are not members of themselves. Russell's plan, then, was to prove that all classes are normal. That is to say, he tried to prove that there is no non-normal class: no class exists that is a member of itself. He then asked the fatal question: is the normal class itself normal or not? To put all this symbolically,

$$x \notin x \leftrightarrow x \in N$$

to read, x is normal, if and only if it is not a member of itself, and the name for N is "the class of normal classes" or, briefly, "the normal class". Russell wanted to prove that all classes are normal. To ask whether the normal class itself is normal is to ask if the following formula is true that obtains from the previous one by substituting "N" for "x":

$$N \notin N \leftrightarrow N \in N.$$

Suppose the formula is true. It reads,
N is in N if and only if N is not in N.
Let us abbreviate "N $\in$ N" as "R".
Then, it is possible to write the formula as "R $\leftrightarrow$ ~R", and to decompose to two:

$$(1)\ R \rightarrow \sim R;\ (2) \sim R \rightarrow R.$$

This is very embarrassing. For, assume R; from it and (1) it is possible to deduce ~R by the *modus ponens*, thus proving ~R by *reduction ad absurdum*. Yet, likewise, assume ~R, and from it and (2) it is possible to deduce R by the *modus ponens*, thus proving R by *reduction ad absurdum*. This is a proven absurdity, a paradox. In logic, this is a catastrophe: in a system that includes a paradox all statements are demonstrably true.

Russell's solution was to outlaw the paradoxical sentence R. That made Russell see clearly that it is possible to assert contradictions, since they are well formed (though false); outlawed expressions are not false: negations of true sentences are false and *vice versa*, but negation of outlawed expressions are likewise outlawed. The logic of meaning got thus divorced from psychology. This way Russell could develop a formal language, one that includes inferences like.

$$(a)\ c/a$$

to mean every statement follows from a contradiction.

Wittgenstein's rejection of Russell's elimination of the paradox as artificial reunites the logic and the psychology of meaning. Their disagreement defined the two branches of the analytic school, the formal and the ordinary-language. For some historical reason the two analytic schools are known as the "Vienna Circle" logical positivism and as Oxford ordinary-language philosophy. They shared the Frege-Russell idea of meaning as truth-value, as well as the idea that metaphysics is meaningless, and that there a meaningful question most have a meaningful (namely, verifiable) answer. Until recently, tradition ascribed these two ideas to Wittgenstein as a matter of course (even though the name "verification principle" is the invention of Waismann). Now it is becoming fashionable to contest this. There is even a suggestion, believe it or not, published in a leading philosophical periodical, that this allegedly false ascription is the invention of Popper (Smit 2015).

The idea that a question with no (verifiable) answer has no meaning is still in occasional use. Now some questions in physics that had no answer for millennia did find answers. (Examples: Why is the sky blue? Why is the grass green?) Hence, the idea that a question without an answer is meaningless is refuted or too vague. The logic of questions and answers should help adjudicate this. This logic takes a question to be the task of choosing between alternative answers to it, where a statement qualifies as an answer when it satisfies some adequacy conditions or *desiderata*. Otherwise, all questions of fact are answerable merely by reference to God's will ("x is the case because God wills that x be so"). If a language analyst will protest, saying that this answer is metaphysical, and then it is very easy to meet this objection by replacing God's will with, say, Wittgenstein's will ("x is the case because Wittgenstein wills that x be so"). This will be palpably false and thus decidedly meaningful. It is thus explanation by the deductive theory of explanation denuded of its *desiderata*. The question is, then, what explanation is scientific, or what explanation is satisfactory? Now the wish for a satisfactory answer may go too far: we all want our explanations to be true. It is the elusive character of the truth (Democritus) that is the very rationale of the present discussion. Therefore, we make our *desideratum* not the truth but the mere option of searching for it. Any *desideratum* such as the demand that an explanation should be verifiable (Hempel), testable (Popper), offering a mechanism (Bunge), or fit a given metaphysical system (Einstein), any of these may be a candidate that will definitely disqualify this falsehood as an explanation. This is clear enough an example for adequacy conditions for a statement to qualify as an answer. Questions usually have tacit adequacy conditions, and many jokes illustrate this. (When asked, why do you rob banks? Dillinger answered, because that is where the money is.) Now *desiderata* are usually tacit because the people who use them take them for granted. Often the same people who take them for granted will find it hard to articulate them. They are nevertheless effective enough. When it matters much and yet remains untreated, then the people for whom it matters may invite lawyers to express clearly the *desiderata* involved. When this will not do, when lawyers cannot render clear what is too unclear, then they may invite law courts to adjudicate. If they too fail, then some lawmakers have to step in. When legislatures fail to do so – for any reason whatsoever – then the vagueness remains within the system. This is an inescapable part of any way of life: things are worse in societies not governed by the rule of law.

All this may prove the philosophy of young Wittgenstein most inadequate: here are cases where something must be said, and even systematically, but without the ability to say it clearly. It seems he was fully aware of it: he noted the ambiguity of the copula that Russell had discussed at length (*TLP*, §3.323) and that "language disguises thought" by sowing confusion (*op. cit.*, §4.002) and that propositions in ordinary language are "perfect as they stand" (*op. cit.*, §5.5563). To this last sentence he added enigmatically, "(Our problems are not abstract, but perhaps the most concrete that there are"). It seems that here a few ideas coalesce. First, Wittgenstein did not mean his language to replace the vernacular. This is why Russell's critique of its limitation came from mathematics and not from ordinary experience, as is done in this paragraph. Second, ordinary parlance can be subject for scientific research. There the study of metaphysics must belong. Peter Strawson took this to heart when he gave his book on metaphysics the subtitle "*An Essay in Descriptive Metaphysics*". Oddly, his own audience failed to understand his neologism "descriptive metaphysics". Third, Wittgenstein meant his first book to be useful for everyday affairs; it puts science on a par with applied science. This was the tough-and-no-nonsense ideal of many a classical positivist, and Wittgenstein proudly claimed to have achieved it. He did not. In any case, the claim that the early philosophy of Wittgenstein is hard to comprehend is usually due to his brevity and to the refusal to ascribe to him false ideas. Otherwise, he is not so difficult to comprehend.

The case of mature Wittgenstein is different: the same question raised for young Wittgenstein arises equally for mature Wittgenstein: what adequacy condition renders an answer non-metaphysical? He never answered it. He denied that he had a general view and was sufficiently satisfied with offering instances. Does this reluctance to generalize find any roots in the ideas of his predecessors within logic? How much does this idea owe to extensionalism?

Chapter 6
Frege

Extensionalism makes it possible to study logic independently of any theory of meaning. Frege's logic was not fully extensional, however: he developed a theory of meaning in order to have classes uniquely determined. That theory is also flexible enough to allow statements of identity to be at times analytic, at times not. He rejected the traditional empiricist epistemology as an obstacle since it is psychologistic – even though he could not replace it. (A few years after Frege died Popper proposed the first non-psychologistic theory of science.)

6.1 Anti-Essentialism Again

This study is still stuck in the history of logic, because its topic is Wittgenstein's *Philosophical Investigations*. Mature Wittgenstein intended to deprived philosophy of all of its explicit informative content. This might have caused the utter neglect of the study of the acquisition of knowledge (including the acquisition of language). The new field of cognitive psychology could and did take care of these, as a separate field of study. Moving learning theory from philosophy to psychology caused excitement; it resulted from the separation of logic from the theory of judgment; this was the result of Frege's rejection of what he called psychologism, namely the mixture of logic and psychology. That mixture resulted from the conjunction of two views: of logic as the logic of science and of meaning as conceptual. The logic of science did not disappear. The distinction between the psychology of learning and the logic of learning is due to William Whewell (Wettersten 1995), a generation before Frege. It became unavoidable once Frege separated logic from psychology as a task that his refutation of the classical theory of meaning rendered urgent. (Whewell still identified the logic of learning with the logic of science; the distinction between the two could not appear before Einstein destroyed the view that improving upon Newton's theory is impossible. Whewell's tremendous achievement was limited, Wettersten observes (2005, 78–9): his explanation of the infallibility of

J. Agassi, *Ludwig Wittgenstein's Philosophical Investigations*, Synthese Library 401,
https://doi.org/10.1007/978-3-030-00117-9_6

science was not satisfactory. Thus, his theory never won public assent despite of his tremendous prestige.)

Cognitive psychology appeared as a separate field of study in Russell's *The Problems of Philosophy* (1912). The later recognition of seminal works in that field entrenched the recognition of the field itself only towards the end of the twentieth century. No less significant was the development of mathematical logic, which likewise separated from philosophy and became partly mathematical and partly the new fields of computer science and of information science. This development was more spectacular and exciting. Russell presented it (1912, Ch. 15) as a general process: "as soon as definite knowledge concerning any subject becomes possible, this subject ceases to be called philosophy, and becomes a separate science".[1] What else, then, does philosophy include in a seminal form that can become a separate science? For this, the field of philosophy itself had to undergo analysis; Wittgenstein's *Tractatus Logico-Philosophicus* that appeared soon after World War I performed such an analysis for the first time. The verdict it offered is disappointing: articulated philosophy has no informative content; to expect more of it is therefore unreasonable. We therefore must acquire familiarity with the new logic as a tool of analysis, and with the view of logic as the formal theory of language and of mathematics, not of science.

This development assumes the downfall of essentialism, according to which essential definitions are both informative and logically true. Kant had said, informative truths are not analytic. Wittgenstein agreed, and added that they are not *a priori* valid either. At least this is the usual reading of his texts. There is an exception, though, in his last word (*On Certainty*, §105):

> All testing, all strengthening and weakening of a hypothesis, takes place already within a system. And, indeed, this system is not so much a more-or-less arbitrary and doubtful starting point for all of our arguments, as it belongs to the essence of that which we call an argument. The system is not so much the starting point as the vital element of the argument.

This observation should interest all those who study the problem of rationality. It is the surprising refusal to pay the price of fideism even while admitting its force: admittedly, a presupposition stands behind every argument; it is essential for all argument; it is thus not the arbitrary presupposition that fideism recommends. It is the proof of fideism by a transcendental argument. Except that Wittgenstein did not explain his view: what is the presupposition that stands behind every argument in a system, and what is a system.

Descartes' radical program to start *de novo* is impossible: it cannot avoid the necessity for some universe of discourse. Wittgenstein added to this in the above-cited passage a system, a universe of discourse whose basic suppositions must be

---

[1]How developed is a field when it ceases to be a part of philosophy? This problem of demarcation differs from the one that Popper studied, though the two obviously overlap. Raising different problems of demarcation should appeal to analytic philosophers and perhaps it will when they will become serious again.

certainly true. The alternative to this is considering the universe of discourse putatively true and trying different ones. Open-minded people may consider this better than the demand for certainty. To allow for it, we have to relinquish the demand for certitude. The above passage from Wittgenstein's *On Certainty, incidentally,* shows that one may reject essentialism and use the noun "essence".[2]

The systematic error of traditional essentialism is in taking the status of statements about essences to be that of definitions (as logical truths). Kant noted that. His most impressive disciple, William Whewell, rightly advised to consider them hypotheses and test them empirically if possible. It may be impossible, however. As the passage just quoted from Wittgenstein says, a test may involve an intellectual system as a whole. That system is irrefutable, and so possibly we are stuck with it. This, however, need not always be so. Somehow, research does help improve intellectual frameworks: our present ones, all of their faults notwithstanding, are by far superior to older ones. At times even conservative thinkers admit that.

At least two different distinctions appear here, between the informative and the tautological, and between the empirical and the metaphysical. They differ. Language analysts took it for granted that the informative and the empirical are coextensive. They are not. The word "metaphysics" is a jargon term for first philosophy or first principles. Metaphysical systems are principles, assertions concerning the nature of things, concerning the basic furniture of the universe, the kind of which Wittgenstein recognized in the above-cited passage (Watkins 1957a).

All this is hopefully a useful introduction to Frege's anti-essentialism. His idiosyncrasy lies elsewhere; he refused to view metaphysics as hypothetical, as doubtful. Following Kant, he declared metaphysics certain but not by definition. Its certitude, in his view, rests on the intuitive obviousness of its axioms. More than that, he insisted that though the axioms of mathematics are intuitively certain, they are also, contrary to Kant, in part logical truths: he deemed arithmetic logically true, but not geometry. His hostility to essential definitions made him an extensionalist in matters of classes. His demand for intuitive certitude of axioms with no reference to psychology, this demand he considered the demand to choose as axioms the most transparent options. This made him reject extensionalism regarding concepts, and study their meanings. This is a central point about him, and it does not obtain for logic today. As to the axioms of the calculus of statement composition, logic today proves them by the truth-table method (invented by Peirce and by Wittgenstein) and by natural deduction. As to the axioms of geometry, as Einstein did replace them, they have lost all claim for being certainly true. (In pure mathematics, they are implicit definitions, acceptable, if at all, only for some very limited ends.)

This does not exhaust anti-essentialism. The idea that knowledge is deductively based on intuitively obvious principles is traditional; in certain contexts, this justifies

---

[2]Wittgenstein rejected the Cartesian solutions the problem of induction and only hinted at an adequate solution to it. (Will 1974 elaborated on this passage.) Wittgenstein suggested that the framework for research is given and its very use solves the problem. It does not: researchers choose systems tentatively; if they improve them, they have to revise their questions and this they usually do.

viewing tradition as essentialist. Even without essential definitions, there is a choice between essentialism and nominalism, and within essentialism between the versions of it of Plato and of Aristotle. As Russell read Frege, he sided on this with Plato, not with Aristotle; Russell tried to construct nominalism of sorts but admitted failure. This is not to say that Frege sided with Plato on the question, what items reside outside space and time, nor even on what is that region that is outside space and time; only that thoughts and truths do exist "out there". In other words, some abstract entities do exist. This Russell tried to do without. He admitted failure (1940, final two pages).

Frege offered three new items: the calculus of statement composition, the functional calculus and the theory of meaning. The root of these three items is the chief point of departure of his view, his discovery that logic concerns statements. Contrary to the whole tradition of logic as the logic of terms [or classes], Russell insisted that words gain meanings only in the context of statements in which they occur. Dummett, Frege's leading scholar, has called this "the context principle". It is important for the calculus of statement composition, as its key words, known as "the logical constants" (as they are not substitutable, since they determine the structures of sentences), the words "... and ...", "... or ...", "not ..." and "if ... then ...", acquire their meanings (not separately as tradition requires, but) together, in context. The context principle has puzzled many commentators, including Dummett, since Frege had a context-independent theory of meaning: the meaning of a term, he said, is its sense plus its reference. The literature includes a detailed discussion of Frege's distinction between sense and reference, and this is the pitfall: it is not the distinction between them but the theory behind them that matters. Every distinction is valid (since essentialism is false); it is the usefulness of a distinction that matters, not the distinction itself: the distinction at hand helps one explain how names that do not designate can have meanings, and this explanation is a theory. The problem that that theory solves gained attention *apropos* of the empty class and apropos of identity. This is why the observation is so important that words, even meaningful ones, do not create sentences (notice that here elipsis is disallowed), that sentences have structures in addition to words, so that the meaning of a sentence is not derivative from the meaning of its words; hence, in a sentence a word has meaning depending on the sentence-structure (Wright 1980, Part One). This is why words have meanings only within sentences. This invites the study of the structure of statements – of simple ones separately and of complex ones, those that can be split into simple ones. The existence of one-word sentences in ordinary parlance is especially intriguing. Indeed.

6.2 What Is a Calculus?

Frege's path-breaking work found little general interest, and the little that it did find was rather unappreciative. The spokesperson for Boolean algebra, Ernst Schröder, declared Frege's functional calculus more-or-less Boolean algebra (since $x \in C$ is the same as $P(x)$ where C is the class whose elements are all and only the objects that have the property P). Now Frege's system includes not only one-place predicates. Yet he responded to Schröder more ambitiously: his system is not only a calculus but also a language, a universal language Leibniz-style. Today we can see with ease the difference between the two: a language has to include contradictions, or else we cannot describe disagreements in it, not to mention debates about them; a calculus must be consistent or else it is valueless since in it all possible assertions are theorems. Thus, a language is a set of rules for sentence formation and transformation plus a vocabulary that fits these rules, whereas a calculus is more restricted: it is a set of theorems – its axioms – and rules for generating more theorems plus a much smaller vocabulary – of terms needed for the wording of them. Thus, the proper presentation of a calculus requires a formal language. Hilbert offered a formal version of Euclidean geometry in 1899 whereas the proper idea of a formal language (that include contradictions) appeared first in 1905, and the first formal language was PMia, 1911–13, and even that not fully. The reason for this is simple: Frege's idea of a formal language sufficed to enable Hilbert (who followed him) to present a formalization of Euclidean geometry.

The calculus of statement composition and the functional calculus differ from all that Frege's predecessors knew about composite statements, as it was a calculus, and one that he developed for his logicism. Thus, logicism changed its meaning with the change of logic. As Wang Hao has noted, we may question Frege's calling his calculus a logic.

The idea of the functional calculus rests on Frege's extensionalism plus his introduction of relations as basic. The formula, $x \in C \equiv P(x)$, where C is a class and P is a property opens new vistas: it allows the addition of any relation R:

$$\{x, y\} \in C \equiv R(x, y)$$

where the class C is a class containing couples related as specified. The formal calculus in which these statements appear is the functional calculus, so called as Frege borrowed from mathematics the concept of a relation as a function. In mathematics the relation R(x, y) is usually written as $y = f(x)$. The presence of functions of more than one variable – such as $z = f(x, y)$ – it was easier to move from f(x) to f(x, y) in order to introduce relations. To complete the process, let us notice that whereas mathematical functions are equations, logical functions are not. We can, however, altogether let go of the equation, or rather of "=", and, say, replace the (false) formula (3 + 5 = 7) or the (true) formula (3 + 5 = 8) with the formula Plus (3, 5, 7) or Plus (3, 5, 8), or, if you wish, + (3, 5, 7) or + (3, 5, 8) so that it is a sentence of three-place predicate. Peculiar about this presentation is the disappearance of the copula (the "a" of SaP, namely, "all – are – "; likewise, the "e" of SeP, namely, "all – are not – "; the "i" of SiP, namely, "some – are – ", and the "o" of SoP, namely, "some – are not – "). Funny: it takes effort to get used to the copula, and then it takes more effort to let go of it. This Frege achieved, as he wished to separate the quality (affirmative or negative)[3] from the quantity (all or some) of the classical categorical statement that tradition condensed together in copulas. To have quality and quantity separate helped him to axiomatize them neatly: he introduced symbols for all x "(Ax)" or (∀x) or (x) and for some x "(Ex)" or (∃x), so that the most basic aspect of classical logic, the square of opposition, is summed up in the functional calculus by the schema, "(Ex)" is short for the schema "~ (x) ~". Thus, for example, "some humans are mortal" is short for "not all humans are not-mortal". To take a mathematical example, concerning prime numbers, in the class calculus,

$$(\mathrm{Ex})\, \mathrm{x} \in \mathrm{Prime} \equiv \sim (\mathrm{Ax}) \sim \mathrm{x} \in \mathrm{Prime}.$$

Or, in the functional calculus,

$$(\mathrm{Ex})\, \mathrm{Prime}(\mathrm{x}) \equiv \sim (\mathrm{Ax}) \sim \mathrm{Prime}(\mathrm{x}).$$

Now the logical constants or connectives of the calculus of statement composition (like "and" and "or") and the expressions of the functional calculus are all that is needed. Those are of the forms f(x), f(x, y), etc., where f( ), g( ) etc., stand for some (one-place) predicates and f(,) , g(,) etc., stand for some two-place predicate, namely some relation, etc. Proper names a, b, etc., as well as the universal and existential quantifiers for all (x) or (Ax) or (∀x) and for some (Ex) or (∃x), complete the list of forms of the language. One can add classes by proper definitions. For this one has to add some meta-linguistic items: the rules of inference and the predicates "is true", "is false", nowadays written as "∈ T" and "∈ F", or alternatively as "= 1" and "= 0". Similarly, "is demonstrated" and "is refuted", nowadays written as the assertion symbol "⊢" placed before or after the name of the assertion in question respectively and "is a tautology" and "is a contradiction" with the sign "⊨" before or after the name of a formula. This is a mere rounding up of what is already in the works of Frege except for his reading of the assertion symbol as designating judgment which was an error (in his own light). He created a system that is more-or-less Leibniz' *characteristica universalis*, and he did so more-or-less single-handed, as became clear in retrospect, after Russell recast his system in Peano's notation and after generations of followers added clarifications and improvements. Amazingly, Frege had the great ambition from the start; the major steps were his and Russell's, perhaps also of Tarski and of Gentzen (see below).

The axioms of the calculus of statement composition are well known and of no import in the present study. Not so the axioms of the functional calculus. They usually appear together with an axiom or a definition that reads thus:

[3] Modern logic has no sign of affirmation akin to the sign of negation. Wittgenstein suggested that each sentence should have yes/no rubrics – to tick off at most one of them.

$$(Ex)F \equiv \sim (Ax) \sim F \text{where F is any formula;}$$

The axioms then read

$$(Ax)P(x) \supset P(a);$$
$$P(a) \supset (Ex)P(x);$$

where "a" is any proper name. Hence (since a ⊃ b and b ⊃ c entail a ⊃ c),

$$(Ax)P(x) \supset (Ex)P(x).$$

To be precise, all of these formulas are schemas for formulas, not proper formulas: whatever holds universally holds for any given individual; whatever holds for a given individual, holds for some individual. The conclusion above means that the universe is not empty, which is innocuous. The axioms may look innocuous too, but they are not, since they imply that every name has a bearer, in disregard for Mr. Samuel Pickwick. This is innocuous for mathematics, where Mr. Pickwick has no business, but not in natural languages. (The word "natural" is in contrast with "artificial", where both Esperanto and PMia are artificial or in contrast with "formal" where Esperanto is not formal but PMia is, together with the many computer languages that have entered the digital world in recent decades.)

When adding a given formula into the construction of a more complex one, it is vital to check that a term is not introduces more than once; if it does, then to avoid confusion it is essential to avoid using the same term in more than one way, as it may then cause confusion. In that case, it is required that different terms have different names. Obviously, the formula "if f(a) then (Ex) f(x)" holds only because we have agreed to name only an existing object. Otherwise, the meaning of (Ex), namely, "exists", is not the ordinary sense of the word. This is a constant source of trouble, and it is the reason the axioms of the functional calculus enter this discussion. Let me conclude this brief note with the observation that some variants of the functional calculus are not limited to names of existing objects. Such a variant is a free logic. Frege's functional calculus is first order functional calculus or the lower functional calculus: in it all functions are set; if we want to discuss functions in general, that is to say, if we use the word function as a variable, then the functional calculus becomes second order. Discussions of it have to be limited to wordings that block Russell's paradox.

Frege had the idea of a formal calculus for logic, even though he did not fully succeed to construct it. Later on, different variants of it appeared. The most important change is the addition (to any variant) of the idea of implicit definitions. It is the innovation of Duhem and Poincaré that Hilbert rendered part-and-parcel of logic (despite Quine's objection). The idea is of the permission to add to a formal language any axiom-system as implicit definitions. Its importance lies in its enforcement of the demand to have all inferences fully made with no skipping of premises, no matter how obvious they may be. In line with this, mathematicians found later axioms previously not noticed; it was then necessary to add them just because they were so obviously true. This turned out to be profitable because mathematicians are no longer averse to the idea that however obvious a statement is, we may wish to see what its import is and what may be the gain due to its omission or to its replacement by its negation or by any other alternative to it. This is the effective way to supersede the reliance on intuition, the idea that what we cannot consider false must be true: our intuition can evolve. This is easier to prove when the intuition is about commonsense matter than about mathematical matters. Whereas philosophical tradition declares some intuitions (which?) undeniable, a newer mathematical tradition denounces all intuition; fallibilism recommends the use if intuitions with the readiness to improve it (Lakatos). This leaves both Moore and Wittgenstein well behind.

It all began with Frege's insistence on stating the obvious. He insisted on it because, he thought, Kant was right about geometry but not about arithmetic. Since he considered both obviously true, and since the non-analytic mathematical truths are as obvious as the analytic ones, he had to follow formally Kant's idea of the analytically true. Hence, on some occasions, only formal analysis helps distinguish between the analytic and the synthetic. Hence, Frege reached his great results by following doggedly Kant's dogged presentation of analytic truths. Kant's dogged presentation came to justify his introduction of transcendental logic. Frege dismissed this idea out of hand.

Kant took it for granted that the analytically true is the tautological. (Gentzen proved this after Frege died.) Second, Kant said, roughly, all logically true statements are of the form of the law of identity (a = a) or variants of it. A sentence of this form is famously a tautology – literally, a repeated word: *auto* (the same) *logos* (word or sentence). As an extension of this he considered of statements of the form "ab is a": what is both a and b is a. Finally, if by definition c is ab, then also "c is a" is a tautology. (His example was "balls are round".) His valuable contribution was the idea that nothing else a tautology. All that belongs to logic but is not one of the three forms mentioned here, then, belongs to transcendental logic. This is Kant's proof that any knowledge, both arithmetic and geometry, and, of course, Newtonian physics, is synthetic. By Hume's elimination of all empirical theoretical knowledge, scientific theories must comprise synthetic *a priori* knowledge.

Frege found all this shoddy. Kant could not prove the obvious "all a-and-b are b". Boole could. Such proofs comprise a feature that, unlike classical logic, modern logic shares with mathematics. This feature may be unimportant, and, indeed, it is hardly ever noted. Yet it is a great relief to have the right to ignore classical logic by considering it marginal and incidental.

Frege invented the two calculi of logic – the calculus of statement composition and the lower functional calculus – together with his theory of meaning that transformed logic from a theory of science to a theory of language. (Analysis is just this, the logical approach to philosophy from the viewpoint of logic, which has become the viewpoint of language, where the rules of sentence formation and transformation comprise the grammar of the language in question. Rules and a vocabulary together comprise a language. PMia has only grammar and no vocabulary. This sounds strange until we realize that the construction of the vocabulary of

PMia introduces it as its grammar.) Frege's research program evolved out of the old logic by the very transformation of logic from the laws of thought or of science to grammar. Despite the tremendous progress due to the correction of the basic errors and confusions of classical logic that Boole's streamlining of it brought about, and due to his enormous expansion of its recognized formulas and his clarification of the difference between terms and statements, he still viewed logic as the logic of science. It was Frege's theory of meaning that changed all that and heralded analytic philosophy.

The traditional three laws of logic, so-called, are statements from the logic of statement composition, and their current status is what Frege gave them: they are tautologies like many other. Recently Quine came up with the proposal that even Boolean algebra is no more than a set of tautologies. He is probably right, and, anyway, he did not worry about his deviation from Kant's view that all tautologies are results of conceptual analyses like "balls are round". Frege realized that Kant was in error, as arithmetic truths differ from both geometric truths and from Kantian tautologies: they are tautologies of a different kind. (Wang Hao raised the option of considering the new kind of tautologies semi-transcendental.)

To sum up, Kant's denial of the status of analytic truths from certain assertions previously deemed essential definitions led him to rescue them by declaring them synthetic *a priori* knowledge; he could not rescue them otherwise, as he stuck to the traditional hostility to hypotheses and to a stringent view of what makes a statement analytic. A generation later, Boolean algebra widened the concept of logical truth. This permitted viewing more truths as analytic, and this raised the need for a logic of statements. A generation later, Frege suspected that arithmetic and geometry have different status. Even Kant had noticed that they differ: non-Euclidean geometry is conceivable (even though it is not applicable in principle [which principle?]), he said, but no alternative to arithmetic: prior to the twentieth century, an alternative arithmetic seems impossible. Even then, non-standard arithmetic includes standard arithmetic, unlike non-Euclidean geometry that deviates from the Euclidean one.

The sharp distinction between terms-concepts and statements-propositions is here to stay. Frege first insisted on the difference between a statement and its assertion, and to make this clear he introduced his assertion symbol "⊢". Russell had no use for this distinction. He therefore changed the meaning of Frege's assertion symbol. In his texts it means, "is a tautology" or "is a theorem" or "is provable". Wittgenstein found even this redundant, and he persuaded Russell to omit it from the second edition of *PM*. This was a serious error (even in Wittgenstein's own light.)

Frege also could not avoid confusing assertion with judgment (Bell 1979, Introduction). The concept of judgment has to do with grounds for assertion, but as he set aside the theory of science, he should have allowed for groundless assertions, for sheer hypotheses; he did not. A little later (1902), Alexius Meinong too struggled with this and wrote a book on suppositions, as he too did not know what to do with them. The process of fully legitimizing hypothesis began only in the twentieth century, so strong was the prejudice against them (since they were deemed prejudices).[4] Today, then, Frege's assertion symbol is read as "is a theorem (in the current system)" rather than as "is asserted" or rather "is properly asserted".

Boolean has its formulas properly asserted because they are self-understood. (Venn's diagrams make this understanding graphic.) Similarly, other classical formulas got clear expression, such as, the ones that define complementation:

$$a \vee \sim a = V$$
$$a \cdot \sim a = \varnothing$$
$$a = \sim\sim a.$$

Other, newer formulas identifying classes with other classes appeared, such as De Morgan's laws: the complement of a class product is the union of their complements:

$$(\sim a \vee \sim b) \equiv \sim (a \cdot b)$$
$$(\sim a \cdot \sim b) \equiv \sim (a \vee b).$$

The worst offense to classical-style texts was the correction of the confusion between classes and statements. Negation was then seen as a part of a statement or as an operation on a statement. Thus, adding to a statement the words "it is not the case that" yields another statement. The operation turns a true statement to a false one or *vice versa*: the negation of a statement is one with the opposite truth-value. It is possible to negate a statement, not a class; classes have complements: the class non-x is the class that contains all and only members of the universal class that are not in x. When a is a statement, non-a is also a statement; when x is a class, non-x is a class. In classical-style texts, it was never clear of any object whether it is a class or a statement. Boolean algebra cleared this confusion; Tarski made this obvious with his theory of contents of statements: the content of a statement is a class that includes all and only statements that follow from it. The theory of content explains the similarity between Boolean algebra and the theory of inference, since content is a class determined by the rules of inference.

This similarity explains the confusion regarding the objects of the three laws of classical logic: at times, they were statements, and at times classes. These are as follows.

1. The law of identity: $a = a$: any object, class or statement is self-identical.
2. The law of (non-)contradiction: $\sim (a \cdot \sim a)$: the conjunction of any class with its complement is empty; the conjunction of a statement with its negation is false.
3. The law of excluded middle $a \bigvee \sim a$: every object belongs to any given class or else to its complement; the disjunction of any statement with its negation is true.

[4] An example is Kant, *Critique of Pure Reason*, Preface to the first edition: "In this sphere of thought, opinion is perfectly inadmissible, and ... everything which bears the least semblance of a hypothesis must be excluded, as of no value in such discussions. For it is a necessary condition of every cognition that is to be established upon *a priori* grounds that it shall be held to be absolutely necessary ..."

Modern logic does not grant these laws the prominence they had traditionally enjoyed, and there they appear in a vast set of valid formulas; but this is not the point here. The point here is that the similarity explained by Tarski kept the confusion between classes and statements. Even *Symbolic Logic* by Lewis and Langford of 1932 accommodated for it by its very effort to present formulas systematically as valid both for classes and for statements.

6.3 The Demise of Transcendental Logic

The beginning of change was not in logic but in epistemology (in the philosophy of science). Hume considered induction merely a psychological baseless fact and science uncertain (moderately doubtful). Others considered this unacceptable. Kant went so far as to seek a new kind of logic of science – a transcendental logic – to return the ship its compass, he said. This move has a bad name now due to the (understandable) hostility to it that Einstein and Russell expressed and that analytic philosophy shares. A more historical approach requires noticing that classical logic was much too poor and as Kant was the first to explain, especially after Hume and Kant refrained from relying on essential definitions. Only tautologies were certain, and Kant exposed their poverty. It was natural then to try a new kind of proof. Even young Wittgenstein did all that: he showed the poverty of tautologies even after logic greatly enriched their ranks. He relied on facts to *show* the truth of the statements that mirror them. (He wrote "show" in italics, Black 1971, 164*ff*., which is no more informative than banging a fist on a table.)[5] Waismann, and after him the Vienna circle, called this the verification principle. Kant knew better: he said observations are not provable unless and until they receive scientific wordings. (Here he considered equivalent scientific wording and Newtonian wording.) He looked for a new logic to replace Hume's psychology and failed. The next step followed about one century later. It was Frege's replacement of transcendental logic by a powerful logic – achieved with the aid of a new theory of meaning.

With the demise of transcendental logic, doubt re-emerged. The distinction between logic and science became clear. The very presentation of the status of logic was the result of Kant's analysis. Gauss learned from Kant that geometry is

[5]Ramsey's review of Wittgenstein's *Tractatus Logico-Philosophicus* declares this obscure. Russell clarified it; the more formal/exact a text is, the more important is the preliminary introduction to it. This is the ladder to throw away. Russell considered this invitation to dogmatism: as preliminaries are pre-critical, Wittgenstein considered them immune to criticism. Oddly, Russell criticized Wittgenstein for his habit of applying forceful criticism in preliminary stages of research. This may be explicable by the ascription (McCutcheon 2001, Ch. 1) to Wittgenstein stress on preliminaries as the fervent pursuit of perspicuity (of achieving clarity through preliminary stages). As the concern of McCutcheon was with religion, she examined Wittgenstein's early attention to the central role he ascribed to preliminaries both in formal mathematics and in studies of miracles. This is a coincidence at best, to impress only the superstitious.

open to new options and Frege learned from Kant that arithmetic should be provable by a new kind of logic. Obstacles made his task Herculean; now that the road is clear, the task seems easy. He did not finish even the task of removing psychologism, since he clung to the old theory of judgment. Despite opposition to psychologism, his theory of judgment is psychological. It forced him to reject the validity of the inference from a contradiction to any statement – as there is no judgment of a contradiction (Dummett 1973, 433). It was Russell who first stated something like the law of *reductio ad absurdum* within the logic of statements – ⊢(~p p) p, or ⊢(~p q·~q)p or similar formulas. Modern analysis began with the recognition that, unlike paradoxes, contradictions are meaningful, well formed. It was the end of the theory of judgment. Frege still tried to improve the theory of judgment: he replaced personal judgment with the wonderful requirement for axioms to do that job. When this was successful, the theory of judgment became superfluous.

Frege's sign for judgment, his assertion symbol ("⊢"), is still in use. This may indicate what use he had for it. Today the assertion symbol comes before the name of any statement declared true; the symbol says that the statement whose name appears after it is a theorem of the system within which it appears. Put after the name of a statement, a⊢, the symbol designates not proof but disproof. Gentzen's theorem says, [in any consistent system] all and only tautologies are theorems and all and only contradictions are anti-theorems.

The revolution in mathematics was loss of the traditional view that axioms are self-evident. Frege did not want to give this up in logic. Logic and geometry are thus different in Frege's system: geometry may have alternatives, he explained, not logic. This is not a dogmatic attitude to logic; yet Frege's reasons for it had to go. It became the universal part of the structure of any language. Questioning all this is a new question: can we alter our language the way we can change our geometry? This question is simple and profound. Seemingly, it demands an answer. Wittgenstein denied that: language in action displays its rules as serving well: as well as is reasonable to expect. This is vague. The alternative to it is the view of language as rules of sentence formations and transformations. It brought logic nearer to mathematics. This shows that the view of mature Wittgenstein on analysis was profoundly different from that of young Wittgenstein, even though this difference is observable in action rather than in some specific declarations.

Frege's early effort was to construct arithmetic out of the theory of sets, while using Boolean algebra. Boolean algebra seemed too poor for that. Fortunately, abstract set theory replaced Boolean algebra in the completion of the project. The Frege-Russell theory presents the number 1 as the (infinite) set whose members are all the unit sets, the number 2 as the (infinite) set whose members are all the couples, and so on. Boolean algebra and the calculus of statement composition plus a minimal rule of inference proved sufficient to develop arithmetic. The *modus ponens* served to that end sufficiently well. Frege endorsed it as the only one to use (except for the rules of substitution that goes without saying). Modern logic was born.

Formal arithmetic now has different versions, but the most commonly used is that of Zermelo and Fraenkel of the early twentieth century, and it is a direct descendent from the efforts of Cantor (complemented with the contributions of both Frege and

Dedekind). Most logicians today consider Frege's logic the complete modern system of logic, though in the notation that Russell gave it, one that he had borrowed from Peano, plus his modification of Frege's arithmetic. Alternatively, the current version of the foundation of mathematics is the system of Frege's logic plus the Zermelo-Fraenkel axioms for abstract set theory. (Category theory is possibly another alternative.)

The most important rule that Frege followed is, never confuse a concept with a thing. It is so obvious that one hardly sees the need to state it, much less to defend it. In line with the new analysis, we now speak not of concepts and things but of words and things, and, more importantly, of statements and facts. We should add to this Russell's stress that only statements, not facts, are possibly true or possibly false. All this was insufficient: it lacks a new theory of meaning. Frege gave an example. The following inference is valid. Marcus Tullius is Cicero; Cicero made a speech; hence, Marcus Tullius made a speech. The following inference looks the same: the morning star is the evening star. The morning star is observable in the morning. Therefore, the evening star is observable in the morning. Problem: how is it that the two inferences look the same when only one of them is valid? The reason for the difference is very obvious, and so one hardly notices it: that Tullius is Cicero is a matter of naming: we name the same person by two names, and the rule of substitution says that whatever is true of the one is true of the other as the two denote the same object. Not so with the two stars: it was a discovery, and the first great astronomical discovery, that the bright star visible in the evening and the one visible in the morning are identical. The names given the two stars are not important; what is important is the identity of the two objects, previously not noticed as identical – which is the reason they were given different names. (The plotted orbits of the two planets turned out to be the same. This led to the surmise that the two are identical.) Frege's refutation of the traditional theory of meaning opened a floodgate; it prompted the study of meaning as means for overcoming the frustration of centuries of stale debate. Meaning became a fashionable item (Broekman 2016, Ch. 1). This flood marked the birth of the analytic school of philosophy.

Now the discussion of the identity of two planets belongs to science and so perhaps Frege's discourse is to be limited to science. To prevent this move (following Linsky 1967, 100) Kripke 1976, 259–63 discussed the case of one who learns that London is beautiful and that Londres is ugly without knowing that these two cities are identical. This suffices to prove that London = Londres differs from London = London,[6] a point that is central to the philosophy of his teacher Quine (who admired Russell most but nevertheless clung to the views of Frege as much as he could).

Philosophers can make mistakes that conflict with commonsense. This sounds paradoxical. It is, said Wittgenstein, and its source is metaphysics that philosophers

[6]It often is a tautology, observes Bar-Am 2008, 47 *et passim*; not if it means mediaeval London is modern London: this is intuitively true but not tautologically so. This is hardly ordinary language, though. "I am the Lord your Lord" (*Exodus* 20:2) is more like it.

endorse and the confusion that it propagates. Perhaps not; perhaps it is not paradoxical at all. Most people are not confused about substitution, as they do not think about it. This is true of any field of study and, *pace* Wittgenstein, it is not peculiar to philosophy. The ability to be confused about anything requires some minimal knowledge about it. Any new field of study may introduces confusions to weed out later, even mathematics and even logic. Hence, even if Wittgenstein had a task to perform, it was temporary, not as inherent to metaphysic as he suggested.

Frege's concentration on logic and mathematics to the exclusion of science was a boon: there were too many cases of confusion in every field of science – at the time and throughout history. Wittgenstein saw analysis in the light of Russell's suggestion to remove puzzles, and puzzles as confusions that lead to paradoxes; and so Frege's refutation of the old theory of meaning was due to an attempt at a clearing of a confusion, even though perhaps not fully successfully. Possibly, the matter is not at all a case of confusion; perhaps his critique was of an error rather than of a confusion. Possibly, his effort was not a clarification but the suggestion that formal systems should include a statement of the difference between the two equations, "Tullius is Cicero" and "the morning star is the evening star". If so, then there are two symbols for identity, analytic and synthetic, and for long both commonsense and logic were vague about this.[7]

Frege's purpose in undertaking this exercise was to refute the classical referential theory of meaning. To distinguish between the two senses of identity he observed that unlike analytic identity, synthetic identity of two expressions comes with different senses. His refutation of the classical theory of meaning led him to his new theory of meaning.

Whether Frege has solved the problem he raised is contested. No one contests that he refuted the classical theory. Or did he? Since the morning star and the evening star are one, why can we not conclude that the evening star is observable in the morning? Answer: because the names for the star denotes both an object in the sky plus its being visible in the evening. The identities of the star visible in the morning and of the star visible in the evening are not parts of their names[8]; they are added items of information. As Linsky (1967, Ch. 3) has noticed, the name we now use to denote the star in question, "Venus", now denotes both the evening star and the morning star; it may stand for the second innermost planet without further ado. The logic of the use of that name depends on choice, and it differs from the logic with which any of Venus's old names first appeared. Worse, one object may have two names accidentally unbeknown to their user; as Kripke (1976) has noted, one may learn some stories about London and other stories about Londres and not realize that they are identical. If the stories are incompatible, then one who endorses them holds an

[7]Tarski [1936] 1941, §19 shows that identity sign is synthetic: (x) (y) (x = y) is true of an empty universe and of one that contains only one object but not otherwise. This does not hold for the identity sign in the sentence "Tullius = Cicero", of course.

[8]In Leibniz idea of his ideal language essences reign supreme, so the name "Caesar" includes all the characteristics of Caesar, so that understanding it amounts to knowing all about him. The literature on Wittgenstein ignores this, which is strange; so does the literature on Russell, which is not strange.

inconsistent view. This is disturbing, as it discloses that names change their meaning in different contexts, so that they have not only references but also senses, and this raises the question, what are the requirements from a translation of an ancient text? Duhem and Quine said, translations are never fully adequate. This is an incredible idea and a bombshell. It also raises the question, how do people translate nevertheless? The answer to this is simple: they translate poorly: all translations are wanting. This age-old knowledge is now a part of a theory of language; this Frege has made possible by his having done away with the traditional theory of meaning.[9]

6.4 Mathematical Logic

The presentation of the contribution of Frege with no background information is deeply rooted in courses in logic as taught in most modern universities, following innumerable logic textbooks that are very similar to each other.[10] Philosopher and historian of science Thomas S. Kuhn viewed this multiplicity of textbooks that echo an ideal standard ideal text as a science; the standard ideal text he called a paradigm. The paradigm of logic is the propositional calculus as Frege has invented it, although in a new notation due to Peano who invented it and to Russell and Wittgenstein who popularized it.

Pre-Boole logic was more standardized. Its final version, however, is not always interesting. A syllogism has two premises, although clearly there is no reason to keep it to this minimum. Boole would have no problem with

> a is a subset of b, b is a sunset of c . . . . j is a subset of k;
> therefore, a is a subset of k.

This version appeals to mathematicians. For beginner philosophy students it might be an excessive burden. Frege would have no problem with it, even though he accepted only one rule of inference (except for the rules of substitution), namely the *modus ponens*. For, either form agrees with the vision of Leibniz that guided both Frege and Russell of a fully formal language.

Frege's standardization is partly due to the tradition that evolved with group theory and Boolean algebra, partly due to the new abstract set theory, and partly as his decision to have a formal language and a formal system that proves Kant right on

[9]The logical connection is obvious between diverse theses on the limits of natural languages considered formal: The Tarski-Quine no analytic-synthetic dichotomy, the Duhem-Quine limited inter-translatability thesis and the Bar-Hillel-Catford no-full-machine-translatability thesis. Perhaps it is only the Duhem-Quine no-complete-refutability-in-isolation thesis that is slightly less direct and less obvious. All these impossibilities are due to viewing the informal formally; informally, all of these theses are obvious. We obviously agree to tolerate some (hopefully small) imperfections. See Agassi 2014 Ch. 6.

[10]The model standard introductory logic course seems to me to be the deceptively introductory high-powered text of Hilbert and Ackermann, 1928.

geometry, not on arithmetic. Yet he could not accomplish all this without his critique of the traditional theory of meaning as reference that Boole's extensionalism rendered impossible. By tradition, this belongs not to the introductory logic standard text, but to the introductory analytic philosophy or to the philosophy of language, or to a similar item.

The theory of meaning of Frege, be it acceptable or not, permits the rise of the theory of identity – as a relation that is reflexive ($x = x$), symmetrical ($(x = y) \equiv (y = x)$), and transitive ($(x = y) \bullet (y = z) \supset (x = z)$). Now the definition of identity as any relation that satisfies these three conditions is standard in mathematics. It is, however, problematic, as Frege has shown in his most important contribution to philosophy, logic and mathematics: identity defined this way may be analytic. Such as "Cicero is [identical with] Cicero" and synthetic, such as "the morning star is [identical with] the evening star". Overlooking this renders Frege's propositional calculus extensional. Yet calling a system extensional does not close the issue. Thus, Frege's most appreciated disciple Quine argued that "All bachelors are bachelors" differs from "all bachelors are unmarried" even when ascribing to the latter assertion the status of a (verbal) definition. This sounds odd, until we realize that the status of a statement as analytic or synthetic depends on the possibility of substitution and this depends on the range of the variables in the language concerning which Quine made his assertion. The range of variables is determined within an artificial language as a part of the description of that language. In common parlance the range of variables is negotiable. This means that the very opposite of the chief observation of Wittgenstein is true. He said (*TLP*, §: Preface), "what can be said at all can be said clearly"; and (*TLP*, §4.116) "Everything that can be thought at all can be thought clearly. Everything that can be put into words can be put clearly." Not so: nothing that we can say is clear unless it is said in a formal language, and what we say in a formal language we do not know what it means although we can make a conjecture about it. In offering such a conjecture, however, we enlist the service of an informal language.[11]

Frege had little respect for the vernacular. He said, it would be possible to develop proper epistemology only after the establishment of the formal language of mathematics will be completed. This made him neglect epistemology. Since he was an ardent believer in certainty as the characteristic of science, obviously, had he not neglected epistemology, the problem of induction would have derailed his researches entirely, especially since his initial concern was to dispute Kant's epistemology.

Unlike Frege, Russell and young Wittgenstein had clearly epistemological concerns in mind.[12] Russell centered on mathematics as a stage towards a theory of the natural sciences, and on this Whitehead was of great help. Whitehead himself moved later to the study of science in general and to Einstein's theory in particular. But he helped Russell finish with arithmetic first. Young Wittgenstein was most ambitious.

[11]See *Stanford Encyclopedia of Philosophy,* Art. Lakatos.

[12]Despite all this, Wittgenstein declared he owed more to Frege than to Russell.

He solved to his own satisfaction all the problems that he found in the studies of Frege, Russell and Poincaré. Fortunately, eventually he left the whole of his early work behind, even though he could not decide wholeheartedly what the final verdict on it should be.

Chapter 7
Russell

Russell inaugurated analytic philosophy new-style (as well as logical atomism and "logical" positivism). His initial concern was with rationality. He took its display in science to be best and the clearest. He therefore had two initial aims: to render philosophy scientific and to prove that science is certain (since tradition equates reason with provability). This led him to efforts to improve logic. The first ambiguity in classical logic that he met was of the copula: in "all men are mortal", the copula is the subclass and in "Socrates is a man" it is class membership. He wanted to show clearly the distinction between the two concepts. He argued that whereas every class is a subclass of itself, no class is a member of itself. He wanted to prove this. So he defined "normal" a class is normal iff it is not a member of itself: $x \varepsilon N \leftrightarrow \sim(x \varepsilon x)$ and tried to prove that all classes are normal: $N = V$. This made him wonder: is the normal class itself normal? As we saw, the answer is, yes-and-no; which is absurd. Now language is full of absurdities: $0 = 1$; $0 = 2$; $0 = 3$; and so on. These were never problematic as everybody always pronounced them false. Russell's paradox, like any paradox (or antinomy or puzzle), is demonstrably true and false. A language that contains a paradox has all its statements proven. This is the end of logic.

Russell suggested then that the rules of sentence-formation should eliminate paradoxes as pseudo-sentences. This view of paradoxes amounts to recognizing contradictions as well-formed and thus as meaningful. It also amounts, Russell showed, to the recognition of the possible difference between the apparent and the real (or ideal or deep) structure of a sentence. To avoid a paradox (or a puzzle, as he called it) he found it necessary to view non-existent things as having no names proper, but definite descriptions, and false ones at that. Unlike Frege, he did not disdain natural languages, and unlike mature Wittgenstein, he did not consider adherence to them obligatory. He developed Wittgenstein's logical atomism as the old neutral monism cast in the new analytic style and then he gave it up for a weaker version. This signaled the end of his aspiration of identifying philosophy with analysis – which Wittgenstein refused to give up (Cook 1994, 12; 2000, 10, 23, 72–4).

J. Agassi, *Ludwig Wittgenstein's Philosophical Investigations*, Synthese Library 401,
https://doi.org/10.1007/978-3-030-00117-9_7

7.1 The Deduction Theorem

Russell was the founder of analytic philosophy new-style, although some instances for it preceded his work and although it was Frege who fashioned the wording of some of its principles – not to mention Kant and the medieval logicians. Russell is the founder nevertheless, as he characterized analysis and offered some tools for it: first, puzzles should be eliminated; second, clear solutions should be preferred to problematic ones even at the cost of incompleteness; third, (parts of) natural language should be rectified to come closer to the ideal – formal – language. The culmination of Russell's analytic and logical work combined was the logical atomism that he accepted from Wittgenstein and developed. It is but an analytic version of classical empiricist epistemology: it replaces in classical empiricism concepts with sentences. Russell's logic, however, was not satisfactory: his view of his system of logic as the ideal language gave way to a more pluralist view of language, more in tune with the views of mature Wittgenstein. Russell's replacement of the early version of logical atomism with some more sophisticated version of it amounts to the rejection of the view of analysis as the sole method of philosophy. As Wittgenstein never gave up that view (at least not explicitly), the break between them was inevitable.[1]

Yet they both took as central the basic distinction between analysis old-style and analysis new-style as the distinction between two views of the very subject matter of logic: it was the progress from the laws of thought and of judgment to the laws of language. The very concept of sense, or of intension, signifies to that effect, since the intension of a term is not necessarily what anyone intends; it is not what any particular person claims. (In the works of mature Wittgenstein, sense is apparently presented as a part of a "life-form" – presumably[2] as a part of a social item, perhaps a social institution.) Yet the theory of the sense of a word is only a part of the story. Perhaps Russell's invention of the analysis of statements is more significant than the transition from the analysis of concepts to that of sentences and words.[3] Why and when did this happen? Who made what discovery? What is its import? What was the rationale for the change, how did it develop, and what made it crucial?

Kant was the first innovator of ways to analyze statements as he examined knowledge. This was a radical change that Bolzano implemented as he proposed a

---

[1]Alam 199, 135, 155. Landini 2007 describes all this in admirable detail. See also Agassi 2009. In the 1959 preface of Popper's *magnum opus*, he conceded to the analytic school what was Russell's contribution and rejected the rest as due to misconceptions (alluding to Wittgenstein's contribution, p. 17). See Glock 2008, 15 and 219. Glock places Popper in the analytic tradition, but while laudably mentioning his protest.

[2]The term that Wittgenstein employed systematically is "forms of life". The reading of this as social institutions is the thesis of Winch 1958, although the expression "institution" appears there sparingly. See Jarvie 1972, Ch. 2 and Gellner 1973, Ch. 4.

[3]A string of words is judged a pseudo-sentence in a language if its words are not placed in accord with the rules of the grammar of that language, or else if some of the words in it do not refer but should (*TLP*, §5.4733).

move from Kant's study of known statements to the study of statements as such. Frege then viewed all logic as the handling of statements rather than of concepts or of terms. More radical was the change from concepts to language. What was the rationale for this change?[4] Dummett, the leading Frege scholar, declared (1991a, xi) that the honor for having made this change goes to Frege as he is "generally perceived as the founder of analytic philosophy" – as the outcome of his critique of psychologism. Stuart Shanker dissented. Since according to Frege logic still was the laws of thought and of judgments (Smith 2009), he said, the honor for having ushered in language analysis goes to Wittgenstein. As Mark Notturno (1985, 1) and more so John Wettersten (1995, 131–2) have argued, the psychologism that Frege was fighting is not a clearly specific doctrine: it is the old analytic approach itself.

This dispute is frustrating: obviously, the transition was difficult and made in steps. Frege developed logic as mathematical logic; he said that he was clarifying the structure of the ideal language, which he took to be the one adequate for mathematics – adequate for viewing arithmetic as a part of logic proper. As to natural languages, he only observed them in order to eliminate from the ideal language any item that is not common to all of them. When he discussed the logic of science, however, he had to speak differently. He then stuck to the view of the logic of science as the laws of thought and of judgments – even though he contrasted Boole's idea of logic as the laws of thought with his own view of logic as the laws of truth. He said (1984, 351), "logic has much the same relation to truth as physics has to weight or heat". Even there he was not very clear. Wittgenstein and Russell clarified the status of tautologies as truths by virtue of their forms. Although this would not have surprised him, he did not say it. Later on, Tarski and Gentzen explained the status of valid inference by reference to sentence forms and the transmission of truth. Sadly, the situation was (and alas still is) worse in methodology than in logic: when Popper rejected psychologism in methodology (1935, §2), he did not impress peers; many of them are still confused on this issue. This is odd, since anti-psychologism is a central tenet of analytic philosophy (and the view of Popper as an analytic philosopher is still popular). The most significant contribution to our understanding of scientific method is his bold announcement that scientific theories are true-or-false (rather than true, as tradition would have it, or as meaningless). This was his view of meaning as a truth-value in his anti-psychologism; this idea did not look as revolutionary as it was, since

[4]This historical question invites a historical answer. Some writers have offered answers that include streamlined versions of Frege's work. The streamlining requires of them to overlook the concerns that engaged the historical Frege. This is sad. Here is the place to stress that he worked as a mathematician. Berkeley's critique of the calculus was conceptual; all effort to answer it were mathematical (Wisdom 1939). Frege and Peano approached the situation differently (Segre 1994). Kreisel 1978, 88 reports Wittgenstein's reference to the "disastrous invasion of mathematics by mathematical logic". This is strange, considering that Frege was the father of symbolic logic and the very first to perform this "invasion" and considering that Wittgenstein admired his work. This was "an exaggeration"; but since "his public performances which were always tense and often incoherent", presumably Kreisel found it unsurprising. See also Rodych 2018.

Frege clung to his idea of judgment. It was Russell's total rejection of the theory of judgment and his total recognition of contradictions as proper statements (though analytically false, of course).

This is the view of Russell's major task as the removal of rubbish – a task that John Locke proudly considered the task of epistemologists. Yet he had to remove more: the classical empiricist conception of the concrete (as inconsistent, no less). To see how revolutionary this was we may notice that Wittgenstein's peers and heirs are still ignorant of this achievement. Indeed, the realization that any description of the concrete is itself an abstract shakes even a strong sense of reality. Russell insisted, and by reference to knowledge of both empirical information and some abstract, non-empirical knowledge. Yet he was not sure. (He left the 1913 manuscript unfinished and unpublished.) In effort to take stock of what he had, he went repeatedly back to basic questions: what does a sentence mean? How do we word an observation report? What is the performance of induction? Frege had ignored the problem of induction, and he did not say what fact a judgment denotes, as he said statements denote nothing in particular (they denote "the true", he said). Russell was interested in induction. When he wrote the preface to the 1927 German translation of his *The Problems of Philosophy* of 1912, he referred approvingly to Keynes's 1920 treatment of induction. Now inductive support is a relation between statements; hence, it is an epistemological study, not a methodological one. Likewise, it pertains to judgment; it is not divorced from it. Russell turned to a full study of induction only decades later, but never as mere methodology. This is a systematic error in his admirable philosophy. Yet, after Frege, it was Russell who paved the way to a better methodology: he worded and applied it with more awareness of its pitfalls.

That Frege was in a halfway position merits stress, since it is there that work is particularly hard. To exemplify this, let me repeat the point made in the previous Chapter: Frege's view of contradictions was a serious error. It was a leftover from the traditional view of logic as involving judgments. Dummett observes this (1991b, Ch. 10) but refers to Frege's theory of judgment elsewhere in the same work (258–9). According to Frege, he says, one can understand a statement without judging what its truth-value is, but one cannot understand any statement without knowing what judgment is, since this means not knowing what truth is. Dummett complains there that other commentators on Frege are not clear about the situation. He blames them unjustly. For, the situation itself is unclear; clarity is desirable, yet historians have to admit clearly the presence of ambiguity. David Bell rightly says (1979, 91), "It is difficult to know what Frege's attitude to indirect proof is" – as Frege was unable to agree [or disagree] with the theorem that every statement follows from a contradiction. Although Dummett has devotes much discussion to the description of Frege's conception of logic, he was not sufficiently explicit on Frege's trouble with contradictions. Had Frege added to his assertion symbol a denial symbol, both ⊢t and c⊢ (for tautology and for contradiction respectively), he would have an easier time.

How much was Russell indebted to Frege? The standard answer is, he was not indebted to him at all, as he did not know of his work prior to having developed his views to quite an extent. This is possible but it is very far from self-evident. In particular, Frege has created an agenda and it is quite possible that Russell developed his agenda in the shadow of Frege's, not knowing that it was Frege's. Peano's familiarity with Frege and Russell's familiarity with Peano renders Russell's ignorance of Frege partial.

It is hard to reconstruct Frege's theory of inference, partly because of the complication of his notation, which is in extreme contrast with the clarity of Russell's notation – a notation first developed by Peano, whose ideas Russell became acquainted with in 1900. (Frege was aware of Peano's notation too; he rejected it and he explained why in their correspondence.) It must have felt odd to discuss notation, as everybody admits that it is merely ancillary. Yet Frege realized that in times of fluidity of research different notations draw attention to different aspects of a situation as it unfolds.[5] Frege's notation is notoriously difficult[6]; commentators add to Frege's wording of his ideas their Peano-Russell rewording. This helps, but it does not facilitate understanding of Frege's view of the rules of inference; they are still particularly difficult to follow. Usually, commentators gloss over this fact fast, on the assumption that there is no difference between Frege's and Russell's views on inference. This is under dispute (Dummett 1973, 83).

Things are even more complex. Both Frege's and Russell's theories of inference are unsatisfactory; it is a high hurdle not easy to overcome and easy to overlook or circumvent. In some context, it does not matter: as long as the context is chiefly of theorems and of their deduction from axioms, there is no need to examine the difference between analytic and synthetic statements. For, in that context questions of validity and of consistency matter, but not questions of fact. Similarly, mathematical discourses prove the existence of objects under discussion. Is the introduction of references to the real world into language possible? For this, one needs a logic of science that is not psychological and that permits the introduction of new hypotheses. Tarski and Popper achieved this much later; they added to the efforts of Frege and Russell, to which they owed much. They had to meet in a new way the purpose that historically the theory of judgment came to meet. These were defects in the traditional theory of inference. To avoid all reference to judgment, the rules of inference should employ no reference to specific truth-values: valid rules hold regardless of whether the statements they apply to are true or not. This does not matter for discussions limited to analytic truths.

The rule for introducing synthetic statements into the theory of inference must refer to falsehoods, of course. It says this. An inference is valid if it belongs to a valid inference scheme.

---

[5]Variance in notation alone need not alter content. This is true by definition. Rewording a problematic differential equation, however, may facilitate its solution. Hence, changing notations may help perform a task. Even the facilitation of reading a formula may help this way, The notations of Oliver Heaviside for both the equations for electric circuits and for electromagnetic equations were every helpful, as was the lovely simplifications that Einstein introduced into tensor calculus.

[6]Landini 2012, Preface. Landini argues that simplifying Frege's notation has raised many new problems for the reading of his ideas.

An inference scheme is valid iff it is not invalid.

An inference that belongs to an invalid inference scheme is invalid; an inference scheme with a counter example is invalid; a counter example to an inference scheme is an inference exhibiting that scheme whose premises are true and whose conclusion is false. To repeat, an inference is valid if and only if the form it conforms to has no counter-example. Hence, the logical form of an inference is necessary and sufficient for its validation. In mathematical logic, where all statements considered are tautologies or contradictions, the difference between an asserted statement and a statement that has not been asserted does not matter: the assertion symbol that Frege introduced to indicate a judgment, was meant to indicate both that a judgment was made and that the object of judgment was found to be (necessarily) true. Russell interpreted the assertion symbol otherwise, to mean, it is assertable: it is a theorem of the system. It makes no difference then that the rules of inference hold also for non-theorems. For the logic of science, where science is a set of hypotheses, this required clear statement; at the time, it was not clear at all. If anything, it was clearer in Frege's system than in Russell's system, since Frege was contemptuous of the vernacular. Russell was more agreeable to the view that the rules of logic hold for natural languages. (On this, oddly, young Wittgenstein followed Frege and mature Wittgenstein followed Russell!)

Frege and Russell were able to make sophisticated deductions despite their differences as to inference. This emerged from their subject matter being mathematics and logic and from their logical abilities. These require attention. When people reason with no concern for the rules of logic they usually apply their intuitions both of logical truth, whatever it is, and of deduction. The move back and forth between these two items is permissible by the deduction theorem that says,

$$a, b/c \leftrightarrow a/(b \supset c).$$

This means, if statement c follows from statements a and b then statement "if b then c" follows from statement a.

$$\text{Hence, } \vdash a, \ \ \vdash b/\vdash c \leftrightarrow \vdash a/\vdash (b \supset c).$$

This means, if theorem c follows from theorems a and b then theorem "if b then c" follows from theorem a. Yet here the concept of inference is different, since in the one just presented, we will see, every theorem follows from every theorem, which makes the deduction theorem between tautologies trivial.

The situation may become problematic differently when a is a theorem but not b or c. in that case, a may be omitted, it looks as if we have, a / b $\leftrightarrow$ a $\supset$ b. Here we have omitted the symbol "/" on the right og the double arrow, and this is a cause for confusion: whereas the formula on the left of the double arrow is of the deducibility relation that is a relation between statements, the formula on the right describes a relation between situations, one described by a and one by b. We may rewrite the formula as, $a / b \leftrightarrow (a \supset b) \in T$, where "$\in T$" is "is true".[7] This holds always within logic or mathematics but not otherwise since the consequent of a synthetic conditional does not follow from its antecedent. It is thus advisable to rewrite the formula as, $a / b \leftrightarrow / a \supset b$, where "/ a" means, the inference with no premise and a as its conclusion is valid, which, to repeat, means that a is a theorem. For, if / a is valid, so is b / a where b is any statement, so that "/" means the same as "$\vdash$" or "$\models$". In mathematics, the predicates "is true", "is logically true" and "is a theorem" may be interchangeable, and by logicism this is always so. In other contexts, this inaccuracy has raised much confusion. It was the logicians of the next generation, Tarski and Quine, who cleared this confusion. Since Russell used the word "implication" for $a \supset b$, Quine stipulated that this word should be replaced either by the word "inference" or by the word "conditional statement" ("conditional" for short), as required.

This is why in Russell's system the *reductio ad absurdum* does not appear as a rule. In the *Principia Mathematica* (*201), the *reductio ad absurdum* appears thus:

$$\vdash (p \supset \sim p) \supset \sim p$$

It is better to word it thus:

$$p / \sim p \rightarrow \vdash \sim p.$$

As a proof-procedure, the general form of the *reductio ad absurdum* is this.

$$\textit{Reductio} : p/q \sim p/q \rightarrow \vdash q,$$

to read, any statement that follows from a statement and also from the negation of that statement is a theorem (is demonstrable). In case p and q are the same, this formula reads

$$p/p \& \sim p/p \rightarrow \vdash p;$$

and since p / p is trivial, we may omit it to say,

$$\sim p/p \rightarrow \vdash p,$$

which is the way we learn it in school in elementary geometry classes.

Let us return then to the comparison between the two versions of the deduction theorem, $a, b / c \leftrightarrow a / (b \supset c)$ and $\vdash a, \vdash b / \vdash c \leftrightarrow \vdash a / \vdash (b \supset c)$. The difference between the two depends on the meaning of the assertion symbol. A statement is assertable in a system when it is a theorem in the system. The best illustration of it is this. Under Wittgenstein's influence, the third edition of the *Principia* has no assertion symbol. This is odd. "$\vdash p \supset p$" means, any conditional statement whose

[7]More precisely, T is the set that contains all and only the true sentences in the language in use.

antecedent is the same as its consequent is a theorem. What does "p⊃ p" possibly mean, then? Perhaps it means that if it snows then it snows, and perhaps it means that if it rains then it rains, but it should mean both: it should be a token, just as in classical logic "Socrates" is a token for any individual's name. If this is so, then the story displays Wittgenstein's preference for assertions over assertions of assertability,[8] such as the assertability of any statement of a given form. If so, then his philosophy precludes altogether the assertion of the rules of logic. This is no news, since in Wittgenstein's first book he stated this explicitly as he claimed that the *modus ponens* and the assertion-sign are senseless!

This shows that Russell was not sufficiently clear on central issues – perhaps not even in his own mind – and even on so central an issue! This suggests that his having gone so far despite this obstacle is testimony to the key role that his tremendous common sense played in his most technical investigations, logical as well as mathematical, and to draw attention that finding the innovations that seem in retrospect obvious required much labor at the time. Admittedly, Russell's motive was philosophical from the start, and the obstacle here discussed is the lack of explicit clarity of the difference between deduction and proof or between deduction from tautological premises and from synthetic ones. Oddly, the philosophical motive and the obstacle were similar: they both concern the limits to provability![9]

7.2 On Denoting

With this correction, the theory of inference became sufficiently formal, with the rules of substitution and the rule of detachment (the *modus ponens*). Since the rule of substitution is a part of the set of rules of sentence-formation, it is often implicit. The conclusion from this is that for inferences proper Frege and Russell allowed only the rule of detachment. This is rather cumbersome. Take the valid rule, a-and-b entail a. Frege and Russell state it thus: (if a-and-b then a), a-and-b; therefore a. The asset of having one rule only is that it reduces the ways to commit error. Yet there is a need

[8]*TLP,* §3.334: "The rules of logical syntax must go without saying, once we know how each sign signifies." §5.132: "'Laws of inference' ... are senseless and would be superfluous". [Note: they *would* be superfluous; they are not superfluous *tout court.*] §6.1264: "(... one cannot express the *modus ponens* by means of a proposition.)" This is a truism: a rule is not a proposition; since early Wittgenstein limited meaning to propositions, his §6.1264 is trivial.

[9]Russell was never clear about the deduction theorem. His "Replies to Critics" (Russell 1944, 969) has a repetition of the theorem and a discussion of the value of any conditional statement. In the case of the dependence of our knowledge of the truth of the conditional on our knowledge of the truth of its antecedent or of the falsity of its consequence, Russell says there, the conditional is useless; it is useful in the case in which our knowledge of its truth is prior to such knowledge. He said this in oversight of conditionals that appear as hypotheticals – in the deduction theorem and in science. They are unknown yet most useful. More obvious is the *reductio ad absurdum* that here Russell overlooked (as inessential?). This also explains his tern "the contradiction" for his paradox (rather than "the proven contradiction"). See also Whitehead and Russell, 1911, I, 62.

for a proof that the rule of detachment suffices. This requires a more general theory of inference, and this was available only when Tarski and Gentzen supplied it.

Russell's use of Peano's notation and his own removal of the theory of judgment altogether, made logic what it still is more or less: beyond this, the received improvements are technical and marginal. Russell may thus have caused some confusion by glossing over certain disagreements he had with Frege, and this may be the reason why the increase of Russell's popularity is also the increase of Frege's. This also led to the continued neglect of Frege's writings. The renewed interest in Frege is post World War II. Russell's review of Frege's work in his *Principles of Mathematics* of 1902, which appeared as Volume I, lost its force as *Principia Mathematica* by Russell's teacher Whitehead and Russell replaced that book, while Frege remained the venerable ancestor. The first, perhaps the only, clear disagreement Russell had with Frege was on Frege's theory of meaning; his paper on it, "On Denoting" of 1905, contains a note to say that in his *Principles of Mathematics* he had agreed with Frege but in 1905 he dissents from him on this matter. The paper was under the shadow of the *Principia Mathematica* for long, since its main points reappear there (*14). Only with the rise of the analytic school did it become as famous as it should be: it is the first piece of modern analysis, of the analysis of statements. Still, it is hard to deny the judgment of Quine 1966, 663 that it is "wholly mysterious" and "hard to follow". Detailed study of it is rewarding, especially for Wittgenstein scholars. Linsky (1967) made this obvious; his (Linsky's) having had difficulty to publish it suggests that it was path breaking.

Having discovered his antinomy, Russell was stuck. He narrates ([1959a] 1993, 60 *ff.*) that he was badly stuck with the problem until he developed his theory of denotation. It was the analysis of proper names offered there that intertwines with his idea that seemingly well formed – seemingly meaningful – sentences are possibly not. He discussed that theory very proudly and in *A History of Western Philosophy* (1945): he expounds it as a significant recent development. As Linsky notes (1967, xvii), though Russell was able to change his mind with ease, he never changed his mind on denoting. It is worthwhile to see why he ascribed to it that great importance. In his *A History of Western Philosophy* he chose it as what he liked posterity to remember him for.

Russell criticized Frege's theory[10]; Alonzo Church tried to fix it.[11] Quine offered a theory in which proper names are "purely designating".[12] Others offered different

---

[10]The criticism of Frege's theory of denotation in Russell 1905, 484–5, is that it is hit by the paradox of empty terms ("the present king of France is bald") and of referential opacity ("the king wanted to know who was the author of *Waverley*"). Linsky 1967, 72–3 and Hintikka 1981, 170 have complained that Russell had not cleared the paradox of referential opacity, and that it does not involve definite descriptions. The two complaints may neutralize each other. Yet referential opacity remains an obstacle – not for PMia that is extensional, not for other purposes. Students of intensional logic may do better if the specify them.

[11]Martin 1963; Salmon 1993.

[12]Quine 1943, conclusion. This is the program for developing a free logic, one with no existential import.

solutions, including Linsky, Lejewski, Hintikka[13] and Bunge. A fashionable novelty in logic today is possible-worlds semantics; it presents different things as possessing different names in different possible worlds. For this, that theory describes names in a way that differs from the way Frege and Russell did. The most widespread attitude, however, is implicit in the works of both Wittgenstein and Popper, who strangely agree on it. Considered local, the matter is open to all sorts of ways to handle proper names. For example, we may replace them with names of classes that have at most one member each; since the empty class is a subclass of every class, this view reverses the truth-value of assertions about non-existent entities. It is possible to replace them with predicates or with special variables. When the question receives local treatment, its presentation in all its generality has lost its urgency, and in different settings it may receive different solutions with no fear of inconsistency.

We have no need to go into the study of meaning of terms, since our concern is with mature Wittgenstein who insisted on the meaninglessness of sentences. We have to discuss Russell's "On Denoting" as it led Russell and thus Wittgenstein to the new idea of the seeming-meaningful meaningless. Otherwise, we need not discuss the meaning of words. Even the difference between analysis of words old-style and new-style need not be discussed, so let me only notice it briefly. The theory of meaning that is a part of classical logic discussed being a logic of terms could make do with meaning of nouns. As the rise of modern logic began in earnest when Frege refuted that theory, he naturally wanted a broader theory of meaning. We need not discuss his theory of meaning of functions and of the logical constants such as "and" and "not"; rather, we may center on Frege's theory of the meanings of sentences. His most powerful and significant assertion here is, a sentence is meaningful iff it has a truth-value. He went further and declared that a true sentence denotes the true and a false sentence denotes the false. It is not easy to see the difference between his view and the view that sentences have senses but no references. The same goes for functions and connectives. Still, it is easy to consider the reference of a one-place predicate to be the same as the class to which it applies (as $x \in C \equiv P(x)$ holds already in classical logic) and so also the reference of a relation is the class of the related couples. It is harder to apply the same reasoning to logical constants. We need not discuss them here. Back to "On Denoting", then.

Why was denoting so urgent a general problem early in the day? This is the central question for all historians of analysis of this situation. Linsky has noted that Wittgenstein's analytic technique altered from the study of the structure of sentences to the study of their proper use (Linsky 1977, xxi). This is the major change in the style of analysis: from Russellian large-scale to Wittgenstein-style case study.[14] This change rested on the questionable supposition that the characteristics of small-scale

[13]Hintikka 1981. (His oversight of Linsky here is baffling.)

[14]"Use" is a social term. Modern logic and modern methodology are sociological. Traditional (Locke-style) conceptual analysis is intentionally psychological: classical philosophy was individualist and demanded the reduction of sociology to psychology. Though Wittgenstein's "use" is clearly sociological rather than psychological, many members of the analytic school prefer psychology to sociology.

analyses are easier to treat than of large-scale ones.[15] To the large-scale problem of Russell's study in his "On Denoting" of 1905, then, and to its urgency.

The paper raises a number of questions and poses to its current readers more. The current literature suggests this. The theories of Frege and of Russell differ in terminology and in scope. What Frege called "sense" and "reference" Russell also called "meaning" and "reference" or "connotation" and "denotation". A variance in nomenclature is not significant; the disagreement is. As to scope, whereas Frege discussed only nouns, Russell discussed mainly descriptive phrases. This, too, is not significant, as both discussed the meanings of words that denote simple entities: proper names and definite descriptions. What is the meaning of "Mr. Pickwick"? This question classical logic ignored or worse. Boole and his followers solved it in a manner that both Frege and Russell (as well as Meinong and others) found wanting. Russell suggested that the very act of naming individuals amounts to the supposition that they exist. In addition, he clarified the case of the allegedly non-existent entity introduced by a description without naming a non-existent entity.

"On Denoting" of 1905 is about a dozen pages long. Russell states there that denotation is significant for logic, mathematics and science. The significance of the matter within science is far from obvious: science generalizes; hence, it seems, it ignores individuals as well as individual events, and so, in principle, it has no need for proper names. (This point won forceful presentation later, in Collingwood's *Speculum Mentis* of 1924.) It is puzzling, as astronomy is unthinkable without the names that it employs, and as chemistry names the elements and physics names elementary particles. Somehow, however, we all recognize the difference between the name of a star and that of a chemical element: names, thus, are no simple matter. Moreover, from the point of view of science, were there many solar systems available, we would be less dependent on the names of the members of our system than we are. True, in astronomical maps, as in any map, names signify. (The best way to name them is by the use of their co-ordinates. Hence, their names are reducible to one name, said Russell, the name of the co-ordinate system used to depict them.) Words like "Man", "Horse" and "Hydrogen" are class names that differ from words for unique entities, be they proper names like "Sir Walter Scott" or definite descriptions like "the author of Waverley".[16] Russell found troubling statements like "the present king of France is bald" and like "Scott was the author of Waverley" as they generate paradoxes. They invite analyses; and then the objects of analysis turned out to be statements, not words. This makes "On Denoting" pioneering.[17]

The starting point is Russell's critique of Frege's (and Meinong's) views. John Searle (1958, 138, 141) defends Frege's theory, claiming that Russell's presentation

[15]Popper 1959, Preface to the English Translation is devoted to a thorough criticism of this idea.

[16]Moore noticed (Schilpp 1944, 214) that that "the Whale" differs from "Man" and "Horse" in its possession of a definite article. Russell dismissed this (*op. cit.*, 690) as a minor irregularity.

[17]Hylton 1989; Linsky 1967, xvi, stresses that Russell repeatedly offered variants of that theory.

of it is too confused. It is perhaps worth repeating that Russell himself stuck to his own theory to the end of his active life (and Wittgenstein always respected it, though he could not possibly endorse it with no qualifications as he rejected the quantifiers). We need not quibble. The question here is not the question of the correctness of that paper but its great significance – for the whole output of Wittgenstein.

The reference of an expression is, we remember, its extension, so that the reference of a name is the individual so named, the reference of a class is all of its members and so on. (The trouble usually hides behind the "and so on".) If names refer to things and their senses are Platonic items, we may just as well say, at least as a *façon de parler*, that meanings are references to two items, the physical and the Platonic. This kind of expression may be too loose, too inaccurate, but we may accept it now for a while. It is suspect because it puts Meinong and Frege in the same boat. Russell did so in 1905, and he did so explicitly. What exactly are Platonic items? This question is baffling, but Russell managed to put it in Frege's terminology, forcing an answer to it out of his theory: what are the Platonic objects to which Platonic names refer? Platonic names refer not only to Platonic items; they also have to refer to other items, as each meaning has two constituents. So what are the sense and reference of a sense of an expression? An obvious answer to the question may be, the sense and reference of the sense of an expression is the reference and sense of that expression. Another answer to the question may be, the sense of a sense is another sense, that the sense of one Platonic item is another; but then what is the relation between the sense of a thing and the sense of the sense of that thing? Will that inflict on logic an infinite hierarchy of senses of senses of senses of...? (Carnap discussed this suggestion seriously.)[18] In Frege's opinion the world inhabits different objects and their qualia, be they properties or relations; language, then, populates names that refer to objects, and functions that refer not to the qualia of things but to the things that possess them. Now Frege's system also allows for functions of functions, and they refer not to functions but, again, to things that have the qualia or the qualia of the qualia. Similarly, the sense of a reference and the reference of a sense have to move within the same universe of sense and reference. How? No possible way charted by Russell turns out to be satisfactory. The empty name, in particular, has no reference, so that an assertion containing it should be meaningless even if it seems obviously false. Again, we see how vital Russell's theory of the meaningless is and how important then is his theory of definite description (that allows for meaningful discussions of the present king of France, since – unlike Wittgenstein – Russell denied that assertions about the non-existent entity are meaningless: he declared them false).

Mathematics has no use for ordinary proper names. Whatever name mathematics employs, it is hopefully replaceable with a description. Russell tried to realize this hope. Try hard as he did, he could not reduce all names. He settled for "this" and "that" as irreducible proper names; he did not need his theory of definite description for his logicism. Our concern in the present study is historical.

---

[18]Carnap 1947, §30; Linsky 1967, 44n, 46, 48.

Russell's theory of definite description is more general than the theory of naming. "The present English monarch", thus, refers to the same individual as the one referred to by the monarch's proper name. This is problematic: we assume that a monarch rules England and this may be an error. Now since we assume that this is not true of France, the expression "the present king of France" should behave differently. This is a remarkable fact: we have here an expression whose structure depends on facts. Hence, a rule of grammar may depend on facts! This is disturbing, since we want to structure language to enable it to serve as means to describe (possible) facts, not for the choice of a rule for describing a fact that depends on factual assertions before we have debated their truth-value. This is a wish to avoid unnoticed dogmatism; it is not obvious, however, that we can meet this wish. Russell said, we could meet it only after we complete the job of analysis: when we understand that an expression – be it a name or a proper description – stands for a thing, we will not allow this expression unless we assume that the object exists.[19] The structure of language does not depend on facts, but if we realize that naming involves the assertion that the named entity exists, we will agree not to name a thing unless we assume that it exists. Consequently, when we have to speak of a fictitious entity like Mr. Pickwick, we will note that the word "Mr. Pickwick" is a name only in a Pickwickian sense. Similarly, once we realize that the expression "the present king of France" represents the hypothesis that France is a monarchy, we will not use this expression unless we admit that hypothesis. Assuming a hypothesis supposes language and should not depend on language, but we have to say whether we endorse that hypothesis – provisionally or not – and the use of certain expressions amounts to doing just this.[20]

Why analyze? Why not read a sentence of this kind literally? Answer: some sentences are obviously ambiguous and they require analysis to prevent obvious error. Thus, Russell's example, "I thought your yacht was larger than it is" may mean different things. To confess having thought that a and non-a is the case, may mean different things. It may mean that a is true only under some conditions / in some circumstances, or that one has changed one's mind, or that one believes a contradiction. Simple analysis can show this. Other statements invite deeper analysis, perhaps. Russell provided an example for that too, one that has simple words not in need of analysis, put in a statement that is not easy to comprehend and so may

[19]Limiting this discussion for unique entities leaves open the question, is it necessary to assume that a class is non-empty before naming it? This is a quaint variant of the problem of induction. For, the assumption that atoms exist is a corollary to the statement that atomic theory that describes them is a true explanation.

[20]Some comments seem obviously exaggerated, such as the refutation of Russell's view that definite descriptions are unique. The alleged refutation is with the expression "the men who . . . ." This is an unintended insult. It stems from understanding Russell's "On Denoting" as a theory of all the diverse usages of the definite article in familiar English rather than as a solution to a problem within the task of formalizing parts of a natural language into a language adequate for mathematics. Russell expressly referred to the definite article in the singular. (In some languages, singular and plural expressions may differ widely.)

invite analysis all the same: "what is is by virtue of what it is" is his example. The example is problematic, but not any word in it!

Analysis suggests that the definite article "the" means the one and only. In mathematics, descriptive phrases occur regularly, often as depicting objects. Mathematicians may then prove that the entity so named exists, often also that it is unique. Otherwise, they may have to withdraw the name in question or take it to be a class name. (When the expression, "unique up to the addition of a constant" refers to a function it also refers to any other function such that the difference between them is a constant. In this case, the name is of a class of functions.) Hence, it is not the structure of language that sanctifies a name, but the meaning of the definite article: it forces us to use it only when we want to ascribe to something existence and uniqueness.

This is not always so. At times the definite article introduces existence ("the son of x", "the way from x to y") with no uniqueness implied and at times it introduces uniqueness too ("the eldest son of x", "the best way from x to y"). Yet our intuitions may falter here, and examples may sway it one way and another. The reason, to repeat, is that we talk knowingly of the non-existent only in research and such; when we do so, we do so in the subjunctive. All this is unimportant, since we have given up the theory of judgment and then it does not seem necessary to refer to our intuitions at all. Intuition did enter analytic philosophy nonetheless. Sir Peter Strawson has rejected Russell's analysis that suits mathematics in preference for ordinary parlance.[21] He has a point: commentators are now agreed (Linsky 1967, 98) that the dispute is a standoff: Russell was concerned to include the empty case, the case of a name or a descriptive phrase referring to nothing. Ordinary parlance is vague about this, or else the nothing receives existence in some Pickwickian sense.[22] When ignorance of the existence or non-existence of an item is at issue – say, the item is a secret weapon in the hands of the enemy – then one ascribes to it a hypothetical existence – to be withdrawn if need be. How does common parlance take care of this situation? This may be an interesting question, but one that Russell did not care for. He cared for mathematics that does speak of nothings like the greatest prime number; it proves their non-existence.

Without adjudicating between Russell and Strawson, one can see that analysis may refer to judgments too: even after old-style psychologism (that takes logic and language as mainly psychological) is dead, the psychology of language is still alive and legitimate. The psychology in old-style psychologism is erroneous only because it poses as analysis, not otherwise.

---

[21]Strawson 1959; Russell 1950; Linsky 1967, 99 considers Strawson's argument exceptionally poor. See also Rorty 1971.

[22]Bunge uses the universe of discourse freely to solve the problem at hand the easy way.

7.3 The Puzzle

Consider the analysis of a definite description. Unless a unique thing fits the expression, any sentence that includes it is false. A statement and its negation then are both false. Which is absurd. This is the puzzle.

Russell discussed this because he had no theory of judgment to fall back on. He was analyzing then a case open to different readings[23]: the reading he offers is not obligatory. The aim of discussion is to prevent a statement and its negation both being false. He resolves the puzzle about the present king of France by analysis: the negation of that sentence considered a conjunction is very different from what looks like a negation at first sight.

So much for the puzzle of non-existence. It is of no concern to Frege. It is referential opacity that Russell troubled Frege with. When we speak of the desire of King George IV to know about that author, who happens to be Scott, it is not clear what is at stake. Perhaps the king knew that Scott was that author but did not know who that fellow is, perhaps he wanted to know who is Scott, not even knowing about his authorship, and perhaps he knew who Scott was but not who is the author of the novels (since, as it happens, they were published anonymously). Now, as it happens, only the third case is true. Yet, since the individual called Scott and the author in question are identical, and since logic allows one to replace any expression by its equivalent, it turns out that by logic the three possibilities just mentioned are all true (since they have the same subject) whereas in truth only the third is. Hence, there is an ambiguity in our understanding of a statement with a definite article in it, and so we do not quite know what its negation is. Russell is very decided on the matter. The statement in which there is a definite description of a non-existent entity has some ambiguity of meaning. It is nevertheless untrue because the definite article in it promises a reference and does not deliver the goods. Its meaning thus differs from its reference, as evidenced by the fact that the meaning of the expression "the center of gravity of the solar system" is complex but the point it refers to is as simple as any point is. Similarly, when we say "Scott is the author of Waverley", then we assert the identity of reference of two terms that differ in meaning, just as the different terms do in the statement "the morning star is the evening star" that Frege discussed in detail. A few commentators – Searle, Linsky, Strawson and Hintikka – criticized Russell. He did show awareness of their objections: he used "meaning" for "sense" and said, his thesis is, "denoting phrases express a meaning and denote a denotation". Indeed, Hintikka spoke of the Frege-Russell view, despite Russell's dissent from Frege's postulate of the third realm (in addition to mind and matter) that he (Russell) identified with Plato's Heaven that he considered non-existent (rather than a meaningless noun).

[23]This is a disclaimer: the theory is no complete account of the diverse acceptable usages of the definite article. Russell ends his paper by asserting a smaller claim as more general: the problem has no easy solution. All his commentators agree with him, but few pay him the courtesy of saying so explicitly.

To dispute Frege's solutions of puzzles Russell presents some new puzzles. The following is such. Consider "Scott is the author of *Waverley*; the king wanted to know who the author of Waverley is; hence the king wanted to know who Scott is". It is a valid inference with true premises and a false conclusion. Russell showed a way out by rendering the inference invalid.[24] This is the meaning of Wittgenstein's famous enigmatic observation (*TLP*, §6.5), "the puzzle does not exist": all puzzles must be fully unpacked; "resolved" is the term for it; to resolve is to analyze away. The solution of the puzzle indicates its urgency: the technique of resolving away a difficulty was new. It was bread and butter for Wittgenstein.

Alas, this is confusing. It is puzzling that so simple an obstacle resists removal over a century. Part of the reason is that Wittgenstein's technique is popular yet useless here. The end of analysis, said Wittgenstein, is to expose some expressions as pseudo-statements. Russell's theory of definite description is the paradigm of language analysis. It helped him to distinguish correctly between the meaningless and the meaningful (in the sense of Frege, who astutely declared meaning as truth-value). Russell resolved a few puzzles, but expressly not by dismissing them as pseudo-statements.[25] The confusion of two ways to resolve a paradox may explain the embarrassing papers against Russell that are insults to his intelligence.[26]

Russell made it quite clear why he presented some puzzles. Unlike Locke or Wittgenstein, he was not in the least concerned with clearing any rubbish. The puzzles (that indeed must be removed and even urgently) were of his own making, not popular errors (Locke) or confusions (Wittgenstein). He produced puzzles, he said, to illustrate the superiority of his own theory of meaning over that of Frege. Unlike the theory of Frege, his resolves his puzzles. (He considered puzzles thought experiments in logic, akin to empirical tests in science.)[27] All this was his progress towards a fully formal system.

[24]To render an opaque problematic inference invalid one needs to use a technique similar to the classical fallacy of four terms (*quaternio terminorum*).

[25]Russell 1905, 484: "one would suppose that 'the King of France is bald' ought to be nonsense; but it is not nonsense, since it is plainly false." Here we see the importance of not calling a falsehood "nonsense".

[26]Russell stressed repeatedly that propositions that contain expressions with no meanings may nevertheless possess clear and obvious meanings. Wittgenstein agreed, even though not clearly: *TLP*, §4.064 says, "Every proposition must *already* have sense; assertion cannot give it a sense, for what it asserts is the sense itself." Nevertheless, Wittgenstein often spoke of the meanings of words, not of statements ("and then whenever someone else wanted to say something metaphysical, to demonstrate to him that he had failed to give a meaning to certain signs in his propositions"; *TLP*, 6.53). This is but a slip of his pen: he proudly presented his atheism as a part of modern logic. Popper 1963, 373 has corrected Wittgenstein here: the theory of definite description allows describing the deity in many ways, all of them traditional. Putnam 2008 repeated the claim that Wittgenstein's philosophy supports religion as a practice devoid of any doctrine.

[27]Russell 1905, 484–5: "A logical theory may be tested by its capacity for dealing with puzzles, and it is a wholesome plan, in thinking about logic, to stock the mind with as many puzzles as possible, since these serve much the same purpose as is served by experiments in physical science." This then is a partial success of his plan to render philosophy scientific.

The difference between Frege and Russell is complex. The pioneer Frege took his sentences unanalyzed; Russell analyzed them. This is no disagreement yet. Frege demanded that every name (and thus every definite description like "the morning star is...") should have both sense and reference. The triumph of Russell's work is not in its presentation of a new theory of meaning or of naming, since as such it is much less comprehensive than that of Frege. The triumph of his work is in the force of the analysis that imposes them: the demand that some expressions should have reference means that names and definite descriptions denote something. Hence, whenever sense and reference are agenda, they have to be unpacked first, no matter what exactly sense is. In his 1905 paper, Russell did not speak of sense except to say that a proposition conveys it.

Historically, the triumph of this discussion was not that it clarified certain expressions in ordinary discourse; Wittgenstein's fans later did present it this way, but in error. The triumph was in the proof that analysis is a must-do for certain ends of admitted significance. This enabled Russell to seek a theory of knowledge. Analysis was his starting point. Here is a representative quote from a response to a contemporary formulation (1914, 885; 1956a, 149): the true and the false are respectively the real and the unreal, considered as objects of a possible belief or judgment. In response to this wording he says,

> This assumes that such phrases as 'the present King of France,' which do not denote a real individual, do, nevertheless, denote an individual, but an unreal one. This is essentially Meinong's theory, which we have seen reason to reject because it conflicts with the law of contradiction. With our theory of denoting, we are able to hold that there are no unreal individuals; so that the null-class is the class containing no members, not the class containing as members all unreal individuals.

7.4 Russell and Young Wittgenstein

Perhaps the hardest question regarding Wittgenstein is, how do his ideas relate to traditional philosophy? Later, Anscombe identified the riddle that Parmenides had posed as what troubled Wittgenstein throughout his career (Bar-Am and Agassi 2014): as there is no unreal, how can falsehood exist? Frege assumed that the reference of true and of false statements are the true and the false respectively. These entities are obscure; they conceal this riddle.

Russell said (1912, Ch. 12), "Our theory of truth must be such as to admit of its opposite, falsehood." Thus, the riddle was the first and most obvious obstacle. Young Wittgenstein took it to be very seriously and declared that it does not exist (TLP, §8.5), treating metaphysics the way Russell had treated riddles. The road to his mature theories was the road away from Russell.

There is no need to go along with Russell's analysis of the text he chose to analyze. Today self-respecting commentators are careful not to speak as loosely as his contemporaries used to do, for which he deserves gratitude (together with Wittgenstein). The passage cited here is a token for evidence that simple matters of presentation of statements and their negations are not easy to accomplish and that

Russell found that the new logic is a tremendous tool for clarifying discussions. So much so, that not much later he declared philosophy to be essentially logic.[28] This is not far from the philosophy of young Wittgenstein; its novelty is in the detail, and Wittgenstein went into detail much more than Russell. This is why Russell appreciated it even though he judged it an utter failure – so much so that two decades later he gave up the whole project.

It is here that young Wittgenstein entered his life and made a huge difference. It is hard to imagine that the most dry-as-dust field of all learning, formal logic, was the passion of two immensely sensitive and highly vulnerable researchers: they viewed logic passionately, as if it was a sort of music (Monk 1996, 251, 253, 259, 567). Russell reported: getting to know Wittgenstein was one of the most exciting intellectual adventures in his life. They were soul mates. Just imagine an undergraduate student coming to his professor's rooms at around midnight and pacing up and down in silence for over a quarter of an hour. Russell finally asks him: do you think about logic or about your sins? Both, he answers. Russell obviously understood.[29] By contrast, Wittgenstein tried a similar exhibition of sincerity with placid Whitehead and it failed. Later on, he tried a different ploy with Carnap and it failed too. Nevertheless, most people who met him, once or frequently, reported that he was a most impressive presence (Bartley 1982–1983, IV).

Russell described the chief aspect of his immense delight at his encounter with Wittgenstein: he was intellectually very lonely and here he found an heir apparent.[30] He was then developing his logical atomism, which he presumably meant to present as the analytic version of traditional neutral monism, the theory that knowledge is dual organization (physical and mental) of our sensations. In line with his newly discovered analytic technique, he argued that neutral monism has to give way to logical atomism. He acknowledged the invention of the idea at least in part to Wittgenstein, though he was a young student whom he had first met but 2 years earlier. Clearly, the idea was only in diapers then. No one ever tried to present a fully developed version of it. The nearest to this is Russell's famous 1918–1919 series of lectures *The Philosophy of Logical Atomism* and Wittgenstein's 1922 book. It was never developed; it is a non-starter. Yet this is its merit: it is clearer than empiricism in general and neutral monism in particular. The superiority of this variant is the ability to prove it hopeless. Russell managed that feat.

Wittgenstein came to Cambridge in 1912 as a student and left in 1914 for the war. In this short time, Wittgenstein impressed Russell so much that, Wittgenstein's sister reports (1984, 2), he told her, "we expect the next big step in philosophy to be taken by your brother." Russell announced logical atomism in 1914, developed it in 1918, and advocated ever better versions of it to the last – yet as a hypothesis to reject if it turns out to be in conflict with commonsense (Russell 1944, 707). Wittgenstein seems to have given up his published version, never to replace it by an explicit

[28]Russell 1914, title to Chapter 2: Logic as the Essence of Philosophy.

[29]Russell 1956b, 27; Monk 1990, 2012, 64.

[30]Monk 1996, 241, 250–2272, 280, 289, 282: "No one except Wittgenstein understood it all".

alternative.[31] In a famous letter of 1916 Russell says (Wittgenstein 1974, 24) of Wittgenstein,

> His criticism ... was an event of first-rate importance in my life, and affected everything I have done since I saw he was right, and I saw that I could not hope ever again to do fundamental work in philosophy. My impulse was shattered.

These are strong words. Russell reported that he went to logic in order to find what makes for a scientific proof, and in this junction, clearly, Wittgenstein both encouraged him to develop logical atomism and shattered his high hopes that this theory, or any other, will deliver the goods.

The dream of reviving logical atomism is hopeless: perception theory has developed by leaps and bounds since, and sensationalism is gone forever and with it all possible versions of logical atomism. Nearest to it is Popper's theory of relatively atomic sentences. When analytic philosophers will have learned to treat Popper properly, they will notice it. Russell's greatest achievement, his mathematical logic, is valuable regardless of questions of their status. His distinction between the declared meaningless and the self-contradictory does not need a full-fledged theory of meaning. His withdrawal of his view of the language of mathematics as the ideal, taking for granted the imperfection of the English language (Schilpp 1944, 690; Russell 1953) leaves meaning still further from the center of action. He never withdrew his theory of definite descriptions; its status is still not clear (Linsky 1967, 116–119). This is a part of a more general question: what is the status of empty names? (In Frege's view, a definite description is a name.) This is a part of a more general question: what is the limit of analysis? Commentators unwisely shun this question. Russell and Wittgenstein answered it quite differently. Meaning has thus lost its relevance to logic and mathematics.

In hindsight, the story of Russell and Wittgenstein has a sad ending. Their meeting after the war, still was joyful and cooperative (Monk 1996, 567), going over the work that Wittgenstein had written under dire conditions. That was very exciting, but it was the end. Russell questioned Wittgenstein's ideas, such as the claim that existential statements are meaningless. (He allowed for them as disjunctions. This idea does not work, as Russell noted.) Wittgenstein questioned the need for them. Wittgenstein said he had no objection to their prevalence in ordinary parlance (*TLP*. §§3.323, 4.002, 5.5563). Later Russell explained why Wittgenstein was adamant: he did not allow speaking of the whole world, as we do yet when we say that a set has exactly three members (Russell 1959a, b). Russell reports that Wittgenstein was adamant. This was new to Russell. Noticing that Wittgenstein was ready to discuss only his own ideas and only rigidly, Russell gave up (Monk 1990, 2012, 73–4).

Russell learned from Wittgenstein even then: he acknowledged that in the later edition of the *Principia* (1925). His criticism of Wittgenstein made him appreciate the distinction between object language and meta-language and its far-reaching

[31] This may be a mistake of mine: as commentators are reluctant to discuss this, I may be missing a point.

consequence: it amounts to giving up the search for the ideal language (Monk 1996, 594).[32] He then omitted the assertion symbol from a later edition of the *Principia* as a concession to young Wittgenstein.

Russell helped young Wittgenstein by writing a foreword to his book and recommending it as a doctoral dissertation, and by recommending him for a position in Cambridge University. But the thrill was gone. Wittgenstein had "grown tired of serious thinking", said Russell ([1959a] 1993, 216).

Young Wittgenstein's philosophy comprises two stages, his collaboration with Russell and his *Tractatus Logico-Philosophicus* that separated them. They shared the wish to develop a precise theory of science, that renders philosophy scientific – with no dependence on any metaphysics. The great manuscript that Russell was working on he left unpublished. Russell finally gave up; he declared defeat only in 1940. In 1919, when they met after the war, Russell reports, Wittgenstein spoke with him only about logic, day and night, in great excitement. That was their last forceful intellectual interaction.

[32]Forster 2004, 230. Engel 2012, 48 denies that Wittgenstein's *Tractatus* relates to the ideal language. "He was attempting, rather, to reveal the essential structure" of any language. As Forster refers to *TLP*, §§3.325 and 4.0031 here, it is hard to ascribe to Engel ignorance rather than defensiveness. Dummett 1991c , 83–5 argues for the impossibility of the ideal language, as all language is social.

Chapter 8
Young Wittgenstein

The discussion still rages over the question, to what extent did mature Wittgenstein disagree with young Wittgenstein? What exactly did he withdraw? How significant was what he withdrew to his philosophy as a whole? These questions are under dispute because he left them open, and intentionally so. Commentators must agree that some of the continuity between young and mature Wittgenstein was unavoidable. The idea of New Wittgenstein (2000) is largely the idea that already the young Wittgenstein had in mind the therapy that his anti-philosophy is supposed to provide. It is very hard to offer a balanced judgment of it. In a sense, it is trivially true, and it does not conflict with an older reading of him except perhaps with that of his "Vienna Circle" followers that he did not particularly approved of. In a sense (and in agreement with the official "Vienna Circle" reading) it is obviously false: Wittgenstein began as a student of and collaborator (!) with Russell, he criticized Russell to the point of making him abandon his major research project. This included the search for certainty. This search expressed itself in two major ideas, logicism on mathematics and logical atomism on knowledge (Bar-Elli 2002, 170). Both failed and Russell admitted them nonetheless, except that they ceased serving his quest for certitude that gave way to his admission of commonsense as much as science permits. Wittgenstein did not share Russell's despair. His logical atomism was an austere image of science and of the universe (Jacquette 1997, 37) as a version of neutral monism cast in the language of modern logic: every atomic proposition is a series of items, beginning with space-time co-ordinates and continuing with a list of possible sensations, vision, hearing, etc. All the rest is excessively sophisticated elaborations (Bar-Elli 2005) – still largely incomplete (Proops 2017, §3.2)!

Wittgenstein never questioned his early, unbearably narrow idea that the undecided questions of metaphysics are undecidable and that undecidable questions are meaningless, even though later on he declared his early philosophy too limited (*PI*, §23). It is this narrow idea that renders all theology meaningless. Young Wittgenstein showed respect for science and indifference to religion; mature Wittgenstein went the other way, without, however, withdrawing his early respect for science and his early view that theology is meaningless. What this means is not that

J. Agassi, *Ludwig Wittgenstein's Philosophical Investigations*, Synthese Library 401,
https://doi.org/10.1007/978-3-030-00117-9_8

theological assertions are not understandable; to some extent we know that they are, but only to some extent; theologians traditionally and generally agree upon this. They are meaningless in a different sense: they are ungrammatical. What is grammar? Young Wittgenstein had a very sharp and clear answer to this question: he developed a precise grammar and thus a language that competes with that of Russell. It is the language of his *Tractatus Logico-Philosophicus*.

8.1 *Tractatus Logico-Philosophicus*

Wittgenstein's 64-page long *Tractatus Logico-Philosophicus*, the only book of his published in his lifetime, is brief and enigmatic. Russell's judgment of it was highly appreciative, yet in utter dissent. Whatever our judgment of it is, we should never overlook the fact that it excited most of his readers, including Russell. Its impact was incredibly powerful. What was its quality that gave it such a great impact?

The booklet offers a complete philosophy, aiming to answer all central problems of philosophy in short trim adages. To begin with the spoiler, its major thesis is, true philosophy cannot have verbal expression. It is disappointing, since it is the standard mystic thesis that appears in all cultures. Rationalists from Maimonides to Russell viewed it as trivially true yet of no value, since critical discussion is limited to what we can utter. (Russell 1917, end of first chapter)

Yet Wittgenstein arrived at his mystic thesis in a very unusual manner: he argued that all efforts to utter any philosophical idea is bound to fail. To prove this, the book begins with a development of a formal language that is a trimmed version of the *Principia Mathematica* of Whitehead and Russell. The *Principia* appeared just before the War, in 1910–1913; The *Tractatus* was finished afterards. It appeared only in 1921 in German and in 1922 in English, due to difficulty to find a publisher. It found one only because Russell wrote an introduction to it.

At first the book was a failure. in 1927 only 27 copies of it were sold. In the post-World War II era it is an undisputed classic. The success claimed for it is in the claim that it offers proof that the austere language it offers suffices to serve as the logic of mathematics and of science, while ousting metaphysical assertions. This claim is as old as positivism, but now it is (allegedly) a part of logic. Except that the proof is invalid, even feeble: the book is a colossal failure, even though commentators on it are almost unanimous in praising it as a major feat – and rightly so.

Much of the reputation of the book was due to its convincing power, the severe criticism of Russell's Introduction to it notwithstanding. Carnap repeatedly declared its thesis proven beyond reasonable doubt; Neurath demanded the suppression of criticism of its doctrine, in disregard for his own poor view of it. In the fifties of the twentieth century concise versions of Wittgenstein's proof were repeated in may publications and much more so orally: there is no room for metaphysics since all that can be said is a part of logic, or science, or commonsense – and metaphysics is neither. This is already the new Wittgenstein: the early version of his philosophy, his *Tractatus Logico-Philosophicus*, had allowed for commonsense but left no room for it. It likewise allowed for but left no room for unscientific expressions, commonsensical or artistic or any other. The vehicle of commonsense is ordinary parlance: ignoring whatever it shares with science, its contents appears in ordinary speech rather than in scientific discourse. Wittgenstein explicitly disregarded it (Black 1971, Ch. xx).

This idea, the idea of the language of science, is the contribution of Wittgenstein and his trademark. When members of the Vienna Circle spoke of Popper's idea as relating to the language of science, they distorted his views unrecognizably. They did not see this, his repeated protests notwithstanding: Wittgenstein covered their whole field of vision.

The idea that science has a language is but the positivist version of the traditional view of logic as the idealized version the structure of rational discourse; almost imperceptibly, Frege and Russell turned the idea of logic as the idealization of language to the idea of logic as the very ideal language. In the philosophy of young Wittgenstein, this idea plays a central role.

He thus came to the view that the revolution in philosophy was incomplete: philosophy is the clarity of language-construction that amounts to the removal of all ambiguity. Mature Wittgenstein simply extended his old idea to natural languages, concepts, thinking, and judgements. Ambiguity then is necessary for the assertion of metaphysical views.

Ideal language is obsolete. Gödel showed that PMia is limited; Wittgenstein refused to admit Gödel's result as he rejected the view that logic requires the services of a meta-language. Tarski showed that the meta-language is indispensable for formal systems; Wittgenstein refused to admit it. Turing developed the idea of formal languages as artificially structured systems for computations. Subsequently, a multitude of formal languages appeared in the market. Wittgenstein ignored it (Shanker 1987a).

In addition to the defects of Wittgenstein's early philosophy as a system of logic and of mathematics, as a philosophy of science, it is a version of conventionalism and thus it shares its defects. In addition, it is a theory of logical atomism, where an atom reports the smallest possible element of experience. Russell first tried to incorporate it in his logic and then he refuted it. Wittgenstein's *Tractatus* represents a very ambitious effort to rescue logical atomism by the use of logic to remove all puzzles once and for all, through the claim that logic leaves no place for the articulation of metaphysics, as it is neither in logic nor in empirical science as described by the doctrine of logical atomism, by now defunct. Here the discussion of it serves as background material for the work of mature Wittgenstein to which this study is devoted.

8.2 Russellian Background

Analysis is always trivial, and there is always the difficulty of seeing why it should engage serious attention. This is precisely the difficulty that Moore spoke of when he discussed the paradox of analysis. The difficulty is rooted in the assumption that trivialities can bring about only more trivialities. This is not so. Russell never shared this idea or else he would not use analysis as the basis of his works in logic.

Wittgenstein was Russell's star pupil and for a time his heir apparent – to the extent that Russell could or wished to plan such things. A book that attacked Russell's new logic appeared in 1912, and he offered it to Wittgenstein to review. The book was lightweight: Russell was so much ahead of the crowd that hardly anyone was able to criticize him forcefully. (He valued Wittgenstein as someone to talk with and to expect some work from: intellectually he was very lonely.) The response to that book was political, not intellectual; the assignment of the task to Wittgenstein, then, was an act of recognition, hardly a challenge. Wittgenstein's status was growing. The review is a charming exercise in emulation of Russell. Later on, Wittgenstein repudiated that review, thereby obviously repudiating his status as heir apparent. By that time, Russell had broken-off with him – intellectually: personally, he still referred to him as a friend and helped his career.

In the published version of Russell's lecture series that he gave in the United States in 1914 and of those that he gave in Cambridge in 1919, he offered logical atomism as a new theory that he ascribed to Wittgenstein, though, evidently, it is the product of a close collaboration. Moreover, the idea itself is not as forceful as its elaboration, and even that is hardly a great feat (since it was soon aborted as a failure – to be resuscitated in ever renewed versions). When Wittgenstein's book appeared, only a few years after Russell delivered his lectures on logical atomism, the two collaborators parted company. They remained estranged for life. Nevertheless, Russell wrote an introduction to Wittgenstein's book in order to help get it published, and he wrote a letter recommending to Cambridge University authorities to grant Wittgenstein a doctor's degree for his book and to invite him to join their faculty. Whatever Russell wrote on Wittgenstein contains some critical remarks. Wittgenstein did not conceal his view that Russell's remarks were the product of a profound misunderstanding of his ideas. For his part, Russell judged Wittgenstein's response to his criticism unacceptable. This led to a total breakdown in communication between the two close collaborators. Russell's obituary on Wittgenstein in *Mind* almost thirty years later, in 1951, ends with the typically English assertion, "of the development of his opinions after 1919 I cannot speak."[1] This was not to say that he did not know: As Landini has shown, Russell constantly checked every new idea of Wittgenstein.[2] Thus, the letter of recommendation that Russell wrote to the authorities of Cambridge University describes Wittgenstein's research program very clearly.[3] Though in that letter he did not conceal his disagreement, he wrote in sincere appreciation. Early in the fifties, the picture changed: soon after Wittgenstein died, in 1953, he attacked the cult of common usage, as he called it, and he

[1]This alludes to the meeting Russell and Wittgenstein had after the first world-war. It ended in total miscommunication.

[2]Landini 2007, 70, 74, 97, 104, 111, 198–9. Russell 1940, 196 and 1959a, b, 249 refer to his failed efforts of two decades of work on Wittgenstein's ideas. The end of his 1951 obituary on Wittgenstein is his expression of final disappointment.

[3]Russell 1998, 439, report to the Council of Trinity of 8th May 1930 on Wittgenstein's work. This document is significant as it presents Russell then as still hoping to see Wittgenstein discuss mathematics usefully.

showed that he knew that the target of his attack was a philosophy that Wittgenstein had initiated. For Wittgenstein's later output, he showed profound disapproval and a total lack of appreciation. At the time, this was common knowledge.[4]

Russell's 1905 theory of meaning is the natural starting point for the study of the early philosophy of Wittgenstein. The current literature conveys the feeling that the paper is of tremendous importance, yet it says little to explain it. Let me try to spell things out.

Russell attacked Frege's view that some names denote nothing, though what he was unhappy about was Frege's view of sense more than his view of reference. Russell always viewed names as denoting entities, in accord with the axiom of the functional calculus that allows replacing a name by the claim that something exists (= it allows replacing a name by a variable bound by the existential operator). His main concern was with sense, however, and his argument comes not from a theory of denoting but from his theory of definite descriptions. This is an innovation, and it is quite a detour. A definite description employs the definite article: "the x...", and its form is presented as a statement with a definite description of the form "the x such that a" where a is a statement. That means, he said, there is exactly one x such that a holds. The remarkable thing is that Russell did not have then a full theory of meaning (to replace that of Frege). He only analyzed the meaning (sense and reference) of terms in statements that include definite descriptions, and he did so in preference for the analysis of names (or nouns). He added three kinds of activity that herald the advent of analysis new-style; the most important among them Wittgenstein and his disciples overlooked; it is that proper analysis is of sentences, not of words. For example, he said that Plato's "only beauty is beautiful" is a pseudo-sentence. Second, he analyzed sentences from ordinary parlance. He defended his analysis by the observation that ordinary sentences (like "I thought your yacht was larger than it is") do naturally invite analysis, and even obtain it quite regularly, though it is not easy to follow the analysis that ordinarily people make in ordinary circumstances and without paying any attention to this conduct. Third, he offered a criterion for the task of analysis – that presents his analysis as preferable to that of Frege. Ordinary discourse, he suggested, includes some puzzles, some seemingly illogical items; analysis should remove them. Unlike Wittgenstein, Russell did not ascribe puzzles to philosophers; nor did he ever say that all philosophy is puzzles; he showed that some theories generate puzzles and so they are unacceptable as they stand. These theories may arise from commonsense or from traditional philosophy or just be one's own inventions. The point is, removing puzzles may enlighten.

Wittgenstein's first innovation is his "the puzzle does not exist" (*TLP*, §6.5): all puzzles are grammatical errors. Russell had no theory of analysis or of puzzles; he simply said that his theory of meaning resolves some puzzles. Wittgenstein generalized Russell's approach. What matters here is the rules of a language, not its vocabulary and pseudo-sentences come from the misuse of grammar, from breaking

[4] Russell's preface to Gellner [1959] 2005 was no surprise; nor was his letter to the *Times* of 9 November 1959 that responds to Ryle's refusal to have Gellner's book reviewed in *Mind*.

its rules. To this he added his central idea: all metaphysics is pseudo-sentences (*PI*, §§5.633-1; 5.641; 6.53).

Russell was of the opinion that the structure of the *Principia Mathematica* was not arbitrary: the truth of its propositions is mostly truth by nature, not truth by convention. He said that the *Principia Mathematica* reveals the deep structure of all natural languages.[5] This directed Russell to the search for the place of philosophy in his evolving system of thought and it sent Wittgenstein to seek in that system proof of the positivist theory that metaphysics is empty. Russell, however, was already committed to a metaphysics. After Moore convinced him to abandon his youthful idealism, he had no system of philosophy beyond realism and commonsense; at least not yet. He first hoped to develop one, and then proved that this is impossible. He began with following the empiricist tradition and rejecting the claim that there are synthetic *a priori* truths (except for the principle of induction). He could not be satisfied with classical analysis old-style. Boole's attention to the empty class and Russell's attention to the non-existent present king of France are not due to any importance that these nothings might have, but due to the challenges that they raised. Finding that logic must permit terms that do not denote is the beginning of analysis new-style. This was a great improvement over analysis old-style, as this way the analysis of concepts gave way to the analysis of words, but this did not suffice, especially in view of the role of old-style theory of meaning in the analysis. Frege discovered that this is absurd, we remember, as he showed that it led to the puzzle of viewing as a tautology the greatest discovery of ancient astronomy (that the morning star and the evening star are identical). Russell went further and showed that the sentence "Scott was the author of Waverley" demanded a similar analysis, except that this was such a great innovation because it was the first analysis of a statement rather than of a mere term. As language mediates between humans and the world, this leads us to discuss the correctness of our comprehension. When we fail in this, we naturally tend to discuss language, the mediator. This shift leads logic to the shift from the search for truth to the search for truth-values, and only then for the question, which option is true? Logic leaves it to empirical science to discuss the answer to this question. How exactly does science accomplish this feat?

The hope was that the new logic would help just here. Yet Russell was stuck. He conceded that words like "blue" are universals, less abstract than concrete empirical observations; yet he also held the opposite view; indeed, as he approached the question from the logical point of view, he inclined to the one answer, and when he approached it from the scientific point of view, he was inclined to the other. This called for a new theory of empirical observations on which scientific theory can rest.

Frege and Russell had at their disposal different kinds of verbal entities for composing simple statements: matrices, namely, n-place predicates[6] and tools for

---

[5]The dichotomy between truth by nature and truth by convention is historically most significant; since 1905 it is outdated. Many works on language take it for granted; they are thus outdated too. Quine stressed this fact repeatedly.

[6]In detail, things are much more complex than they look. The matrices comprise x, y, z, etc. F, G, H, etc., 1, 2, 3, etc.: $F_1(x)$, F_1 (y), etc., $G_1(x)$, $G_1(y)$, etc.; etc. $F_2(x,y)$, etc.; etc. $F_3(x,y,z)$, etc. a, b, c, etc. (x), E(x), and the connectives of the statement composition calculus. As Paul Benacerraf (1973) has

turning these matrices into sentences: names, to replace the variables or the universal and existential operators to bind them; and the operators of the calculus of statement composition. The universal quantifier was the most troublesome: if it denotes, then it seems to denote everything and its meaning then seems to be Platonic. Predicates are troublesome too, since they are universals proper. Names seem least troublesome as they refer to particular entities. Now particular entities (singulars) can have names. What exactly makes them particular or singular? This is the problem of individuation. This Russell did not presume to know: the details of what makes a thing a thing is a matter for empirical science, not for logic. His theory of definite description does not tell us how we pin down a description. Whatever it is, it will be a sort of definite description. Hence, proper names in a natural language may be nothing but definite descriptions (Kripke). If so, then the identity that a description offers is between the references of different definite descriptions: "the wisest man in town and the ugliest man in town are one and the same". Now call the wisest man in town "Socrates"; call the ugliest man in town by the same name. The two need not be the same person.[7] Hence the equation "Socrates is Socrates" may be false; hence, its truth is not analytic. Confusion between the analytic and the non-analytic version of "Socrates is Socrates" is confusion between the two senses of names. This has made classical logicians demand that every name should be unique. Is it better to relinquish names altogether and replace them with definite descriptions? What is the need for names? At some time, Russell was ready to entertain the idea that names proper are pointers. If so, then the only basic – irreducible – names are pointers proper: "this" and "that".

Thus, Russell needed a better handle both on language and on facts. He then postulated bravely – and erroneously – that sentences are divisible to the smallest items, to atoms, and so is the world divisible too: there are atomic facts and atomic propositions that are thus true or false. This raises the question, why should facts imitate language? Decades later Popper's theory of relative atomicity answered this question (Popper [1935] 1959, §29): a given task to test theory divides the world artificially to atoms of possible tests for if; this leaves to scientific theory (and in a pinch to metaphysical one; Agassi 1971) the question, what facts are atomic? Russell did not have this solution. The problem troubled him. He repeatedly altered the theory.[8] This is where Wittgenstein stepped in.

---

noted, this should be available prior to the construction of abstract set theory as the foundation of mathematics, yet it refers to (infinite) sets that require abstract set theory for their proper management.

[7]Two Lord Russells declared jointly that the one of them is not the other. Joint letter, *The Times of London*, February 25, 1959.

[8]Russell [1959a] 1995, 9 dated his earliest logical atomism as 1899–1900 (as a variant on Locke's empiricism). He could then not develop it for want of tools and a clear problem. That crude version served his cooperation with Moore (against idealism) and then his refusal of Moore's theory of judgment. Almost every discussion of Russell of logical atomism refers to a different version of it; see the impressive Bostock 2012 and Hager 2012, 9, 56. All this testifies to Russell's generosity to Wittgenstein (Bostock 2012, vii, 72).

8.3 Logical Atomism

The next move was logical atomism, the theory of atomic propositions. It is not clear who invented it. Russell, who finished work on the first volume of *Principia Mathematica* in 1911, spoke of logical atomism first in 1914. In 1919 he lectured on it extensively. He declared that Wittgenstein had initiated the idea, or that they developed it jointly. At times, commentators ascribed it to Wittgenstein; after Russell rejected it, their ascription changed. Authorship is not important, at least not by comparison. What is important is logical atomism, although it is false. We have to show this.

Objects exist; referring to an item as an object is assuming that it exists. This act looks differently in mathematics than in empirical science. In mathematics and in *Principia Mathematica*-style logic, the assumption that an object exists is the default option, unless it leads to inconsistency: self-identity is universal there, so that everything is self-identical, so that of course everything that is self-identical exists, so that everything exists. If an assertion of the existence of some item (say, the greatest prime) leads to inconsistency, then that refutes the assertion. Even this result Russell achieved only after long struggle. First David Hilbert said, the mere assertion of existence (of a solution of some differential equation) is metaphysical, to be ignored. He then conceded that, though unsatisfactory, the sheer assertion of existence signifies, as it attests to consistency (to the possibility of a solution to a given differential equation). Existence in the real world is different. Even existence in mathematics turns out to be different. Russell admitted that his logicist program had failed: there is no logical reason to assume the existence of infinite sets, yet their existence is crucial for mathematics. Hilbert and Robinson, who hesitated whether to declare that infinity exists, conceded that assuming its existence is crucial for mathematics and they refused to give up abstract set theory. (Hilbert called it Cantor's paradise). Yet they were inclined to consider it meaningless.

Although Frege invented formalism, Hilbert put it to mathematical use. He wished to prove the consistency of mathematical systems. He suggested that since there is a finite number of mathematical signs, we can look at mathematics formally as a set of meaningless axioms which are transformed according to purely formal rules – of substitution and the *modus ponens sans* any meaning – and ask, is there a set of permitted moves from the axioms to the formula "$1 = 0$"? The system in question is finite, and so we may hope to discover that no permitted move achieves such a result. With the return of viewing signs as symbols, this amounts to the proof of the consistency of the system in question.

Russell raised the problem, in what sense does the infinite set exist? Assuming its existence in any sense is the reopening a new kind of Platonic heaven. This raises a more pressing question: how does physical existence differ from mathematical existence? How do we assert it in a manner that distinguishes it from the Platonic existence? There is more to it than that: there are social institutions and societies, and a question about them comes up repeatedly: what kind of existence do they possess?[9] This question Russell and young Wittgenstein ignored. As to mature Wittgenstein, his attitude on this is under dispute, and so I will postpone it.

We usually take for granted the empiricist claim that we attain information from the senses and that this information is basic and so most certain. Russell could not accept that: for a time, he considered "blue" a universal and he refused to endorse the traditional psychological empiricist theory as it stands. He did not know how to rely on the senses without it. He fell back on his theory of descriptions, but that refers to individuals, yet observations are of facts, not of individuals. Facts are not blueness as such but a fact like the one that this patch is blue. This should be the answer, of course, since to be a blue patch is universal and a definite blue patch is not. Suppose we begin with facts, then; we may seek a general theory of facts, or more than one: logical, mathematical, and physical. Where do logical facts reside? What do statements of logical facts designate? He did not know.

Russell's view of what we observe is not obvious: what individuals take as facts, as the basic experience, etc., is hard to decide: it requires very careful studies. The way science rests on empirical evidence is different. The scientific community agrees upon very few methodological ideas: usually different people hold different theories about science. Unlike inductivist philosophers of science, all practicing scientists endorse one and only one rule about science. Indeed, the scientific community considers outsiders those who do not follow this rule strictly. The rule is, *all and only those observations are scientific that are repeatedly reported by independent eyewitnesses and whose generalizations are endorsed* (Agassi 2013, Ch. 14 §§14–16). In particular, they need not be true: some scientific observations, it turns out in retrospect, are untrue; only unrepeatable ones are not scientific. Russell still took singular observations to serve as the basic units of empirical science. This forced him to concede that knowledge by acquaintance is private; science, being public, rests on knowledge by description. Still, he insisted: the latter somehow rests on the former. Here we are now concerned with Wittgenstein's early career. On the point at issue, concerning knowledge by acquaintance, at the time he agreed with Russell: the most central – and controversial – item in his early philosophy is his claim that facts *show* themselves. It even let him reject the principle of repeatability (*PI,* §265; Agassi and Laor 2000, 529), though it is the only rule that the scientific tradition is unanimous and adamant about ever since the scientific revolution.

The reading of Wittgenstein's *Tractatus Logico-Philosophicus* offered here is of necessity controversial, since there is no consensus on its reading. The general strategy adopted here is the suggestion to avoid scholasticism: although some interpretative discussion is unavoidable, try to avoid it as much as possible. The

[9]Quine considered the existence of sets the same as the existence of pieces of furniture. Bunge said the same about collectives and institutions. As distinctions are never objectionable, it is easy to postulate diverse kinds of existence. Already tremendously precise Lejewski 1954 has invited such a move.

interpretation offered here has a special status, as leading Wittgenstein scholar Rudolf Haller shares it with Popper.[10] Approving of this, Dale Jacquette says (1997, 37),

> Wittgenstein tells us that the forms of simple objects are space, time, and color (the latter of which I believe he intends in the broad ... sense including any phenomenal property).

This is a sophisticated version of neutral monism: every atomic proposition is a multiple of space-time co-ordinates and of the diverse possible sensations with decisions of the degree or the yes-or-no kind. This is the same fantasy as the *Characteristica Universalis* of Leibniz and that of "The Library of Babylon" of Jorge Luis Borges. Young Wittgenstein recognized only events, not things.[11] We carve out of the set of events a continuous subset and declare it belonging to that object. How do we do that? No answer. His concern here, however, was to develop his later philosophy. This includes his assessment of his earlier theory, and this assessment is of his theory as the theory of logical atomism and thus as a solution to Russell's problems as they emerged out of his mathematical logic in general and his theory of definite descriptions in particular.

Viewing Wittgenstein's ideas as coping with the problems that beset Russell's philosophy is an especially great compliment to him; there can hardly be a higher compliment to him. Yet his disciples seldom find this compliment pleasing.[12] They wish to read him in a way that somehow puts a good face on their own philosophies,

---

[10]Neutral monism looks to positivists attractive as it looks ruthlessly empiricist. Less ruthless versions of empiricism lose this asset by allowing continuity and thus allowing what Reichenbach 1944, 21 called inter-phenomena and outlawed – due to uncertainty or to the uncertainty principle. (He was vague on this, the central point of his book.) Ruthless empiricism admits as given only sense experiences. Young Wittgenstein advocated a version of it. This was brave but doomed to fail. He gave it up after he returned to Cambridge. His thesis, metaphysics is meaningless, lost its positivist background and thus became baffling. Followers then declared it sheer commonsense. It is what Popper presented as Wittgenstein's neutral monism. (Anscombe 1971 rejected Popper's ascription of neutral monism to Wittgenstein as more definitive than warranted. Later, Hintikka 1996, 127 refuted her argument. Popper's ascription prevails. See Jacquette 1997, 37 and Craig 1998, Art. Neutral Monism.) Here mature Wittgenstein differed from young Wittgenstein. (*PI,* §654 presents as *Urphänomene* what TLP, §3.262 presents as what *shows* itself). In 1929, Wittgenstein published his paper that showed how far we are from solving the problem of induction. He then realized that the number and characters of the co-ordinates that he postulated for allowing science to store its data are neither *a priori* valid nor *a posteriori* valid: they are themselves products of observation. This Borges elaborated on in his "New Refutation of Time" (Agassi 1970).

[11]*TLP,* §1.1: "The world is the totality of facts, not of things"; §4.1272: "Wherever the word 'object' ('thing' etc.) is correctly used, it is expressed ... by a variable name. ... [Otherwise] nonsensical pseudo-propositions are the result." Wittgenstein said a few times that one cannot say that x. This usually means, one cannot say "x" in truth; here it means one cannot say "x" meaningfully: it is meaningless. This, as Popper has observed, illustrates Wittgenstein's deviation from the ordinary meaning of "meaning".

[12]Monk 1990, Chapter 4 said, the master-apprentice roles of Russell and Wittgenstein reversed early in the day. This hagiographic claim rests on the information that Russell aaccpeted criticism of Wittgenstein. It has conter-examples from other critically-minded teachers. It nonetheless won great popularity. The hagiographic character of any commentary is apparent in its disregard for serious, honest, obvious criticism.

whereas he would never have approved of these philosophies, as they are only partly analytic. He demanded of philosophy to be as purely analytic as possible. The philosophy of Russell, incidentally, at the time also went in the direction of the purely analytic. Except that he never went all the way; Wittgenstein did. In 1922, Russell went out of his way to express disassociation from Wittgenstein on this very point. Let me backtrack.

In 1914, Russell tried to render philosophy essentially nothing but logic. Wittgenstein realized that in its austerity this effort might lead to the proof of positivism (of anti-metaphysics). This was a daring extension (perhaps also a modification) of logicism as the rejection of Kant's claim for synthetic *a priori* knowledge. Now traditional positivism leaves all legitimate metaphysical questions to science: it advocates scientific realism as a matter of course. Not so the positivism of the neutral monists: it is (counter-intentionally) anti-realist. Russell could not stomach it. Wittgenstein found it congenial to his mysticism.

The prime question is, then, what kinds of objects does Russell or Wittgenstein permit? Russell repeatedly settled for the answer that he found in commonsense or in science, and perticularly in their overlap. Wittgenstein's answer is, points in space-time. Every point has a proper name, all of them derived from the one name of the origin of the co-ordinate system. The point of origin has the name "this" and the mathematics that the system requires will be available as a part of logic proper. This makes Wittgenstein a standard Russellian; even more Russellian than Russell, and in two ways. He had no use of the concept "object", "thing" or "entity". (It is a pseudo-concept.)[13] For, logical atomism considers things compounds; the things that Wittgenstein allowed were logical simples: only space-time points. He also claimed to have done away with infinity (*TLP, §4.1272*). This, the item that Wittgenstein presented proudly, Russell had admitted as a defeat of his logicism. Each point has certain properties and bears certain relations to other points. A purely logical system or framework within which these facts are describable it presents as simples; all other facts are composed of simple facts. This, then, is the leanest system scientifically possible. It has no room for metaphysics or even for laws of nature.

It is also the leanest program in the Frege-Russell tradition. To succeed in executing it is to secure fame as the greatest philosopher since Kant. No wonder Wittgenstein did not want to let go of it. It was most important for him to outdo Russell. The most significant, perhaps even central success he could claim was his success – so he declared – in reducing arithmetic to finitist logic. Does infinity exist? He blocked this question. Thus, Russell's rejection of this idea (Introduction to *TLP*) was a tremendous blow. He remained resolute. Russell condemned his blocking

---

[13]TLP, §1.1: "The world is the totality of facts, not of things"; §4.1272: "Wherever the word 'object' ('thing' etc.) is correctly used, it is expressed . . . by a variable name." "The world we live in is the world of sense-data, but the world we talk about is the world of physical objects" (Wittgenstein b, 82, cited in Hintikka 1996, 71–2 and by others). "We primarily observe things, not observations. But the act of observation may be inquired into and form a subject of study". Dewey 1929, 12 rightly denied this: "We primarily observe things, not observations", he observed. "But the act of observation may be inquired into and form a subject of study". Thus, neutral monism is dead.

some questions; he took pride in it, forbidding the very asking of certain unpleasant questions. Its justification was this. Though it is more austere than Russell's system, it purports to have solved all the problems Russell had solved and more. The loss of the ability to articulate a metaphysics then is no loss at all. Clearly, this claim invites careful analysis of the situation. This takes us into details of mathematical studies that have no room in the present study. Still, a few observations may highlight the situation.

8.4 The Significant Meaningless

The Preface to *Principia Mathematica* notes that the *modus ponens* is not a part of its system: to speak anachronitically, it is meta-linguistic. In efforts to stay in the object language Wittgenstein said, rules of inference are "senseless and superfluous" (*TLP*, §5.132); one cannot assert them (*TLP*, §5.132).[14] Yet the theory of inference is most important. Metaphysics, then, like the theory of inference, is *unutterable yet significant*. Wittgenstein opposed not metaphysics but its utterance. This way, the rationale of traditional anti-metaphysics is lost: it turns from error avoidance to a limitation on speech.

Consider, then, Russell's resolution to his own paradox. Its strength is in its claim that we are not clear about what name can stand for what variable. Example from ordinary parlance for this: at times "Socrates", "Tom", "Dick", and "Harry" are not names but variables like "somebody". (Choice between reading "Socrates" as a name or as a variable is context-dependent.) Boolean algebra permits altering the universe of discourse, thereby altering radically the complement of any class; the universe of discourse is thus the range of the variables for class membership. Russell suggested that membership should be of objects in classes, of classes in classes of classes and so on. This is his principle of stratification: all classes have to be stratified. Classes whose members are from different strata are not legitimate. Now this principle is unassertable, as it refers to classes from all strata. Wittgenstein pushed the prohibition further: he recognized the leanest apparatus that suffices for describing neutral monism and developed a version of arithmetic with it.

This, then, permits us to speak of data in space, where different sensations receive their names from their functions, where different points in space-time have proper names and nothing else. This is the world as depicted by Hume and by Mach, presented with the aid of the logical apparatus to create a sophisticated version of it. It suited Wittgenstein's temperament, as he was skeptical about everything but sense data that (like Hume and unlike Russell) he deemed certain: true atomic propositions and their combinations *show* their own truth. This describes not

[14]Wittgenstein did endorse the rules, even while declaring them meaningless and thus unassertable. Here his concept of ineffability differs from mysticism. It is amazing how many confusions this prompted.

everything but everything describable in *the* language: solipsism then is true[15] in that system although, he added, as it is unassertable, it is not clear whether it is true also in God's world[16]: Kant's theory of the limits of reason that prevent knowing this appears now as Wittgenstein's theory of the limits of language (as such).[17] Whereas Kant spoke of the limits of reason but admitted the possibility (possibly, not advisability) of going beyond these limits, Wittgenstein spoke of the limits of language, declaring these the limits of "my world",[18] whatever this means. Wittgenstein spoke in the name of clarity, and yet his book is obscure on major points. He gave up hope that Frege or Russell would understand it, and hoped to meet one reader who would. Commentators on it found it impossible to understand it. On the strength of this obscurity and taking his greatness for granted, Anscombe compared him to some leading ancient philosophers.[19]

8.5 The Dawn of a New Era

Wittgenstein declared metaphysics dead; more precisely, he declared all articulated philosophy dead. We may speak of philosophy as short for articulated philosophy, since we all recognize and put aside whatever we know but cannot articulate. This is what Michael Polanyi has christened personal knowledge. He said, it is what a master – a master artist, scientist or any other – teaches disciples but cannot

[15]Russell narrates that Wittgenstein refused to admit what he could not see rhinos in Russell's chambers, and he even refused to admit that there was no rhinos in Russell's chambers. Russell adds that he suggested Wittgenstein should look under the bed. (In different versions, the animal is a rhino or a hippo, the place is a residence or a lecture room, and the place to look under is a bed or a desk.) This is a bit unkind, of course, as is the parallel story of the refusal of Dean Swift to open the door to Bishop Berkeley as a response to his anti-materialism. This conduct is unkind, yet it is the commonsense criticism of the strict empiricist demand to ignore the unobservable as unserious.

[16]*TLP,* §5.62: "... what solipsism *means* is quite correct; only it cannot be *said*, but makes itself manifest." In a discussion of this paragraph where commentators read it to mean the same as *TLP,* §5.6: "*The limits of my language* mean the limits of my world" Hintikka 1958 makes Wittgenstein commonsense rather than puzzling.

[17]Wittgenstein spoke of the ideal language, of the colloquial or the vernacular, and of language. Norman Malcolm added shared language and denied that Wittgenstein had any concern for it: his concern was the characterization of language. It is, he said, following rules (Malcolm 1989, 12). This is insufficient.

[18]This is a new, idiosyncratic meaning of the expression "my world". I will not try to fathom its meaning.

[19]Anscombe 1990; 2011, 18*. Wittgenstein said his ambiguity and inability to communicate his thoughts was a curse (Scharfstein 1980, 330). It was the inevitable outcome of his demanded not to try to say the unsayable (*TLP,* §7). This demand is not his invention: it is ubiquitous: all gurus who sanctioned it referred to the limitation of language; they did not discuss it, however, much less the barring of metaphysics. The central innovation of Wittgenstein is his claim that the western scientific traditional demand to bar metaphysics is demonstrable. In his mature writings, he took back or at least ignored logic yet he sought certitude.

articulate. Like Wittgenstein, Polanyi advised not to try the impossible. They were in error: we repeatedly attempt the impossible, and though we never fully succeed, we are not always fully frustrated: miraculously, we have progressed. Although there is no recipe for success, hope is always permissible and often advisable. Both Wittgenstein and Polanyi were reasonable; Shaw responded wisely to them (1903): "The reasonable man adapts himself to the world: the unreasonable one persists in trying to adapt the world to himself. Therefore all progress depends on the unreasonable man."

Like hope, upholding the critical attitude does not require guarantees. The fruit of the critical attitude is uncontestable: examining a general theory and finding a counter-example to it is the end of a game of a classical dialogue. Here the intuitionist philosophers of mathematics are right: there is no list of possibly successfully completed dialogues. When a dialogue is completed, it is added to the list of completed dialogues known as the history of science. Not otherwise. Imre Lakatos added to this: possible counter-examples suffice in mathematics. In science, a counter-example counts only when reports exist that it was observed repeatedly, by different observes (Popper). The claim that something is impossible – philosophy in the cases of Wittgenstein and methodology in the case of Polanyi – invites the critical response to try to break some barriers. We may then respond to the claim of Wittgenstein thus: not philosophy is dead, but traditional philosophy is (Hintikka 1996, 10). The first properly philosophical theory was that of Thales, Aristotle has reported. It was, "all is water". His followers disagreed about the question, what is the substance of the universe, not that there is one. That there is one is the theory of substance. Einstein superseded in 1905, suggesting that a more abstract theory of science as laws of nature should replace it. Russell accomplished this same task in his own different way.

Einstein's famous equation of mass with energy translates to Aristotelian as the equation of the actual with the potential. For the Aristotelians this equation is absurd. Russell's destruction of the theory of the substance began with his claim that statements that describe relations – two-place predicates – can be as real as categorical (subject-predicate) statements – one-place predicates – and the former are not reducible to the latter. He continued with the observation that by the definition of substance the universe cannot possibly possess more than one substance at most. The attraction of neutral monism is that it does not postulate a substance of the universe. Russell then endorsed commonsense-realism-sans-substance. He still felt that some version of logical atomism is necessary to maintain this new realism with the aid of empiricism. We may agree and offer Popper's relative logical atomism (Popper [1935] 1959, §38). What is atomic statement is still an observation report, and its atomicity depends on the theory to be tested (and on the background knowledge that comes with it). This keeps the new theory, the realism-sans-substance depend on what we consider natural laws, and this is metaphysical; to be satisfactory it should be a part of a metaphysical system proper (*à la* Einstein). Quite a few thinkers have contributed to this image, including Hintikka who always considered himself indebted to Wittgenstein, I do not know why.

The claim is popular that there can be no universal history of philosophy, since every specific history of philosophy reflects the philosophy of its author. This is not quite true. It is true that irrationalist philosophers are at liberty to write the history of philosophy any which way they like. The paradigm case here is Martin Heidegger who used arguments to support his philosophy from ancient Greek philosophy and while using philological case studies to help him reach his conclusions. Except that his philology does not receive the blessing of run-of-the-mill philologists.

Rationalist philosophers share much of the discussion about the history of philosophy. This is the best and most interesting argument for the clearly anti-positivist thesis that philosophy does progress (although not as satisfactorily as science does). Many classical histories of philosophy display some clearly stated philosophical viewpoints that repeatedly influence the specific historical judgments that they offer. Nevertheless, the locus of agreement – in philosophy as in science – is its history. Not that this comes automatically, but that forging a history is rendering a greater portion of it universally admitted. My hope is to add to the history of philosophy the still unwritten canonic presentation of the place of the mature (anti-)philosophy of Wittgenstein in it.

Chapter 9
Interim Period: Carnap Versus Popper

Self-proclaimed Wittgenstein followers repeatedly faced one philosophical problem that they found pressing: it is the problem of induction. They considered it the problem, how are [scientific] beliefs justified? They could not solve it. Young Wittgenstein claimed to have solved it (*TLP*, §§6.31, 6.363). Mature Wittgenstein ignored it.[1] Popper viewed beliefs as private; he said, science has no need for justification, as it progresses through refutations. His peers in the "Vienna Circle" took his solution to be, belief is justified by failure to refute it. Does this proposal of Popper conflict with young Wittgenstein's identification of language with the language of science? Carnap did not think so. He combined the two. He was in error. Wittgenstein rejected the traditional view of the negation of a scientific theory as unscientific: science is not closed under negation as language is. (Traditionally, a string of words is or is not a wff together with its negation, but not all wffs are scientific. Wittgenstein declared scientific all the synthetic statements in the system of his *Tractatus*.) Popper endorsed the traditional view as he declared the negations of scientific theories unscientific. The views of Wittgenstein and of Popper are thus as clearly incompatible as any two views can ever be; Carnap and Hempel managed to ignore this fact.

In a sense, Popper was a skeptic. Wittgenstein and his followers were interested in the justification of knowledge in search for an answer to the skeptic. The text that

[1]When Wittgenstein demanded of Popper to present a philosophical problem, he mentioned this problem; Wittgenstein dismissed it. Late in life Ayer 1982, 18 disagreed: "For my own part, I think that if one were looking for a single phrase to capture the stage to which philosophy has progressed, 'the study of evidence' would be a better choice than 'the study of language'." Notice that Ayer did not say what philosophers should study; he only noted that philosophy has progressed in a direction that Wittgenstein said has its road forever closed. Why then did the problem of induction trouble so many of Wittgenstein self-proclaimed followers? I do not know. Young Wittgenstein did and Mature Wittgenstein did not discuss the problem of induction; he responded to Popper saying that the problem was not philosophical. As he said this in an awkward situation, it need not count.

J. Agassi, *Ludwig Wittgenstein's Philosophical Investigations*, Synthese Library 401,
https://doi.org/10.1007/978-3-030-00117-9_9

Wittgenstein worked on in the eve of his life is "On Certainty". He too could not let go of it, try hard as he did.

9.1 Background Again

This study is about Wittgenstein's *Philosophical Investigations*, the book that was published as the first of his posthumous output, as the most comprehensive representation of his later philosophy. Previous Chapters here were devoted to an outline of the background to and the central ideas of Wittgenstein's first book. There is one more excuse for the postponement of the discussion of his second, posthumous book. The postponement here is for a discussion of the philosophical community's reception of his first book. This topic invites a separate study, but here let me center only on two aspects of the matter, namely, the evolution of the analytic method and the growth of the hostility to metaphysics. For, the chief question on the analytic school's agenda today is, does analysis expel all metaphysics? Or, was this only a passing phase of excess zeal? Even the discussion of those restricted matters requires strong measures to limit this detour: this is achievable by stressing the views of the "Vienna Circle", and, more specifically, of those of its most influential member: Rudolf Carnap.

Let me present some conclusions of the earlier discussions, to conclude with the general picture of philosophy and of analysis at the time when Wittgenstein returned to Cambridge after a period of self-imposed exile to follow his philosophical pursuits and develop his new variant or variants of the method of analysis. My main concern here is the contrast between the methods of the "Vienna Circle" and those of mature Wittgenstein, better known as Oxbridge philosophy or as ordinary-language philosophy or as analytic philosophy.

The new system of logic constituted a far-reaching revolution; even those who could scarcely comprehend it found its general ideas greatly impressive and appreciated them as eminently revolutionary. They all took for granted that the new logic should help forge a new philosophy. No one knew how to do that. Russell himself tried and developed the method of analysis and of its applications. He said quite explicitly that he had one aim in developing both the analytic tools and formal logic: if he ever neglected epistemology, it was only in order to return to it triumphantly: he always identified rationality with science; since he initially identified science with certitude, his hope was that his investigations would lead to clear scientific certitude. He thus identified his analysis and his logic, and hoped that the new tools will somehow help surmount the problem of induction. On this he used the help of Wittgenstein, whom he praised sky-high. Wittgenstein was thus a leading figure even for people who hardly followed the reasoning behind the new logic and/or the new philosophy that it had signaled (allegedly or in truth). They wanted results, however; hence, when analysis disappointed, they abandoned it or they lost interest in induction. The empirical basis of science, induction, was the focus of analysis after Wittgenstein had his first book published (1922) and after he had realized that it did not deliver the goods. This however did not stop the development; rather, it served as the basis for a new revolution, the one associated with his *Philosophical Investigations* (1953). Let us then consider the stage in his thinking in the interim period, before he wrote the draft of his *Philosophical Investigations*. Revolutionary philosophers then were still trying to get the most out of the first revolution, the one his first book had heralded. Though mastering its techniques looked to them all too easy, the application of these techniques to solve the burning problems was baffling.[2] Some things, however, were clear. Whereas the classical empiricists had developed their theory in a psychological mode, Frege, Russell and Wittgenstein did it in an analytic mode – as a part of logic. This sounded like re-founding empiricism on logic, and thus raising its authority to the limit. Did Wittgenstein solve the problem of induction, then? If so, how? What was his solution? What should a solution be if it is to be analytic?

Should we seek the answer in commonsense? Would that answer be obligatory? Whereas Frege took commonsense in his stride, Moore and Russell took it seriously and declared allegiance to it. Russell then discovered (Wittgenstein 1974, 24) that he was identifying commonsense with *naïve* realism, and that *naïve* realism and science clash. He thus found problematic his allegiance to both science and commonsense, and particularly so to his wish for induction to serve as a link between them. This rendered the problem of induction very urgent, yet he did not think it soluble. Later on, he said, a reservation in Hell awaits those who claim to have solved it. This is odd, since he explicitly endorsed the view of Einstein that scientific theories comprise series of approximations to the truth. Unlike Einstein (and Popper), he did not regard this idea as license to dispense with induction.

This is how things stood. A great problem-shift took place then. All commentators have noticed this shift. Wittgenstein's *Philosophical Investigations* exhibiting a lack of concern for epistemology made Rorty, a leading analytic authority, declare that mature Wittgenstein opposed all attempts to put knowledge on firm foundations. Putting Wittgenstein and Heidegger in one philosophical school with no comment as to their own preferences, he violated basic rules of scholarship. Nevertheless, he noted a significant kernel of truth: before Wittgenstein changed from his early to his later philosophy, he did lose interest in all attempt to put scientific knowledge on a firm basis. Yet he did not say anything like what Rorty has ascribed to him; he just lost interest (in science, not in certainty): all science ceased to interest him but he took care not to oppose it. When he did refer to any science, it was only to say – quite rightly, of course – that, as it needs clearing of its confusion and paradox, it too can benefit from analysis. Yet his grave concern was with the old question, how do I know that you are in pain? The true answer he ignored. It is, I cannot prove it and I cannot explain why it looks obvious to me. I cannot prove that your exhibition of

[2]Hintikka 1991, 180, viewed *Philosophical Investigation* as an unreliable pastiche that reflects only the views of its editors. They were the literary executors of Wittgenstein. Hintikka called them derisively "Wittgenstein's literary heirs", among whom was his *Doktorvater* von Wright, whom he appreciated and who denied the view that Hintikka ascribed to him (Wright 1969, 488).

pain is due to your suffering pain, but it looks to me sound. All contemporary commonsense philosophers and all their students cannot put all this together again.

9.2 The Interim Period

The members of the "Vienna Circle", especially Carnap, were both ardent followers of Wittgenstein and deeply pro science. Their insistence on the problem of induction prevented them from following him to the very end. This is also true of Russell, who, in the preface to his last significant analytic work, his An *Inquiry into Meaning and Truth* of 1940, declared his position nearest to that of Carnap. Few members of the "Vienna Circle" have published comments on the epistemological works of Russell; still fewer have published comments on the problem-shift in Wittgenstein's philosophy, for want of admission that he left a job unfinished. In consideration of all this, the present discussion has as its thesis the following statement. *The root of the transition of Wittgenstein's philosophy was his inability to cope with the problem of induction; he should have admitted that inability; he did not.* This is no censure: such an admission would amount to the assertion that the analytic technique of Frege, Russell and young Wittgenstein is limited the way the analytic technique of Moore that mature Wittgenstein partly emulated is not – if only because this technique aspires to achieve less (Moore) or nothing (Wittgenstein). This amounts to the admission that not all philosophy is analysis and not all philosophical problems are to analyze away. To admit this truth was not possible for Wittgenstein, much less for his fans.[3]

Leaving a job unfinished is not culpable. The very fact that the analytic technique of mature Wittgenstein was not supposed to solve the problem of induction makes it more reasonable,[4] but staying within it displays some indifference to science. It looks like a desperate move rather than a development; and if this impression is erroneous, then the error merits open discussion. It never started. Ryle (1996) even considered Wittgenstein a philosopher of science (even though he could not ascribe to him any interesting idea about science). Commentators ignore this claim, although Ryle was a highly respected, leading analytic philosopher.

Comparison of the views of Russell and of young Wittgenstein is a popular topic of discussion. Some commentators follow David Pears, the prime Oxford Wittgenstein apologist. He presents (1985, xii, xv) Russell's and Wittgenstein's versions of logical atomism as soft and hard, and declares the hard version superior, as only it confines language to atomic statement and their compounds. As he rightly observes,

[3]Monk 1990 is the most popular life of Wittgenstein. It is defensive. This defensiveness is an insult to Wittgenstein's memory.

[4]In response to Popper's solution to the problem of induction that rests on his view that there is no inductive method, a few Wittgenstein fans suggested – in the spirit of mature Wittgenstein – that only a general method does not exist; many detailed (local) ones do exist. This of course, is a true observation. As it is no solution to the problem of induction, it is a red herring.

all versions of logical atomism constitute neutral monism stated in its traditional, psychological versions or in some new, linguistic ones. The hard variant is either unique or impossible. (Proof: were there more than one, then the need to choose between them would refute them all as that very choice is informative yet neither atomic nor compound.) As to soft logical atomism, the first detailed soft variant of it appeared in Carnap's 1928 *The Logical Structure of the World*. As to the less detailed versions, Russell repudiated them all already in 1936, in his "The Limits of Empiricism", where he there gave up all versions of traditional empiricism (and endorsed the principle of induction as a synthetic *a priori* valid truth). He did not relate this idea to logical atomism. He developed another soft version of logical atomism in 1944, and it was still further away from the original version, in that it presented his logical atomism as a hypothesis. This was necessary, since, as a hypothesis the theory cannot possibly serve as the empiricist basis for all science. Russell intended to return to it; he never did.

In 1922, no one wished to adopt any soft version of logical atomism, as the reason for this wish transpired only after much deliberation. Pears' report is thus a-historical. (This no criticism.) In retrospect, the question always is, what merit does any modified version have over the original? The answer to this question depends on the modification in question. It is possible to advocate any idea, soft logical atomism included, without discussing its merit. The only reason for noticing it then is the one that has made Hume suggest it in the first place, namely, the recognition of one's inability to solve the problem of induction. Any new version of it invites examination against the old one: does it solve the problems that the old version had failed to solve? Pears does not say. He does not face the problem of induction squarely, and so we may disregard his discussion without loss. I take this as an example for most of the extant discussions of this issue – for any issue regarding Wittgenstein for that matter. (This too is no criticism: most of the academic output these days is of little value, even in the empirical sciences.) There is no need to complain, especially since modern logic is still very valuable. The only advisable way to go about problems that it has left unsolved is to discuss the situation further.

Back to 1922, then. However complimentary Russell's Introduction to Wittgenstein's book was, it tormented him.[5] Though that Introduction was laudatory, it also declared the project a failure and even a mere delusion. It was indeed a deathblow, as it included a renunciation of the ideal-language project. Things look especially unpleasant with the realization that came only over a decade later, in 1936: only then did Russell declare explicitly that the joint ideal-language project is bankrupt: it was then not only the recognition of the need for a separate meta-

[5]This is my understanding. The only direct evidence on it that I have found goes against it. Hintikka 2006, 37, reports that Wittgenstein said to him, "... let me tell you – if I had come to a university when I was a young man and professors had been such fools as to not understand my ideas – I would not have been unhappy at all. That would have been a marvelous opportunity!" Hintikka added this: "I am not sure that Wittgenstein's counterfactual was true, but it does illustrate the facet of his personality that has not received its due."

language; it was the desertion of the strict empiricism that was the heart of Wittgenstein's book. This is no ground for a complaint; it explains the deep hurt that Wittgenstein felt and the unease that he quickly developed towards Russell and that he nursed for the rest of his life. For a historian, it is always essential to avoid matters of blame as much as reasonably possible, and to try to understand different opinions, attitudes and feelings. Here we have a great unease that seeks explanation. To strike a balance between understanding unease and refusing to endorse it is not always easy. In the present case it is. Wittgenstein biographers overlooked it because very regrettably most of them are defensive for him.[6]

Wittgenstein's fans complained about Russell's conduct: he could and should have diverted much of the unease had he said, as he was criticizing Wittgenstein, that he did not have as comprehensive a view of things as Wittgenstein had. This was contrary to all that Russell was doing.[7] In his "On Denoting", he was criticizing Frege's theory of meaning and offering his own, yet he did not have a theory half as comprehensive as Frege. He took this for granted: this was his analytic style: his views developed as his analysis helped him develop them, and he could not say how far they would go. Incompleteness looked to him an aspect of the analytic style itself. He thus saw in the comprehensiveness of Wittgenstein's view some unpleasant defect, as it was delusional. Presumably, for Wittgenstein this view of Russell's was the most painful part of his Introduction.

The personal aspect of the matter is less important than the philosophical difference. What the difference between Russell and Wittgenstein was is clear enough: Russell took logical atomism as a partial and temporary solution to the problems at hand, since he was still struggling with the problem of induction; in his Author's Introduction, young Wittgenstein declared his own work meaningless, yet final and "unassailable". It included logical achievements, and Russell lavishly praised them. The reduction of the axioms of the calculus of statement composition to truth tables,[8] the reduction of the number of basic operators from Russell's two (negation and disjunction) to Sheffer's one that combines them, the development of a new arithmetic system within his system of atomic propositions, and the elimination of the cumbersome theory of types. (Russell never relinquished that theory but he never liked it, considering it much too artificial.) What Wittgenstein considered the greatest achievement of his first book, the elimination of metaphysics as puzzles that do not

[6]This holds particularly for Ray Monk's popular lives of Wittgenstein and of Russell.

[7]The disagreement between Russell and Wittgenstein as to the possibility of the meta-language is irrelevant here: Russell ascribed to Wittgenstein the mystical view of the totality as ineffable, dismissing him as irrationalist. Incidentally, Wittgenstein's refusal to recognize the meta-language is the view of the vernacular as an ideal-surrogate. Thus, Wittgenstein's view is in conflict with the Sapir-Whorf hypothesis (Harvey 1963, esp. 175). The literature is full of attempts to square them, which is easy, of course: already Maimonides used profusely the technique diffusing metaphysical assertions by considering them metaphorical.

[8]Hintikka too praised Wittgenstein for his truth tables (Hintikka 2006, 35 *ff.*). Otherwise, Hintikka found Wittgenstein's first book no more than gross errors and vague suggestions. Nevertheless, he considered him a "major thinker" in "the entire field of human thought" (Hintikka 1996, 7).

exist, Russell dismissed. Here lies the great difference between them. Russell was critical of the logical material in the book, despite all the praise he lavished on it. He said, despite all of its brilliance, he judged it a failure too: actual infinity is essential for mathematics, and Wittgenstein's system proscribes it.

The most basic difference between Russell and young Wittgenstein is this. Wittgenstein was consistent and Russell was ambivalent. Wittgenstein's denial of the possibility of articulated metaphysics was unwavering. It characterizes his version of analytic philosophy: all philosophy is analytic; there can be no articulated synthetic philosophy. Russell often saw synthetic philosophy as traditional Continental philosophy, and he saw in Hegel's metaphysics (perhaps in the versions that John McTaggart or Francis Herbert Bradley have constructed) the culmination of synthetic philosophy; and it appalled him. The more seriously Russell took the problems that the Hegelians claimed to have solved, the more he was willing to be analytic all the way – like Wittgenstein. When he considered genuine philosophical problems, as when he wrote his classic 1912 *The Problems of Philosophy*, he could not share any simplistic dismissal of all philosophy. So he wavered. It is regrettable that most commentators do not express sympathy with him on this.[9]

The Problems of Philosophy offers a criterion for the proper subject matter of philosophy: empirical and the logical-mathematical assertions aside, whatever assertions remain should undergo analysis, and then, after the conclusion of the analysis, what does not yield to analysis is philosophy proper. Analysis is over only when all is clear. What remained, then, is clear philosophy. Is such a thing possible? Wittgenstein said, no: utterly clear philosophy is utterly silent; Russell wavered.

What Russell found most objectionable in Wittgenstein's early system was the idea of important nonsense (Alam 1990, 6), especially mysticism, Russell's *bête noire*. He evidently considered mysticism unfair as asserting something and then withdrawing it before it can benefit from undergoing examination.[10] In the conclusion of his essay on Bergson, "Mysticism and Logic", he said, the mystic's claim for the infallibility of intuition is false; the intuition that leads to any idea is irrelevant to the demand of rationality to examine ideas; and an idea not explicitly asserted is not available for examination. Hence, it is best to ignore an idea until it acquires an explicit assertion.[11] This goes well within the scientific tradition; Wittgenstein's discussion goes well within the positivist tradition. Yet positivism is traditionally within the scientific tradition. It includes laudable cases of critique of science. The paradigm case here is the (failed) effort of Mach to restate Newtonian mechanics without absolute space and without forces. Wittgenstein lost interest in science; his followers in the "Vienna Circle" were very interested in science. Their appreciation

---

[9]Landini 2007 describes this wavering in fascinating detail, but without saying that this is what he does.

[10]It was a common practice in mid-century Oxford to finish an observation saying, Wittgenstein would not have approved of what I have just said and to take back the observation. This conduct won much appreciation as austere though it was no show of courage.

[11]The idea that before criticizing an idea one should develop it and strengthen it and put it succinctly follow the traditions of Plato and of Galileo. It makes reading them and Russell enjoyable.

of his book flattered him, but he kept his distance from them; when he finally consented to appear before them, he read a text from the Indian poet-novelist Rabindranath Tagore.[12]

This way Wittgenstein expressed his concern with ethics and esthetics. He included religion somewhere within the ethics-esthetics complex.[13] The current name for this complex is "value theory". That it is hard to say exactly what in the view of Wittgenstein on this is matters little, since he said almost nothing about it; nothing, really, except that it is one and that it transcends language – or perhaps it is one in that it transcends language the same way. This transcendence Wittgenstein proved with one swoop. Ethics says, "Thou shalt . . . ", he said, which is meaningless, since, what if I will not? *(TLP, §6.422)* It is not at all clear how this tallies with the insistence of Wittgenstein that the vernacular is all right as it is. Possibly this requires a framework of some general theory of meaning of descriptive and of prescriptive expressions, as imperatives do not describe. The analysis of imperatives requires a theory of language. Young Wittgenstein offered none. Max Black's once popular 1971 *A Companion to Wittgenstein's "Tractatus"* displays this in great detail. John Passmore (1969, 353) says of it, "It is a book, indeed, which one sets out to describe with more than ordinary diffidence". This is a significant general difference between the discussion of Wittgenstein and of those of many disciples, from the religious ones, such as Gertrude Anscombe, Peter Geach, Alasdair MacIntyre and Alvin Plantinga, through the religiously indifferent or mildly religious majority, to the anti-religious like Ayer[14] and Flew (Flew and MacIntyre 1955).

Not that this is what young Wittgenstein said of value theory, since mature Wittgenstein expressed a different view of it. Logic, he learned from Ramsey, is a "normative science" (*PI*, §81). Moreover, he said a never-obeyed imperative does not deserve to be called imperative (*PI*, §345). This is not surprising: most of *Philosophical Investigations* concerns rule following.

The interpretation of Wittgenstein offered here is very coarse, overlooking most of the details that are subject to much commentary these days. This commentary exhibits pseudo-scholarly apologetic and/or self-serving attitudes that are easy to

[12]"When he finally came, instead of answering their questions about his book, he sat facing away from them reading Tagore, the Indian poet-novelist, for over an hour and then got up and silently left the room. Afterward Carnap remarked to Schlick, 'I guess he is not one of us'." Schlick's response is not on record. His politicking still has adverse effect on the philosophical tradition; only the Heidegger scandal tops it. McGuinness 2002, 189 reports Herbert Feigl's explanation for Wittgenstein's conduct; "he did not want to see their expressions as he read". French 1993 presents it as an illustration of his view of life forms, no less.

[13]Politically, Wittgenstein was a conservative (Hayes 2003, 2; Vinten 2015 suggests philosophers should ignore this). The Romantic view of religion appealed to him as Engelmann's memoir on him indicates; I suppose this is why he found Tolstoy appealing.

[14]Griffiths 1991; Ayer 1936, Ch. 4. Ayer recognizes the empirical part of ethics and then he decides that this is all that there is to ethics. Here he definitely disagreed with Wittgenstein. His subjectivist theory oddly conflicts with any case of *akrasia* (moral weakness): victims of kleptomania disapprove of the thefts that they fail to resist. More generally, emotivism fails to notice moral qualms and moral growth. More generally, the reduction of value to taste ignores comparative valuation.

avoid by demanding that authors say how they decide which details to examine and to what end. Rules for the interpretation of Wittgenstein are sorely required. Here suffice it to say a few words about Wittgenstein's general views on the arts and the sciences at large. First ethics. Paul Engelmann, Wittgenstein's friend of the early days, published in 1967 a moving, very sympathetic memoir, seemingly as a footnote to some letters from Wittgenstein to himself and to others. He blamed there members of the "Vienna Circle" for an unintentional but irreparably damaging misreading of Wittgenstein's early work: they were concerned with science and they presented his ideas as a philosophy of science rather than the ethics that it really is. "This book's point is an ethical one", he quotes a letter of Wittgenstein, directing it against the way the "Vienna Circle" read him. This is odd, since Carnap's earliest book, *The Logical Structure of the World*, ends by making more or less the same observation, and Engelmann endorsed the received view that Carnap was the mouthpiece of the "Circle".

9.3 Engelmann

A literature sprang around Engelmann's small book. It does not discuss this difficulty of interpretation – because commentators are unduly defensive. Reference to the view of ethics and aesthetics that Wittgenstein advocated in his first book would then scarcely resolve the conundrum posed here, nor would his conversations on it (Wittgenstein et al. [1967] 2007). The letter that Engelmann cites suggests that the Wittgenstein deemed the moral lesson in question in the sentence that appears twice, in the original Preface and conclusion of the book, the famous Proposition 7: thou shalt not try to say the unsayable! Engelmann finds it a moving indication of the feel that Wittgenstein had for the arts, a certain kind of awe. Not every reader of the relevant text finds it as impressive as Engelmann did, since the book's hard-nosed set of ideas on logic and experience, its extreme tough-and-no-nonsense attitude to science, does not tally with a sensitive attitude towards the arts. Therefore, very simply, articulation aside, Engelmann, who was an artist and who knew young Wittgenstein in artistic settings, took the concern of Wittgenstein to be the arts. Carnap took it to be in science, akin to that of Mach. Engelmann was in error, yet his error is enlightening: though his oversight of Wittgenstein's concern with science is erroneous, he was right in spotting in his book a different interest as well. As to Carnap, he reports in his intellectual autobiography this. The view of science expressed in that book is unacceptable, forcing him and the book's author to part their ways: Wittgenstein gave up all interest in science and his erstwhile follower gave up interest in him.

There is no need to be finicky, though. In all of its versions, the philosophy that goes by the name of young Wittgenstein with this or that degree of accuracy is summed up with his injunction, do not try to say the unsayable! The logic behind it is his observation, "the puzzle does not exist". Needlessly, he adds to this a puzzling remark, be it a rider or an explanation. He says (*TLP*, §6.5), "If a question can be put at all, it can also be answered." This is high-handed and distasteful – even from Wittgenstein's own viewpoint.

The book ignores the logic of questions. That philosophical questions are meaningless Wittgenstein simply decreed. The theory of meaning that he introduced, if there is such a thing, related to statements and to statements only. What is the theory of the meaning of questions? Briefly, the whole theory is this. Meaningful questions must have meaningful answers. Let us examine this theory in action. The most obvious question to ask is, does God exist? We cannot know: as Wittgenstein said, "God does not reveal himself in the world."[15]

This is not how the "Vienna Circle" understood Wittgenstein. They said, the question is meaningless as no answer to it is verifiable. This could become verifiable by viewing "God" as a tacit definite description. As such, they adjudicated, it is wanting, so that "God exists" is meaningless as an incomplete sentence. Alternatively, it is well-articulated (say by identifying Him with the pillar of fire that went before the Children of Israel as they wandered in the desert), and then it can be proved, they said, that He does not exist. Moreover, they said, traditional theology wavers between the two readings and leans towards the view of human language as too poor to describe God. Hence, the answer is not open to empirical test and definitely has no meaning. This is an application of the verification principle[16] of Wittgenstein's first book, on the reading of the sentence just quoted with the expression "can be answered", to mean, can be answered-with-certitude. A meaningful assertion is authoritative, then, and a question is meaningful if it has an authoritative answer.

Here ethics enter the picture, if at all, by the arch-positivist maxim, do not say what you cannot prove! Do not just throw speculations about! This sounds like stern moralizing. All analytic philosophers agree that ethics should not demand the impossible (in their stilted jargon this idea reads, "ought implies can") and traditionally the claim for the possibility of proof rested on no theory of proof but on the allegation that Newton's theory had been proven. This altered after Einstein's brave improvement on Newton's mechanics. Many deemed then the stern moralist attitude out of date. Thus, Russell, for example, stated in his "Scientific Method in Philosophy" of 1914, that it is in the spirit of science to make hypotheses to test and correct later, in the hope that the process of conjecture and correction will prove fruitful.[17]

By contrast, young Wittgenstein propounded a view of science that leaves no place for hypotheses: in it (synthetic) atomic propositions come in pairs, first they are

[15]*TLP,* §6.432. Already Maimonides deemed harmful superstitions all beliefs in divine interventions in the world – except for the miracles reported in Holy Writ.

[16]Hempel, the last of the "Vienna Circle", renounced verification (Linsky 1952, 163–88), yet he approved of the dismissal of theology as meaningless. This renders the demand for verification a mere excuses, and (as Popper [1935] 1957, §6 has observed early in the day) the epithet "meaningless" is a mere swearword.

[17]This view of Einstein is problematic, yet it is the starting point of all serious philosophy of science. Hence, most active philosophers of science today do not count (Agassi 1995a). Incidentally, Einstein endorsed Popper's solution to the problem of induction; Russell did not.

both hypotheses, and evidence imposes one of them on us. Traditional neutral monists had to dismiss abstract scientific theories, but they endorsed them nonetheless, as mere *façon de parler*. Wittgenstein declared (*TLP*, §§6.341–2) theories not hypotheses but a framework within which to organize atomic facts.[18] Even the concept of an atomic proposition receives no explanation, nor even an illustration. So much so, that Anscombe says that he was not decided on the matter and left it to psychology to discover [!] what proposition is atomic. Indeed, he said too little on psychology. She thus makes his philosophy much less analytic than Russell's, or even Frege's, who advocated the move away from psychology. Wittgenstein expanded a bit on his views about physics. He took space and time to be a part of logic, we remember. Which space and which time? Newton's? Einstein's? In 1922, he spoke of Newtonian mechanics as if it was the last word in physics: he said that the fit of atomic facts with Newton's theory says something about the world, but what they say is not open to explicit articulation.[19] (Most of the commentators on this are too defensive to discuss or even to raise the question, did he know of Einstein? He must have, as Russell could scarcely avoid talking about him.) What did Wittgenstein do with causality? Answer: he said, the faith in causality is superstitious (*TLP*, §5.36311), but he also expressed faith in natural laws, saying, if the principle of induction could be stated, it would read as something like, there is a natural law: natural laws exist, but we cannot say so, as this is neither an atomic proposition nor a compound one. Does this square with neutral monism? No one knows, but clearly, Hume shared this view, as he thought Newton's theory is here to stay to the end of days. On this Wittgenstein said less than Hume did – even by the standards of his all too short first book – yet he seems to have agreed with Hume on both points.

There is almost no literature on all of this. What Wittgenstein said does not begin to be satisfactory. In 1922, he spoke of physics as if it still complies with Newton, seemingly having no suspicion that Einstein rendered Newton's mechanics obsolete. Pears, the already mentioned authority on Wittgenstein for the true believer, has alluded (1988, 203) to this fact in passing, insinuating that Wittgenstein wrote advisedly on Newton rather than on Einstein in order to make a point that I for one do not see.[20] Wittgenstein wrote on science under the influence of Heinrich Hertz who, in his early positivist phase, tried to eliminate from Newton's mechanics the concept of force (as idle). The task of elimination of idle words was popular at the time. The influence of Mach on young Wittgenstein is manifest in the assertion that "The totality of true thoughts is a picture of the world" (3.001), which echoes Mach's assertion that the scientific worldview is nothing but the totality of science.

[18]*TLP,* §6.341. The best presentation of this view – conventionalism – still is that of Poincaré.

[19]Regrettably, commentators on Wittgenstein ignore an original and interesting idea. Admitting the conventional character of the fit of facts and theory, Wittgenstein observed that this refutes the conventionalist view of theories as utterly uninformative. This is important.

[20]Possibly references to the likes of Pears are unwise. In this chapter – on the reception of Wittgenstein – his case is an instance of the pervasive hero-worship. Biletzki 2003, 194 agrees; Hacker 2001, 1 does not.

A major critique that Russell had launched against Wittgenstein's book is one that many commentators have discussed: Wittgenstein manages to say much that he declares unsayable. Wittgenstein answered this in his book: these ideas are like the ladder that one throws away once one has reached the roof. This sounds reasonable, since the metaphor presents the ladder as weak but temporarily useful. It is still a dangerous metaphor, as it says, having used Wittgenstein's ladder, reached the roof, and then dropped the ladder, one is irretrievably stuck there in what in 1985 Pears called a false prison.[21]

A soft version of logical atomism in Pears' sense is possible; what recommends it? One can advocate it without discussing its merit. One can adopt scientific theories either without justifying them, or by justifying them as mere *façon de parler*, or by appeal to induction without reference to any specific theory of induction. This is hardly informative or interesting. One must tolerate them for a while nevertheless: they belong to the ladder. Except that we still need the ladder: we have not reached the roof yet and so perhaps we may always need it; perhaps it is here to stay!

9.4 Ideal Language

One final point before getting to discuss Carnap: ideal language.

Russell always took logic to describe – correctly or not – the form or essence of all natural languages.[22] He ascribed this idea to Wittgenstein too. The strict constraint that Wittgenstein put on language made Russell rethink. After all, there was a hint of this in Russell's own work, when, in order to escape his own paradox, he suggested the rule of language that all classes should be stratified, which rule allegedly holds for classes of all strata. This violation can be justified, of course, by placing the proposal in the meta-language while restricting its reference to the object language.[23] Similarly, as rules of inference are meta-linguistic, he realized that as construction-tools they are not a part of the language under construction. Wittgenstein took this as an example of a significant point that is nonetheless

[21]Goethe used a better metaphor: "Hypotheses are only the pieces of scaffolding erected round a building during the course of construction, and taken away as soon as the edifice is completed." Now the removal of the scaffolding is the proof of the hypothesis – that happens to be a false hope. The removal of the ladder is different. The difference between the meta-language and the object language includes the difference between mention and use, and Wittgenstein's throwing away the ladder is then the loss of ability to mention words that he used! The limitation that *TLP* imposes is much wider than it looks: how can one write a history, say, of the rise of writing?

[22]The medieval nominalism-realism dispute ended when Frege destroyed its basis, the reference theory of meaning. The sought-for alternative was – still is – realism with no Platonic Heaven and no Aristotelian definitions. This presentation is not quite adequate; I do not know how to improve it.

[23]This is not the only way to avoid the paradoxes: there is the Zermelo-Fraenkel abstract set theory (ZF) and the von Neumann–Bernays–Gödel set theory (NBG) and there is Quine's modification of Russell's theory. Russell was not pleased with any available solution, as he found all of them too artificial. Although he considered sets fictitious, he still wanted them intuitive, namely, quasi-natural.

unassertable: the rule of inference, he said (*TLP,* §§3.334, 6.1264), is not assertable. Russell remarked on this briefly, saying (Introduction) that perhaps a hierarchy of languages should enable one to assert the many ideas declared in the book unassertable while being asserted well enough. Now an admission of a multiplicity of languages forces one to question the idea of the one language that is the one essence of all natural languages. Russell could not have it both ways. This Carnap recognized: he gave up the idea of the essence of language. When Wittgenstein re-entered the scene, he took it for granted that there is no ideal language; he then made do with the vernacular.

Wittgenstein was Russell's best critic. He shattered Russell's view of science. Russell had hoped that the method of analysis would help him overcome the insoluble problem of induction without recourse to the Kantian solution that Frege and he did so much to avoid. However, Frege did not altogether avoid the *a priori.* Russell tried hard not to accept any view *a priori.* What options are there? If induction is valid, it is valid either *a priori* or *a posteriori.* Otherwise, it is hypothetical; it is then useless, since it cannot validate hypotheses. Now it cannot be valid *a posteriori* without begging the question, as Hume had argued. Russell tried to circumvent this but in the light of Wittgenstein's work he gave it up. In 1936, in his "On the Limits of Empiricism", he said, all our knowledge of the world is *a posteriori,* except for this very statement. It comprises *a priori* knowledge. Later on, he improved his view yet again: the principle, and with it empiricist philosophy as a whole, is hypothetical. Alas, by then Wittgenstein had lost interest in science.

Wittgenstein stayed intermittently in Cambridge between October 1911 and August 1914, when he left for military service. After the war and almost a year of internment as a prisoner of war, he studied in a college for schoolteachers, taught for 6 years in remote villages, and returned to Cambridge in 1929. He stayed there except for holidays, especially in Vienna, for service in London as a hospital porter in World War II, and for some trips abroad. In 1924, the "Vienna Circle" tried to get in touch with him. A meeting finally took place in 1928, a year after Carnap arrived in Vienna. It was a disaster, yet the "Vienna Circle" soon adopted Wittgenstein officially as its patron saint, endorsing with the greatest enthusiasm his idea that logic unmasks metaphysics as sheer confusion, defunct though by then this idea was.[24]

As Carnap gave up the idea of ideal language, he had no reason to assert that metaphysics is all confusion, much less that logic proves it. Yet he asserted it repeatedly. Only in the evening of his life did he alter it somewhat. He said then, he still hoped that a language can be created that allows for no metaphysics. In 1927–8, he wrote two works, *The Logical Structure of the World* and *Pseudo-problems in Philosophy.* The first offers a quasi-Russellian program of translating neutral monism from its traditional psychologistic form into a linguistic form and its execution in an axiomatic version. It incorporates an idea from some psychological theory of learning that some experts then took very seriously and that soon

[24]For the argument for the impossibility of metaphysics to hold universally, it has to employ a general characterization of metaphysics and rules that hold in all languages. Wittgenstein's hostility to "craving for generality" blocks this option. The ideal language obeys only general rules (Frege).

disappeared. The project of that book is by now considered an utter failure, especially since the basic attitude it endorses, methodological solipsism, is by now rejected by all, either on the basis of more modern analysis in the style originated with mature Wittgenstein, or on the basis of modern cognitive psychology. Nevertheless, Quine, for example, deems it a classic of analysis and so a milestone, though his own – Quine's – criticism of Carnap as it appeared in 1950 ("Two Dogmas of Empiricism") was most decisive and was recognized at once as such.

Quine's critique of Carnap's empiricism of 1950 was a turning point of great historical significance. It presents Carnap's empiricism as phenomenological, i.e., either as neutral monism, or as some other view based on Russell's early opinion that sense data are given, are certain, and should play the role of the sole basis – the ideal language – of all empirical science. To repeat, Russell himself rejected this view already in 1936, in his "The Limits of Empiricism", where he said that empirical science rests on both data and the synthetic *a priori* valid principle of induction. Russell himself tried to amplify this in his 1948 *Human Knowledge, Its Scope and Limits*. There is then Russell's defense of empiricism proper in 1914 and in 1921, and his later, constrained empiricism of 1936 and of 1948. The difference between these two is stark. It is no accident that Carnap was engaged in the nineteen-forties with the problem of induction, and that his *magnum opus* is his *The Logical Foundations of Probability* of 1950: the logic of the situation was the same with all pioneering philosophers. Indeed, when Frege ignored science, it was merely taking time out: the success of logic in the field of mathematics naturally led to the examination of what the same tool can do for science. Tradition characterized logic as deductive and science as inductive. Efforts to clarify this all failed because repeatedly the view of knowledge as resting on sensation imposed neutral monism. Analyzing this, Russell and Wittgenstein presented neutral monism as logical atomism. That proved to be a dead end. If science is neither *a priori* nor *a posteriori* valid, then it is not valid at all: it is hypothetical. This violates the verification principle that they all clung to, despite Einstein and Russell.

Carnap clung to the classical theory of knowledge as resting on sensations. He relinquished his initial position of 1928 fast, but only in detail, not in general outline (not its metaphysical foundations). He continued in the vein that is clearly the result of his endorsement of Russell's criticism of Wittgenstein and of subsequent work of mending the fences accordingly. He took seriously the proposal to build the meta-language, and considered formal languages conventional rather than natural; and he admitted, what Wittgenstein denied, that universality and existence are (meaningfully) assertable. This should have made him realize that Wittgenstein was in error, since grammatically correct metaphysical assertions are possible, regardless of whether "God exists" is true or false.

It is hard to see the difference between the ideal language and the sum of Carnap's two languages, the language of sensations (L_0) and the language of [scientific] theories (L_T), except that he recognized the need for a meta-language. On occasion, he did soften his view as he said that he had no desire to oppose any metaphysics like atomism that serves science. Other metaphysical claims, those not rooted in science, he declared confusions and so as meaningless. To that end, he developed a new, surprising theory of the confusion that is, he said, at the root of all historical cases of metaphysics: it is the confusion between the meta-language and the object language. This is a historical hypothesis that the case of atomism refutes. He ignored this and showed that certain statements that seem to us ambiguous can be reformulated more precisely once one decides that they are in the meta-language and once one decides that they are in the object language – but definitely not in both. Quine found this idea fascinating although, of course, he could not endorse it either; in the 30's both Gödel and Tarski studied translations between the object languages and the meta-language; they argued in a manner more in the style of Russell than in that of Wittgenstein, and this showed the poverty of Carnap's mode of argument.

9.5 Wittgenstein and Carnap

This development is of the 30's. Earlier, the central thesis of the "Vienna Circle" was basic Wittgenstein: it was an amalgam of a theory of meaning and a theory of science into one. It was the verification principle, the slogan: "the meaning of a statement is its method of verification". It is questionable whether this has any meaning, as so many admit inability to understand it.[25] How is one to judge such matters? By checking this both against a theory of induction and within a theory of meaning. As to a theory of induction, they had none. Only in 1950 did Carnap come up with a study of induction, his already mentioned *The Logical Foundations of Probability* that soon proved to be a disaster. As to a theory of meaning, which one did the members of the "Vienna Circle" use? No one knows. Unquestionably, it is a modern theory, as it speaks of the meaning of a statement, not of a term. How modern? When exactly did it find its canonic wording? In 1936, Carnap's *Testability and Meaning* offered a new, very problematic theory. Hempel soon published his shamefaced "Problems and Changes in the Empiricist Criterion of Meaning" (Linsky 1952, 163–88) which is a tacit admission of bankruptcy. It ends by declaring the meaning criterion non-arbitrary as it is refutable and thus revisable. This way he glossed over two observations, implicit but important nonetheless. The first smuggled observation is that the criterion in question is subject to a higher criterion: refutability plays here the role of the higher criterion of non-arbitrariness with which every criterion should comply. The second smuggled observation is one that Popper has repeatedly observed: Wittgenstein-style meaning is divorced from comprehensibility, since one must comprehend a lot about refutation and revision before one can decide about a criterion of meaning. (Despite all disclaimers, both Carnap and Hempel ascribed to Popper a theory of meaning. This is highly improper.)

[25]Black 1934. Speaking of the standard wording of the verification principle, Linsky 1952, 6 says, "the vagueness and lack of precision in this formulation are obvious." Since the "Vienna Circle" praised that principle for its securing clarity and precision, this observation condemns the "Vienna Circle" as not serious.

Let us take the matter slowly and a bit more thoroughly: a necessary condition for any statement to be meaningful is that the names and predicates it contains are meaningful. We know what makes a name meaningful in Russell's and in Wittgenstein's systems, but these systems are incomplete. Incompleteness need not be an obstacle, but when one speaks of lack of meaning, one does need a complete theory! The theory of the meaning of predicates is more difficult than that of names, but let us leave this aside. What is the theory of the meaning of a statement? There are two answers to this, both deserving serious considerations. There is, primarily, the formal answer. A well-formed formula with no free variable is a statement; a statement has a truth-value; truth-value is meaning. Explanation: a well-formed formula with no free variables is one that has all its variables properly replaced by constants, or else its universal or existential operator binds its variables. Thus, "Px" is a well-formed formula; it is not a statement as it includes a free variable; it can become a statement in two ways. One way is to put a proper name for the variable: replace "x" with "a" and obtain "Pa". The other way is to put an operator and write "(x) Px" or "(Ex) Px". This will render the formula meaningful, to mean that a is a P, or that everything is P, or that something is a P, respectively. Each of these formulas has a truth-value, and is thus meaningful, provided the terms in the formulas, "P" and "a", have meanings, of course.

Can any of these forms enable one to write a metaphysical formula? Popper said, yes: The classical metaphysical formula "God exists" is tantamount to "something divine exists". That is all that there is to it. The members of the "Vienna Circle" knew of this objection, of course, and did not yield. They said, as one might expect, that the word "divine" has no meaning. Why? Because the sense of the word is not clear. What do we mean when we say of something that it is divine? If we can present the predicate "divine" the way we present any predicate describing sensation, then it would be all right. Can we? Yes: the divine inspires awe, and we know what awe is. No good. The inspiration of awe is not like the inspiration of love, especially as love has a clear object but not awe, which is the root problem with the word "divine". Popper, however, cut things short: the predicates "powerful" and "present" are clear, and it is easy to add meaningfully the qualifier "all" to each of these three predicates, and then, together they represent the traditional theological sense of the divine as omnipotent and omnipresent. The claim that an omnipotent, omnipresent entity exists is theological enough, and is neither verifiable nor refutable. Popper proved it grammatically correct and hence assertable. End of argument.[26]

There is no need to continue in this manner. The "Vienna Circle" fans will be happy with an example only if it has imprimatur from science. They equated the demarcation of sense from nonsense with the demarcation of science from pseudo-science: all and only scientific assertions have meaning. The scientific character of

[26]Not quite. Rorty does not ascribe to Popper a criterion of meaning but he still distorts Popper grossly as he describes (200, 135) Popper's criterion of demarcation as "tracing the border between good science and bad metaphysics". In his most ardent aloofness from metaphysics, Popper never said it was bad *tout court*.

observation statements then is their being observable in principle: we remember that it is not the business of the logician to decide whether a synthetic statement is true or not but to leave it to science to make this decision. How, then, are theoretical statements verified? This is the problem of induction all over again. It was Quine's criticism that emphasized this: his "Two Dogmas of Empiricism" refers to two central and interconnected doctrines, phenomenalism and a rigid theory of meaning. Such a theory of meaning has sense only as phenomenalism, he added. In his view, meaning is different from scientific character, and so he broke away from the Wittgenstein spell over the "Vienna Circle".

Quine's view will not be discussed here, as it is not relevant, since it contains a peculiar kind of ontology (theory of being), and ontology is the heart of metaphysics. Moreover, Quine insisted that his ontology is faithful to science, whereas mature Wittgenstein was indifferent to both. Furthermore, Quine introduced a new contestant into the ring, called "ontological commitment". Whatever this term means, it conveys a regress to the psychologism of the old empiricist tradition. Therefore, it is alien to the present discussion, whose point is that the logic of both Wittgenstein and Quine is the same. Both agreed that the problem of induction forces the discussion into a different direction than the problem of meaning. Indeed, the most significant outcome of the transition from young to mature Wittgenstein is not only the shamefaced disappearance of logical atomism from the philosophical map but also the shamefaced disappearance of the identification of the demarcation of meaning with the demarcation of scientific character. Wittgenstein intensified his discussion of meaning, silently dropping his concern with science as soon as he learned (1929) that his very system of atomic propositions is informative. Russell, the "Vienna Circle" and Quine kept their allegiance to science. The rift was complete. Query: can one retain allegiance to science and maintain the meaning analysis that mature Wittgenstein has offered?

This question is open. Quine is supposed to have an advantage over Popper in that he has demarcated both science and meaning – despite his denial that he had a theory of meaning and his assertion that he was following Popper on the demarcation of science. (To the extent that Quine had a view on meaning, he agreed with Frege, as Alonzo Church observed in 1943.) At least it is clear that, under the leadership of Moritz Schlick, the "Vienna Circle" was scientifically oriented. Its members assumed that Wittgenstein was in the same boat: they admired his new applications of the new logic this way. Neurath, the second leader of the "Vienna Circle", deemed Wittgenstein's doctrine metaphysical and advocated a return to Mach. The current renewed popularity of Neurath despite the recognition of his weakness is a maneuver intended to put Wittgenstein's first book aside surreptitiously. In any case, the question is still open: since science broadens language repeatedly, can one have a theory of meaning while retaining an interest in science? For an affirmative answer to this question, the problem of induction should be solved, and in agreement with the general tendency of philosophical analysis. Such a solution is indeed available: Popper has supplied it. If it is not as satisfactory as is desirable, then it is wise to offer *desiderata* for a satisfactory solution, to discuss the feasibility of any extant answer (including Popper's) in their light, and try to improve upon it. One of these *desiderata* can be an agreement with a theory of meaning. Which one? Perhaps that of mature Wittgenstein. This remains to be explored – by anyone who can explain the import of such a *desideratum*. Thus far, the classical discussions of meaning (of Frege, Russell and Wittgenstein) were significant both as replacements of the defunct classical theory of meaning and as means for the radical transition from the theory of meaning and truth of names to the theory of meaning and truth of statements. Preferably, this should accord with Tarski's semantics (or with any theory that is an improvement on it). Popper's solution of the problem of induction tallies with Tarski's semantics. This should lay to rest the discussion of any future theory of meaning. If so, did mature Wittgenstein have anything to contribute with his new theory of meaning? We will return to this question later.

The canonic study of the analytic tradition is Coffa's *The Semantic Tradition*. He presents Popper as an analytic philosopher. For this to be acceptable, one should notice that Popper rejected this ascription. Coffa did not. To continue discussing this point, a specification of analytic philosophy is required. This holds not only for Popper but also for Russell and Quine and for all other philosophers who disassociate themselves from Wittgenstein. Popper was on the side of Russell here, as Russell viewed the principle of induction as not due to experience; also, (following Einstein) he declared science hypothetical and improvable, which is close enough to Popper's view of science. If Russell's work is a paradigm of analysis, then one may deem analytic Popper's view of science as a set of hypotheses improvable by empirical tests. All that this proposal requires is an official (or quasi-official) divorce between analytic philosophy and the Wittgenstein-style hostility to metaphysics. The hatred of metaphysics by itself, objectionable though it is, is no obstacle for this proposal; the claim that analysis backs hostility is. Young Carnap said, logic proves that metaphysics is nonsense. By contrast, mature Carnap concluded in 1956, saying, he still hoped for a language that disallows metaphysics. What are we to say of this contrast?

9.6 Criticism

This study is an attempt to evaluate the contribution of mature Wittgenstein to philosophy. This attempt already fills a huge library. Yet none of these takes the critical realist viewpoint as its starting point. Moreover, most of the studies that present attempts at an evaluation of the output of the mature Wittgenstein take it for granted that he made an extremely valuable contribution to philosophy and they ask, what is it? This sounds a bit funny, but it need not be. Nor is Wittgenstein the only philosopher whose contribution is difficult to evaluate.

Approaching Wittgenstein from the critical rationalist viewpoint is no help. Any attempt to evaluate his contribution whose starting point is the appreciation of Popper's critique of Wittgenstein and of Gellner's critique of his school, bump into a serious obstacle. The writings of many leading commentators on Wittgenstein are embarrassingly hagiographic, even though, obviously, personal matters are irrelevant to the evaluation of the intellectual contributions of a thinker, even if praise to that thinker may be due to all sorts of other matters. At least one leading disciple of Wittgenstein, G. E. M. Anscombe, has stated this forcefully. She probably found the prevalence of Wittgenstein hagiography embarrassing: she said she wished she could destroy all personal information about him.[27]

The defensive attitude so common among Wittgenstein commentators sadly renders useless much material that is the fruit of love and care and good will. We can scarcely absolve Wittgenstein of this defensiveness. Gilbert Ryle was a leading fan, yet when he went from oxford to Cambridge to participate in the Moral Science Club meeting (that was chaired by Wittgenstein) he found that "veneration for Wittgenstein was so incontinent that mention, for example, my mention, of other philosophers were greeted with jeers" (Ryle 1971, 11). Yet culpable as this conduct of Wittgenstein is, he is long dead and we are not to judge his character or his conduct. By the same token we should ignore the apologetic discussions of his output and the discussions that are presented apropos of his contributions that are frankly presented as quite different from his views, such as the works of Kripke and of Hintikka – regardless of our appreciation of their contributions. Nevertheless, it is quite possible to appreciate his having inspired them, and then we will have to decide how valuable these contributions are. When doing so, we should not make our assessment of the value of their works on our assent to them or on our dissent from them.

Agreement is no condition for approval or appreciation. It is easy to cite some unimportant item from a writer to signal appreciation. To explain the importance of an idea while engaged in its refutation requires some independence. Also, to appreciate say the disagreements between Russell and Frege, requires to appreciate both and to explain this appreciation. For, despite all problems about them and despite the diversity of readings of their works, it is easy to see how important their disagreement still is.

Nevertheless, we understandably consider some error easily avoidable and recommend making some effort in order to avoid them. When a critic finds us having made such a mistake, we are naturally embarrassed. We should therefore make clear the difference between errors that we should avoid and other errors. Even then, we need to maintain our sense of proportion. While discussing, say, the significance of the contributions of Freud, we should take it for granted that many things that he said are embarrassing errors. Yet we have to agree that by our very engagement with his central ideas we are showing respect for them, even when this engagement is mainly critical. For, no matter what area of study we consider, we can take it for granted that it includes too many texts that do not deserve public attention. They may deserve the personal attention of friend, though, since criticism is a favor (Plato, *Gorgias*). Public criticism, however, is a favor to the public, not necessarily to any specific

[27]This step is timely but insufficient: it invites assessment of contributions in oversight of biography. Alas, concerning Wittgenstein such assessments often rest on the ascription to him of old ideas, often Galileo's picture of the way theory influences perception (Niiniluoto 1999, 115).

individual. Mock criticism then is a public nuisance. What to do when the learned literature is loaded with mock criticism? This is a serious problem. The paradigm case is the scholastic literature in the age of the scientific revolution that included critical discussions but stale ones. The contribution of Francis Bacon then was therefore of great value when he argued against criticizing it and for ignoring it instead. Therefore, the default option before anyone who wishes to assess the value of the works of famous thinkers is not condemning these thinkers but to ignore them. Even then, we may examine the question, what make them famous. The paradigm case here is the contribution of Nicolas Malebranche, who was once very famous. It is easy to see why he earned his fame and why we tend to ignore him: his concern was to reconcile the philosophy of Descartes with Catholicism. And so only those who find this still interesting will remember him. If we agree that young Wittgenstein defended neutral monism and that neutral monism was of great concern for classical, Locke-style empiricism, then we cannot appreciate this empiricism and ignore young Wittgenstein. If we consider traditional anti-metaphysics (positivism) a serious concern, then even though we consider it superseded we cannot but appreciate the concern of mature Wittgenstein as the last effort in this direction.

Chapter 10
Ordinary Language Analysis

There is no agreed strategy as to how to approach Wittgenstein's *Philosophical Investigations*. This is regrettable, as one is essential for reaching any consensus regarding it. The rules agreed upon should be commonsensical: comments should be respectful but not defensive: they should follow Wittgenstein's own guidelines whenever possible, but critically, not in blind admiration. They should explicitly discuss the critics of their philosophy and their impact. They seldom heed these rules. What follows is the fruit of my effort to abide by them, as it is my wish that you rate my performance in their light. While paying attention to some simple general ideas regarding the background to this text, let me survey it.

To examine a monograph is to discuss its theses and its methods of argument, namely, the rules of presentation and debate of its author.[1] The natural starting point here is what has achieved general agreement: Wittgenstein's *Philosophical Investigations* says something striking and new about rules. In his *Tractatus Logico-Philosophicus* he says the rules of logic are meaningless (even though they are understood and should definitely be obeyed); in his *Philosophical*

[1]It is advisable to ignore the sub-standard contributions to any given literature. This is not always easy, as leading and highly influential contributions to a literature may be sub-standard. That Schlick, Neurath and Carnap cared more about the politics of the Circle than about the possible validity of Popper's criticism of its major tenets is the paradigm case here. It is time to stop sweeping it under the carpet.

J. Agassi, *Ludwig Wittgenstein's Philosophical Investigations*, Synthese Library 401,
https://doi.org/10.1007/978-3-030-00117-9_10

Investigations he says, following some rules is right, and this makes them (not meaningless, *TLP*, §85, but)[2] empirical (*PI*, §§84-6), an empiricl proposition "hardened" into a rule (Wittgenstein 1956, vi-22)![3]

10.1 *Philosophical Investigations*: Background

Wittgenstein's *Philosophical Investigations* rejects two significant traditional doctrines: the psychological foundation of the old theory of meaning and radicalism. Hence, regardless of any criticism of the philosophy of mature Wittgenstein, these rejections are uncontested if not also incontestable. These rejections are not new, and Wittgenstein was scarcely original in asserting them.[4] Assuming that he had an original message, then, surely these rejections are not it.[5] His rejection of metaphysics is not new either: it goes back to Bacon. The novelty of it in his *Tractatus Logico-Philosophicus* was in its basing of his anti-metaphysics on the new logic. Whether he later repudiated it is under dispute; it is of little interest here, since he ignored it in his *Philosophical Investigations*. There the rejection rests on common parlance, whatever this means. Obviously, he could not offer as simple an argument as his first book did: there he had complete description of what one can and what one cannot say; he decreed[6] that [in his language] only atomic propositions and their combinations are assertable. The view that all and only basic propositions are scientific and that the overlap of science and metaphysics is empty yields Wittgenstein's anti-metaphysics. Nothing like this is possible when the subject matter is a natural language. *Philosophical Investigations* offers few arguments and many examples.

[2]Leading commentators have spent much ink on the question, on what, if on anything, did the later and the young Wittgenstein disagree? Many of them say, he did not change his mind. Hintikka 1996, 47, 52 says, he underwent an essential change, yet he almost never gave up any explicit assertion of his: he only reinterpreted his early philosophy later in life. The great change in Wittgenstein's philosophy, Hintikka added, is his readiness to "involve … human activities". Involve!

[3]*PI*, §84 may allude to Robert Louis Stevenson's *Kidnapped* Ch. 4 that describes the hero standing before a door that opens into an abyss and escaping disaster by sheer luck. In his very last work, *On Certainty*, Wittgenstein spoke of degrees of certainty and of the possibility of erroneous feeling of certainty. Russell had said ("A Free Man's Worship", 1903), "… let us remember that they are fellow-sufferers in the same darkness, actors in the same tragedy with ourselves. And so, when their day is over, … be it ours to feel that, where they suffered, where they failed, no deed of ours was the cause; but wherever a spark of the divine fire kindled in their hearts, we were ready with encouragement, with sympathy, with brave words in which high courage glowed."

[4]Psychologism and radicalism support each other. The reactionary tradition that began soon after the French revolution with Edmund Burke's critique of it that was hostile to both.

[5]It is possible of course to view debates as a language game. This adds very little.

[6]Russell [1959a, b] 1995, 88 was censorious: "He, himself, as usual, is oracular and emits his opinion as if it were a Czar's ukase, but humbler folk can hardly content themselves with this procedure." (Etymologically "ukase" is showing. A Czar's ukase is a decree expressed by showing.)

The central thesis of the book is, there are no properly stated metaphysical assertions. The arguments for it are from theories that Wittgenstein advocated. The main theory is one that Russell initiated and gave up: meaning is use.[7] In England at the time of the book's publication, this attracted more attention than the rest of the book. Nevertheless, it is inessential. A few commentators tacitly agree with this, since they claim that the novelty of the book is in its method of argumentation. It is that of ordinary language analysis, whatever this exactly is. Views on it underwent changes; Wittgenstein might have approved of all of its versions as he insisted on the multiplicity of language games. This multiplicity by itself refutes his anti-metaphysics: we can describe what metaphysicians and theologians do as a language game.[8] This is no defense of what they do, since we can describe what impostors do, too, as a language game, as Wittgenstein has noted.[9] Even were Wittgenstein's view of all metaphysicians as slum-landlords true of all of them, this has nothing to do with meaning. He did create a huge confusion: he never said clearly whether he was denouncing only some[10] contemporary metaphysicians or whether he judged so all of them, ancients and moderns alike. In either case, there being immoral language games[11] makes metaphysics not the morally worst. Wittgenstein's ire was then obviously excessive. His view of meaning is then irrelevant to it: some metaphysicians who talk nonsense are decent about it and some who talk sense are nevertheless irresponsible, Gellner (1991, 33, 107 *ff.*) has rightly observed.

So much for the background material for the topic of the present study of Wittgenstein's major posthumous publication, his *Philosophical Investigations* of 1953. The frustrating historical fact about it is the cloud of vagueness surrounding it since its publication and to date. He was a prolific writer; he wrote until the very end. He published almost nothing after he returned to Cambridge to pursue his renewed career in philosophy. This is the period of about two decades in the thirties and in the forties, including the interruption for about half a decade due to World War II. His

---

[7]PM 30, 66–7, 71, 162. Quine 1981, Chapter 5 disproves it – together with behaviorism.

[8]Copleston 2002, 7 *ff.* Gellner 1992, 47, 84–5 *et passim*. The famous Sapir-Whorf hypothesis takes better care of the difference between languages and structures than Wittgenstein's theory of life forms. The ideas implicit in language are obsolete (often magical). Russell 1923, opening, says, vagueness boosts the unnoticed implications of our assertions and awareness of it improves thinking.

[9]*PI,* § 249: "(Lying is a language game that needs to be learned like any other.)"

[10]Moore's open defense of metaphysics makes it hard to ascribe to Wittgenstein the view that all the metaphysicians he knew were impostors. His condemnation is thus too vague.

[11]Hermann 2015, §2.5.3 takes it for granted that language games are moral when he presents moral disagreements as disagreements about the proper rules of some games. This obviously does not hold for tricksters. O'Connor 2002, 22, states this explicitly and explains it (on the assumption that language games are authoritative by consent, thus presenting Wittgenstein as a conventionalist in ethics). Referring to *PI,* §249, about lying as a language-game, Searle 1969, 324 said, "I think Wittgenstein was wrong when he said that lying is a language game that has to be learned like any other." It may be advisable to view lying as following rules of a language-game akin to the conduct of cheating chess players. We will all agree that to lie is different from to be a liar. Wittgenstein's use of his term "language-game" is highly problematic.

authority grew in leaps and bounds, and his ideas and some of his works circulated among admirers in that period. In 1945, he prepared an early version of his *Philosophical Investigations* for the printer; it had been in circulation earlier – he refers to that fact with displeasure in the Preface to that book, dated January 1945, and as the chief incentive for his preparing it to go to the press (it was towards the end of the War). The publication was posthumous, almost a decade later. It was a major event. The insiders knew why; outsiders could find out why only with great effort. In contrast to the author's short Preface and a shorter Editors' Note that declared the book unfinished: it includes a repudiation: much of it is not final. The situation is still not quite clear, but that much is clear enough: what one does learn from the Editors' Note is that one should not take the letter of the book too seriously; one has to absorb its spirit. The trouble with this counsel is that it invites bias concerning reading it. Those who find good things in the spirit of the text read it correctly; others are ignorant, obtuse, and even hostile. Well, if the choice is between friendly insiders and hostile outsiders, then, outsiders cannot possibly deny their hostility. The flaw of this division is obvious: the splitting of comments to two, inside-friendly and outside-hostile, unquestionably overlooks other possible options. Efforts to comprehend an author from the outside may interest insiders.

In the half-a-century that followed the publication of Wittgenstein's *Philosophical Investigations* loads of exegesis on it appeared. Stuart Shanker has issued five classic volumes of selected essays in this vein. Much material on Wittgenstein's private life also appeared, most impressive and least well received among them was a life by W. W. Bartley (1973). Wittgenstein's executors dismissed it as hostile on the ground that Bartley had dismissed Wittgenstein's philosophy. Right or wrong, this is not a sufficient reason for a dismissal, especially since Bartley was a confessed admirer of Wittgenstein. In any case, I prefer to overlook the private aspect of Wittgenstein's life.[12]

Wittgenstein refers in the two-page Preface to his *Philosophical Investigations* to criticism of his *Tractatus Logico-Philosophicus* that had impressed him: he mentions there Ramsey and Sraffa; his omission of Russell is tantamount to the statement that he found Russell's Introduction to it unimpressive. This made commentators on the matter go a long way in efforts to minimize Russell's tremendous influence on mature Wittgenstein, an influence that he acknowledged even while declaring that Frege had influenced him more (Wittgenstein et al. 1967, §712). The vainness of the comments by Pears on this were already discussed a few pages ago. Anscombe deserves mention too. Together the two should suffice as examples, considering their

[12]Hintikka 2004, 89, said, some idiosyncrasies of Wittgenstein did matter, such as his dyslexia. Now an author's dogmatism or anti-dogmatism may signify, and, as a character-trait, it appears in diverse strategies for responding to criticism. It is often idiosyncratic, especially in writers who serve as objects of cults (Copleston 2002, 7). Hintikka's extensive writing on Wittgenstein, include mention of many and detailed personal characterizations. He avoided referring to the uncontested report that Wittgenstein was regrettably "a thoroughly disagreeable character" (Vinten 2015, note 13; Pascal 1973, 23). The most that gentle Hintikka could bring himself to say is, Wittgenstein was impatient.

prominent positions in the Wittgenstein *entourage*. Anscombe claimed (1971, 91) that Russell's intention "was plainly contrary to the intentions" of Wittgenstein "as is very easily shewn". This is a slight exaggeration, and a slight on Russell, who described the cooperation between him and Wittgenstein in the warmest terms: they had aimed at the characterization of the ideal language (in the sense of the archetypical). Wittgenstein never returned to any critical discussion of formal matters.[13] It is very likely that it was Russell's criticism of Wittgenstein's search for the ideal language that has sent Russell to look for new directions. This is a serious contribution of Wittgenstein, unacknowledged out of narrow partisan considerations.

According to Wittgenstein, the role of the analysis of ordinary language is to dispel metaphysics in a new manner: it should unmask metaphysics as deceitful. Whereas young Wittgenstein described the rules of the ideal language and argued that they render metaphysical sentences impossible, mature Wittgenstein had no rules to fall back on when he spoke of rules; he appealed not to rules but to his reader's intuition.[14] There is no guarantee that any analysis is going to produce any results, much less results that Wittgenstein and his chief disciples would endorse.

To repeat, Russell's distaste for all the synthetic philosophies of his day is rooted in his distaste for some Hegelian philosophy coupled with his view of it as the logical culmination of modern western synthetic philosophy – to the extent that the epithet "logical" is at all applicable to such an anti-logical philosopher as Hegel, of course. (Russell accentuated this with his view of German romantic philosophy as the source of fascism and with his sensible proposal that the lack of interest in politics – such as the one that Wittgenstein displayed – admissible but far from commendable.) This leaves open the question, will analysis ever supplant worthwhile synthesis? Can the future bring about a better synthetic philosophy? Is philosophy doomed to remain analytic to the end of days? This is the prime question for all serious readers of texts of mature Wittgenstein. Most commentators avoid reference to his insistent, relentless anti-metaphysical bent. It is too easy to praise people's views after ignoring their errors. This prevents any reasonable effort to salvage from their errors whatever is worthwhile in them. Wittgenstein having spotted techniques that some sloppy metaphysicians often use did not help in this venture.[15] Our first task in reading Wittgenstein, then, is to decide to try to comprehend his views, warts and all.

---

[13] Although systems are ubiquitous in mathematics, in Wittgenstein's 1956, 1975 and 1978 writings on mathematics, systems enter only very superficially and only when unavoidable.

[14] The explicit appeal to readers' intuition is new with the mature Wittgenstein. Findlay 1984, 1 says, "a comprehensive critique of the thought of Wittgenstein from a standpoint which recognizes him to be, both in his earlier and his later thinking, a systematic philosophical thinker... is not so easy since the mature Wittgenstein professed to have no doctrines and theories, but rather only a method to lay to rest such doctrines and theories." Hintikka seems to have agreed, but he viewed Wittgenstein as a rationalist, and as a matter of course. See also Cavell 1976, Ch. 1 and Nieli 1987, 71, 105–6, 129, 176–9, 183.

[15] Stuart Hampshire undertook this venture: he found that Wittgenstein's condemnation of metaphysicians as slum landlords is unfair to Spinoza; he tried to salvage from this condemnation whatever he could. More liberal is Bar-Hillel, who says (1962), he will not argue with philosophes who do not wish to speak clearly.

Are there rules for philosophical analysis? The rules of synthesis are relatively unproblematic: they are the rules of deductive logic. What are the rules of analysis? This question appeared early in this study, and now it reappears. The "Vienna Circle", that claimed to have based itself on Wittgenstein's first book, took the tool of analysis to be logic and nothing more. Does this make the activity of analysis and of synthesis identical? Not quite: (as Aristotle has noted) it makes the formal tools identical; the use of analytic tools may be casual and unsystematic, whereas the use of synthesis should be systematic: analysis is exploratory, synthesis is (relatively) definitive: it begins with given axioms. It is just because within philosophy there is no justification for any assertion other than tautologies that the paradox of analysis holds that its results must be logical and so in a sense not new. So, we simply have to declare our aim and go freewheeling, try at times this approach, at times that, and see what gets us where – with no guarantee. The prime model for that is logicism, the analytic effort at a grand synthesis.

Some followers of Wittgenstein would strongly disagree with all this. Some of these have presented systematic tools to justify their position; they call these tools "informal logic". This is a folly (*PI*, §108).[16] The very distinction between formal and informal logic is due to ignorance. There are degrees of formality, of course.[17] We can call low degrees of formality informality, of course, and we may trace the rules as to the proper degree of formality. Thus, we do have formal and informal relations between people; they depend on the social situation within which they take place, and rules in that social situation decide the degree of formality that is required or permitted, such as the need to have an agreement to address each other by first names before doing so. Some rules, of course, determine the degree of strictness of the application of other rules. To come to our case, synthesis follows stricter rules than analysis. This includes even the question concerning the aim of an exercise and the power of the tools used in its execution. As analysis is so often groping, this raises the question, how long may one go on groping before showing some results or else drop the exercise as futile? This question holds for all freewheeling activities, of course. As Gellner has forcefully argued, the answer may be much too permissive, as it may be the recommendation to engage in analysis permanently and to no end.[18]

We may then have to trace some of Wittgenstein's moves in order to find his strategy. This is not easy: he is not exactly a reader-friendly writer. His *Tractatus*

[16]Wittgenstein denies there the existence of language in Plato's heavens but does not say that it is an institution: he speaks of life forms rather than of institutions. As to the advocates of informal logic, they treat texts in formal logic as if they were fully formal. Similarly, they often view Newtonian mechanics as formalized. (See Popper 1947a, b, c, 198.) What is a formal system and where it dwells they do not discuss.

[17]Lakatos divided mathematical research to pre-formal, formal, and post-formal; yet by "formal" he meant "axiomatized", in oversight of the difference between Euclid and Hilbert.

[18]Gellner [1959] 2005, 25, 85, 160, 177, 186, 190, 196, 200, 230, 253: the tasks of analytic philosophy, he said, are trite and endless.

Logico-Philosophicus has clear aim and structure, but many of its crucial sentences are enigmatic. His *Philosophical Investigations* is the opposite; its sentences are much less enigmatic, but not his having made them. Even its structure is enigmatic: how its paragraphs keep together is hard to say. In his Preface, Wittgenstein contrasts his early book and the one at hand, and he expresses his intention to publish the two together, so that the contrast between them will be obvious to readers. Obvious! He suggested that putting the two texts side-by-side suffices for the comparison/contrast between them! Heaps of comments about this matter are available, whose very existence easily refutes his suggestion that the result of the exercise is obvious.

Is it advisable to study the first book first in preparation for reading the second? Wittgenstein suggested that his first book was relevant enough to suggest study of it even if in order to be rejected – totally or in small part – at least as a stepping-stone, not to say a ladder. Was he right?

10.2 Young and Mature Wittgenstein

When speaking of Wittgenstein's philosophy, simple decency requires of commentators that they should repeat as often as necessary that he denied having any, as they overrule this denial. They usually violate this rule, thus expressing disrespect for him.[19] He said (*PI*, Preface), his first book is required as means for the understanding of his second. What aspect of his first book is required for that? Perhaps the portion that remains through the transition from his early to the later philosophy. If so, what portion of his early philosophy has survived this transition? We have clear knowledge of something central to the early philosophy that did not survive: he gave up logical atomism (Sluga 2011); as this is the received opinion on it, there is no need to discuss it. He gave up not only the doctrine that every statement is either atomic or compound; he gave up even the claim that some statements are atomic in one sense or another: he said in his later book, we do not know which statement is simple.[20] He retained the empiricist principle: we experience elementary sensations – of sound, of color, or of heat; they are elementary in the sense that they are not open to analysis. That our knowledge of our sensations is not open to analysis is, of course, trivially true in

[19]"... even the most loyal of his disciples (and he exacted very high standards of loyalty) treat his passionate revulsion from the idea of himself as a philosophical theorist as... aberration..." (Pitcher 1968, 10).

[20]*PI,* §§45–7: "We use the word 'composite' (and therefore the word 'simple') in an enormous number of different and differently related ways".

some sense and trivially false in another.[21] To take them as basic, however, is a different act altogether. It is smuggling back in the basic tenet of classical empiricism. It is doubtful that Wittgenstein endorsed the classical empiricist tenet. The only reason he had to say that sensations are not open to analysis is that when he gave up logical atomism what remained of it is that sensations are not open to analysis; that much he could rescue from the withdrawal of his earlier views; and this he did. All quotations from his texts that assert the obvious without explaining why he asserted them have this characteristic: they are results of rescue operations, and hence apologetic, and hence not to take too seriously. Those who take them very seriously in oversight of their apologetic character are doomed to consider deep some trite assertions, and more so some trite problems. In particular, Wittgenstein has posed – to the classical empiricist – the question, how do I know that you are in pain? He poses it in a verbal frame: how do we learn the use of the word "pain"? All serious language acquisition theory includes an answer to this question of Wittgenstein; they all deviate from his answer (Erneling 1993, 198, 208, 235).

What does mature Wittgenstein save from the text of young Wittgenstein after he repudiated it? What exactly did he repudiate?[22] He said, logicians (such as his young self) discussed only one kind of language-games, when in reality there are so many of them.[23] Admittedly, until Wittgenstein appeared on the scene, logicians had discussed only declarative assertions, whereas in reality language includes also other things, like modalities, questions, commands, requests, jokes, rhetoric, and more. Admittedly, Wittgenstein had predecessors. Already Charles Sanders Peirce spoke of the logic of questions and answers. He hardly developed the field, as some logicians do these days. We should grant Wittgenstein credit for his attention to these

[21]Empiricist philosophers still talk of colored patches, in ignorance of the ubiquitous spread effect. The classical empiricist effort to base science on sensations imposes the view of them as unanalyzable. Analyzing them is part of the study of their place in psychology and in physiology and in the study of their place in the arts and much more. This Wittgenstein noted in the only paper that he published (Wittgenstein 1929) and that he denounced with no explanation as soon as it appeared. Perhaps this rejection is due to Wittgenstein's realization that this paper transcends the anti-metaphysical stance of his first book. The parts of human physiology (our sensory system) that serve as background to his system and that he had declared not given to articulation is thus articulated elsewhere. If so, then his refusal to explain his withdrawal of his paper is a suppression of this obvious truth. Incidentally, Pap 1950 presents this obvious truth as synthetic *a priori* knowledge.

[22]*PI*, §23: "It is interesting to compare the multiplicity of the tools in language and of the ways they are used, the multiplicity of kinds of word and sentence, with what logicians have said about the structure of language. (Including the author of the *Tractatus Logico-Philosophicus.*)" This is a repudiation only of the book's narrowness. Commentators understand this as his withdrawal of his picture theory and of his logical atomism!

[23]Mature Wittgenstein did not specify the mistakes in his early work, except perhaps to say that he admitted the criticism in Ramsey 1923 of the picture theory of meaning and the theory of atomic propositions that he (Ramsey) viewed as one. Ramsey also asked, what Wittgenstein's view of negative atomic propositions was. Anscombe's book on this has little success (Bar-Am and Agassi 2014).

matters. Yet this we should notice: the reason for the study of the logic of declarative statements rested on some distinct significant problems, especially within mathematics. The problems that led to the first serious study of modal logic turned out to be solvable with no reference to modalities.[24] Some logicians (Quine in particular) felt that there is no need for the study of modality as a part of logic. Any study of any logic other than the logic of declarative statements requires problems to make it worthwhile.[25] The text of Wittgenstein's *Philosophical Investigations* raises no problems and dismisses all philosophical ones.

10.3 Signposts in *Philosophical Investigations*

The book does not center on any topic or on any problem, and if it has a thesis, it is that meaning is use and that there are no proper metaphysical assertions. Commentators zero in on its terms that represent theories. The following are the most conspicuous: it uses a famous metaphor, showing the fly the way out of the fly-bottle,[26] it discusses at some length family resemblance, forms of life, language games, private language and following rules. Commentators often insist on treating these items as if they were technical terms, although he used them in their normal everyday sense. Much of what they say of Wittgenstein's terminology is either superfluous or contrary to his intention. Thus, discussing his concept of family resemblance some try to present theories of this resemblance, in contrast with what Wittgenstein says: it may be hard to describe resemblance or even to know what exactly it is, suggesting that this does not matter.[27] Indeed, it does not. To prove this Wittgenstein offers many examples of verbal ambiguities in everyday parlance. When it matters, this requires asking for more specifications. Somehow, Wittgenstein's use of the duck-rabbit ambiguity caught the imagination of commentators, although he uses many examples.[28] Why it concerned him he did not say,

---

[24]Lewis and Langford 1932 studied modality to clarify the difference between a $\vdash$ b and a/b that is now a part of logic proper as the properly worded the deduction theorem. In their presentation, "$\vdash$a" is "a is necessary". Their study of modality involves more. For example, it raises the question, is "a is necessarily necessary" the same as "a is necessary"? Why this question signifies is still not clear.

[25]Heijenoort 1967; Agassi 1978.

[26]*PI*, §309. See also *PI*, §123. "A philosophical problem has the form: 'I do not know my way about'". This is another example of a guru asserting an obvious falsehood for followers to interpret.

[27]*PI*, §66: "complicated network of similarities overlapping and criss-crossing: sometimes overall similarities, sometimes similarities of detail." §67: "I can think of no better expression to characterize these similarities than 'family resemblances'".

[28]In TLP, §5.5423 the Necker cube serves to illustrate ambiguity, explained as different facts perceived as one. In PI, §74 the same item serves to illustrate the influence of gestalts – stereotypes – on vision. So now we view (not different atomic facts but) different stereotypes in action. This is a shift away from empiricism (Gattei, 2008, 18); most commentators ignore it. Leinfellner-Rupertsberger, 1990, 856 is an exception: she asks, "How deep is this rift between Wittgenstein's early and late philosophy?"

though he devoted pages (*PI*, II, ix) to it. Ambiguities abound and most people scarcely notice them. Lawyers, mathematicians, and other experts notice the ones that are significant in their normal activities. Frege, Russell and Thoralf Skolem were terrific at spotting ambiguities that matter in mathematics. They have changed the scene so much that consequently it is difficult to find older texts whose logic is satisfactory. Still, Wittgenstein is right in considering normal ambiguities insignificant and using them to erect metaphysical systems reprehensible.

The terms, "forms of life" and "language games" are conspicuous in the book. A large literature is devoted to the question, what exactly do they mean? As Wittgenstein compares his early and late philosophies early in the book (*PI*, §§7–24), he says, there are "*countless* kinds" of language games; he lists a few: describing, giving orders, making jokes and more. He adds,

> It is interesting to compare the multiplicity of the tools in language and the way they are used, the multiplicity of kinds of words and sentences, with what logicians have said about the structure of language.

This reference to "the multiplicity of the tools in language" is odd: do jokes, for example, use any specific verbal tool? Do they display a specific verbal form? Hardly. Wittgenstein allowed for a sufficient flexibility in the use of his term "language games" to take it as unproblematic to begin with, especially since aims abound and games with different aims differ. The question here is, what are these to the grammar of a language? The answer is that Wittgenstein was concerned (*PI*, §75) not with different language games but with the question, "What does it mean to know what a game is?" This is a bit misleading, since his question was more general: What does it mean to know anything? More generally, what is meaning? He seems to have answered it (*PI*, §§43, 138, and 561): meaning is use. More specifically, he said, the meaning of words is manifest in their use: meaning reveals itself in action. Thus, Wittgenstein answered his question, "What does it mean to know what a game is?" by observing that answering this question we compare this game with other games; generally ("and so on", he says), falling back on our knowledge of other games when describing "this game". This of course means that we use meanings to explain [not to define] meanings, and the explanation has a vague aspect, since we fall back on games in general when explaining "this game", yet the concept of a game is vague, as what games share is a mere (family) resemblance.

At least this is Wittgenstein as all and sundry understood him soon after he died. Yet, to repeat, as Linsky has noted, Wittgenstein's analytic technique changed from the study of the structure of sentences to the study of their proper uses. Oddly, this change came into view unannounced. Similarly, Wittgenstein moved from the assumption that meaning is use to the assumption that meaning reveals itself in its use. Perhaps the final assumption was more remote than the one it came to replace: the assumption may be that the function of a sentence determines its very structure (Chomsky-style). It is not clear what assumption Wittgenstein made, and whether he illustrated it. As Russell has noted (1953), even what sentence is ordinary is unclear.

It turns out that all this is not very important: Wittgenstein argued not so much that meaning is use than that we learn the meaning of an expression by watching its operation in its use. This is a theory of the acquisition of meaning, not a theory of meaning, and whatever his study of language acquisition was, newer views on it have since rendered it obsolete.

The term "forms of life" may be no less problematic, seeing that Wittgenstein spoke of diverse sorts of forms, including forms of expression, of grammar, of language, and even of three-dimensional vision (*PI*, §74).[29] Indeed, the term enters here in its most non-technical manner: "to imagine a language means to imagine a form of life" (*PI*, §19). This insight is not different from the insight of foreign-language teachers who convey to their students information about the foreign country. Other appearances of this term are of the same character. Nevertheless, commentators view it as problematic, since the view now dominant is that Wittgenstein's reference is to the multiplicity of forms of life as the multiplicity of cultures, and as an expression of relativism. This is clearly not the case, since we have here a contrast between shared opinions and shared forms of life (*PI*, §241):

> 'So you are saying that human agreement decides what is true and what is false?' – It is what human beings say that is true and false; and they agree in the language they use. That is not agreement in opinions but in form of life.

This way we come to Wittgenstein's solution to the problem of induction. It is only here that things get murky. He asks the question,

> And how are these [calculations] tested in their turn?

And he answers it (*PI*, p. 226):

> What has to be accepted, the given, is – so one could *say* – *forms of life.*

The solution is, people do justify their views, and to have meaning the problem of induction should refer to the ways people do that.[30]

The dominant term in the book, to the extent that there is one, is that of following a rule: follow the rules of language and do not deviate from the straight and narrow; this is the necessary and sufficient rule for achieving immunity to the bewitchment of metaphysics, which was always Wittgenstein's major concern:

> Philosophy is a battle against the bewitchment of our intelligence by means of language.[31]

---

[29]Wittgenstein ignores the language game of courtship. The nearest to it here is in *PI*, §40–1; at the time it won a parody with innuendos in an oxford student publication.

[30]*PI*, §479. "The question: 'On what grounds do you believe this?' might mean: 'From what you are now deducing it (have you just deduced it)?' But it might also mean: 'What grounds can you produce for this assumption on thinking it over?'"

[31]*PI*, §109: "Philosophy is a battle against the bewitchment of our intelligence by means of language" Wittgenstein never explained why he recommended the avoidance of bewitchment. Elsewhere he said, "One cannot guess how a word functions. One has to *look at* its use and learn from that. The difficulty is to remove the prejudice that stands in the way of doing this. It is not a *stupid* prejudice" (*PI*, §340). Here, it seems, we see Wittgenstein's analysis sliding from Wittgenstein-style to Russell-style. Russell 1956b, 77 said that the claim for total freedom from all prejudice is humbug. A leading text on the need to avoid prejudice is Wheeler 200, 238, which is disappointingly slight.

This raises many questions. Wittgenstein makes his reader ask (*PI*, §217), "How am I able to obey a rule?" Wittgenstein admits at once: he does not know (*PI*, §219): "I obey the rule *blindly.*" What rule is under discussion? Wittgenstein showed no interest in standard grammar – traditional or modern. His interest was in the exorcism of metaphysics. This, he declared repeatedly, one achieves by finding the proper rule for the terms that metaphysicians use; this, he further declared, one achieves by going to the root of their applications. This is the innovation of his later years: What we do is to bring words back from their metaphysical to their everyday use (*PI*, §116). Why? He should have left it to its usual language game, since what "the understanding has got by running its head up against the limits of language" is "plain nonsense" (*PI*, §119). The arguments that Wittgenstein offers for this assertion invite scrutiny: it is not to take on faith.

Before arguing with Wittgenstein, we need to comprehend his text. What may help comprehend the text of *Philosophical Investigations,* Wittgenstein suggested in its preface, is its departure from Wittgenstein's early views. He said there, his early views were much too narrow. This is no departure from its contents. The major departure of mature Wittgenstein from early Wittgenstein concerns logical atomism. It is a theory of the essence of language.[32] It leads with necessity though with no intention to one sort of a meta-language or another ("a super-order between … super-concepts"[33]). He found this too pretentious. Ordinary language analysis appealed to him more than his earlier search for the essence of language: it is more "humble" (*PI,* §97). The expression here is rather disturbing: the word "language" is in use in the generic mood, with no qualifier. The word may be qualified – by "a", "some", "few", "many", "any" and "all", or "this", "my", etc. The discussion would go differently for the same case but with different qualifiers[34]; the difference is especially great between the use of the definite and of the indefinite article here. This is the most common confusion that different philosophical literatures suffer, including that of some of the latest Wittgenstein devotees.

---

[32]On Russell, see Kripke 2005. For Wittgenstein see PI, §97:

> Thought is surrounded by a halo. – Its essence, logic, presents an order, in fact the a priori order of the world: that is, the order of possibilities, which must be common to both world and thought.

This is the problem of scope presented not as a problem but as a plain fact. "Asking whether and how a proposition can be verified is only a particular way of asking 'How do you mean?' The answer is a contribution to the grammar of the proposition" (*PI*, §353). This is his refusal to withdraw his verification principle that depends on the admission of the ideal language – the essence of language: it hovers over all of the output of mature Wittgenstein, except that (unlike the scientism of his young phase) it is verification in its ordinary sense. It is thus fallible, unlike Moore's, yet Wittgenstein fell back repeatedly on the finality that is a claim for perfection. This is the last sentence he wrote on this earth (end of *On Certainty*): "To be sure there is justification; but justification comes to an end."

[33]See *PI,* §§38, 68, 89, 92, 94, esp. §92.

[34]This is not to suggest the hypothesis that the same qualifier is tacit in Wittgenstein's texts in every case; he was unsystematic about this, and unclear.

An intriguing allusion in the late book to the early one merits mention. Its context is the claim that utter clarity is as impossible just as the absence of all friction is impossible.[35] The view of the uniqueness of language [language as such? any language? my language?[36]], Wittgenstein continues, is "a superstition (not a mistake) proven itself through grammatical illusion". This is an allusion to Russell's Introduction to Wittgenstein's *Tractatus*. The illusion he alluded to is one due to reading depth into confusion.[37] Such a reading Wittgenstein deemed over-ambitious, the sin that leads to ruination. In his view the confusion that he studied, "problems that arise through misinterpretation of forms of language, have the character of depth". The depth, he said, is manifest in the claim present in his early book. It is the claim that "the general form of a proposition is: The case is so and so" And, he added, this leads to an infinite regress: "x" becomes "the case is x", and then "the case is that the case is x", and so on.

Strange. Wittgenstein's view was and remained to the last the idea that statements are [generally] self-introducing. Why in his later years did he pick on his early wording of this idea? He wished to stress the obvious: it is impossible to escape language when studying language. This sounds trivially true, yet it is either confusing or trivially false. While talking – on anything at all – it is obviously impossible to avoid language as such. Yet it is easy to escape any specific language while discussing it by using another language within which to discuss it. This Wittgenstein refused to recognize. (This sounds odd, but is not impossible: it is permissible to consider the option that the meta-language is a part of the language under study.) He had no argument to support his refusal except from humility, which is a moot point. Indeed, Feyerabend ends his 1955 review of the *Philosophical Investigations* (echoing Russell's Introduction) with the claim that all of the criticism voiced there may be easy to evade by transferring metaphysics to the meta-language, with no possible objection but from artificiality. Yet the notion of the meta-language is not wholly artificial. Why then did Wittgenstein inhibit all traces of this in his analyses (even in mathematical contexts)?

To whom did Wittgenstein address his demand not to use the meta-language? Not those who do so with no concern for philosophy. Among these are students and

[35]This is not to deny that mature Wittgenstein demanded utter clarity. Indeed, he said, (PI, §133): "For the clarity that we are aiming at is indeed *complete* clarity. But this simply means that the philosophical problems should *completely* disappear." As is well known, he always sought perfection (Pascal 1973, 32). Here, however, we may read him to say, the perfection I seek is limited to the perfect exorcism of meatphysics.

[36]Wittgenstein 1961 Notebooks, entry for 29.5.15: "But is language the *only* language?"

[37]*PI*, §110; see Anscombe 1971, §19. Cavell 1962, 86 says, "Wittgenstein speaks of the illusions produced by our employing words in the absence of the language game which provides their comprehensible employment". Hence, meaninglessness is capable of remedy by completion. Wittgenstein declared this none of his concerns (*PI*, §§91–2). His concern, he said repeatedly, was exposing the meaningless talk that comes out of efforts to cross the limits of language, and that is therefore irreparable. It is not that Cavell was ignorant; it is that paper is patient.

teachers of foreign languages; logicians and mathematicians[38]; natural scientists, (cognitive) psychologists and novelists. Even the resolution of paradoxes may be a matter for computer programmers. Wittgenstein did not forbid the use of a meta-language; he merely declared that there is no such thing, meaning, whatever you do right is not the use of the meta-language; at times he possibly explained his own refusal to use it in certain contexts, or else why he viewed his own seemingly meta-linguistic expressions as not meta-linguistic. Admittedly, there is a problem here: the split of the language into the meta-language and the object language comes to resolve the paradoxes, especially the liar: Wittgenstein (rightly) spurned the received way to resolve paradoxes.[39] His rejection of Gödel Hintikka found regrettable.[40]

Wittgenstein said, paradoxes are no cause for worry. To this Ayer responded (1985, 65–6) blithely: "If this is Wittgenstein's standpoint, I am in sympathy with it. But then I am not a mathematician." This is the right response: paradoxes are harmless in normal discourse[41]; they render formal systems useless; these ceased to interest Wittgenstein. This, however, renders it improper to offer his works as guidance to the interested in formal systems. Still, formal systems are formalized living systems, and they can help study natural languages. This wants explanation, though – particularly since there is a middle ground where parts of natural language are partly formalized; the use of computers to solve scientific problems testifies to that (since most scientific theories are not formalized, as Popper 1947a, b, c, 198 noted).

This rests on the later admission of the impossibility of absolute clarity. Much depends on the limit of desired and/or possible clarity. It is, he said, the limit of philosophy: clarity is sufficient when the task of ousting philosophy is completed, he insisted:

> philosophical problems arise when language idles (*PI,* §38d);
>
> philosophical problems . . . are solved, not by giving new information, but by arranging what we have always known (*PI,* §109);

---

[38]Mathematics employs equivalence meta-theorems. The most famous is the meta-theorem in projective geometry: when the words "line" and "point" replace each other in a given theorem, the result is another theorem. Some equivalences are possible to read as within systems and as about them – Carnap's demand to be clear about such choices notwithstandaing.

[39]*PI,* §182. TLP, §3.333 offers a poor resolution of Russell's paradox (Popper 1954, 1963, Ch. 14).

[40]See Russell [1959a, b] 1995, 88. Floyd 1995 and Rodych 1999 against Hintikka, 1996, 101: "In general, he never managed to relate his problems and ideas with actual work in systematic logical theory. As a result, most of his speculations about logic and mathematics remained unfulfilled and frequently superficial." Wittgenstein considered himself primarily a philosopher of mathematics and in this he impressed Georg Kreisel, Alan Turing and Hau Wang. Turing parted company with him. Kreisel expressed great disappointment in him in 1958. He softened this in his 1978, where he compared the views of Wittgenstein on the tradition of the foundations of mathematics favorably with the Bourbaki manifesto, no less.

[41]Bar-Hillel 1966. Mach still played down paradoxes; Russell did the opposite. This should serve as a warning: presentations of paradoxes are usually mere sketches: it is in fully formal systems that they signify.

> the clarity that we seek is . . . complete clarity. But this means only that the philosophical problems should completely disappear. (*PI,* §133.)[42]

Full clarity, he later declared, is not attainable, even not desirable (*PI,* §108): it is as useful as the friction required for walking (*PI,* §107). The degree of clarity sufficient to dispel philosophy, however, is still required. Wittgenstein repeatedly assured his readers that this is possible. His anti-philosophical prejudice that he allegedly proved when he introduced it in his *Tractatus Logico-Philosophicus,* he repeated also in his *Philosophical Investigations,* this time with no shade of proof.

Of course, not all confused assertions are philosophical but all philosophical assertions are confused; thus spoke Wittgenstein. Only philosophical confusions interested him, he said, and these are present even within mathematics and within psychology. Even then, they are philosophical, he explained, since when a philosophical confusion is resolvable, the result is not a new piece of knowledge. He found valuable seeking these clarifications nonetheless, since its aim is to remove some restraints. Now this may be true of all confusion, yet Wittgenstein was concerned with verbal ones; moreover, there are diverse verbal confusions, yet Wittgenstein was concerned with philosophical verbal confusions, those that result from efforts to transcend the limits of language and thus of reason. Query: if ousting confusion helps free mathematics and psychology, can it not also free philosophy? Why can philosophy not have the same privilege as the other fields of inquiry?

The answer of young Wittgenstein to this query is clear: his theory of meaning left no room for philosophy. The theory of meaning of mature Wittgenstein is different. The answer of mature Wittgenstein to the question, why not some meaningful philosophy, is the same as before: philosophy, that is to say metaphysics, is the efforts to transcend the limits of language, and thus it is hopeless. This won elaborations, such as the one that Bernard Williams has offered, we remember: were metaphysical knowledge possible, it would be neither analytic, else it would be logic and mathematics, nor *a posteriori*, else it would be empirical science; hence, it must be synthetic *a priori* – which is impossible. QED.[43] Although this proof is absent from Wittgenstein's writings, it is generally attributed to him. Why then need we cite Williams and his likes?

This alleged proof is of the impossibility of metaphysical statements. It tacitly turns into the proof of the impossibility of metaphysical knowledge. In metaphysics, bridging between statements and knowledge is the exclusion hypotheses from

[42]The completeness of the clarification, Wittgenstein explains, is its having reached its goal, not in the sense that it has reached its limit. Thus, clarity serves diverse purposes, and the one that Wittgenstein cared about is the exclusion of metaphysics. Hence, if clarification has not reached its goal, it should continue with more vigor. How long?

[43]Williams 1968. Rorty 1971 refuted the argument that Williams had used, mentioning only earlier writers who had used it. Rorty also cites Strawson, who had presented skepticism as involving hostility to hypotheses. This holds for Pyrrhonism, not for skeptics like Maimon, Einstein, Russell and Popper.

language. This is the verification principle. Kant declared [philosophical?] hypotheses illegitimate.[44] Wittgenstein and the "Vienna Circle" did that in three steps: postulating rules, endorsing the traditional view of metaphysics as speculative, and declaring it thereby meaningless (Schlick 1936, 341; Waismann 1968, 39–40). Williams did all this tacitly.[45] His celebrated paper is thus but a conjurer's trick.

Wittgenstein stressed: metaphysics does not exist only as well-formed propositions; otherwise it exists and is very important; this is why allowing any ellipsis, any non-standard expression that may be ellipsis, and any religious assertion taken literally rather than as a metaphor (and even "bless you!"), suffices to create counter-examples to Wittgenstein's assertion that ordinary language is metaphysics-free. He knew this well, of course.[46] He catered for it. He said, one should look at an expression in its language game and seek its origin. As an activity, it does exist; and he recommended it as corrective. This rests on the assumption that philosophical questions do not exist. Now, for any question to make sense, he added, there must be an answer to it. This is a wild supposition.[47] Why not have some unanswered questions, even some unanswerable ones? Possibly Wittgenstein might say that such questions are idle, that there is little to do with unanswerable questions are only troublesome. This is false. Moreover, if what characterizes philosophy is the questions that it comes to answer, and if the questions are answerable, then the answers it provides may be untestable or testable. If they are untestable, then they may be debatable in some other way (remember Plato's early dialogues that discuss ethics not empirically). If the answers may be open to empirical studies, then perhaps we should undertake such studies. Why can philosophy not become an empirical

[44]Kant took it for granted that Newton's theory is knowledge, not hypothesis; Maimon disagreed, much to Kant's annoyance. Worse, Maimon said the same of Kant's system contrary to Kant's disclaimer (Preface to first edition of his *Critique of Pure Reason*): "in this sphere of thought, opinion is perfectly inadmissible ... everything which bears the least semblance of a hypothesis must be excluded".

[45]Russell's plan to reduce all philosophy to logic did not include the exorcism of metaphysics. He stated that (with Spinoza as a possible exception) all theories of substance are false. Hence, there is a need to restate Descartes' mind-body problem with no reference to substance. This move replaces one problem with another; it is thus not anti-metaphysics.

[46]Wittgenstein 1938. Malcolm 2001, 59: "Wittgenstein once suggested that a way in which the notion of immortality can acquire a meaning is through one's feeling that one has duties from which one cannot be released, even by death." I do not know what this means.

[47]*PI*, pp. 214–15: "When I read a poem or narrative with feelings ... the sentences have a different *ring*. ... But the figurative employment of the word cannot conflict with the original". Wittgenstein has here a valid argument against Descartes, who in the mood of universal doubt had said that he might be dreaming; not so: dream-life is not as comprehensive as waking-life. This is a strong argument, but it is answerable. Some oriental philosophers said, life itself is but a dream. Gibson and Huemer 2004, 9 rightly argue that Wittgenstein considered art informative but not explicitly so, and this includes the verbal arts. (P. 187 there is an impressive discussion of a Wittgenstein text, refuted by Hintikka 1958, 88 as he has shown mundane "the only language I understand" to have become through sheer mistranslation the intriguing "the language that I alone understand").

science proper?[48] Why do people like Williams take it so much for granted that philosophy cannot be empirical? It is traditional that philosophy should perform a certain task, offer a worldview, offer a sense of the meaning of life, and answer certain central questions. Philosophy seeks the meaning of life. Does Wittgenstein offer a view on meaning to life? Theology does; Wittgenstein said it only seems to do, as it offers only pseudo-assertions, since all assertions are logical, scientific, or commonsense.[49] This his followers do not endorse any more. They may, then, want to demarcate philosophy anew – perhaps in a manner that will not render theology a branch of philosophy. Why not say, then, philosophy is a reasonable kind of religion? Wittgenstein was ready to take philosophy not as propositions but as attitudes. However, he insisted that philosophical questions, like philosophical propositions, are sick,[50] meaningless. Why? Because, he insisted, meaningful questions must have informative answers. He assumed that we know when a question has an answer and when not. This was Popper's strongest objection: in Wittgenstein's system, one has to understand an item in order to decide whether it has meaning. Hence, in his system meaning differs from comprehensibility. This is somehow tolerable for young Wittgenstein; not for mature Wittgenstein, let me add, as its aim was to return thinking to the everyday. Why is the sky blue? The everday has no way to find out whether this question is answerable or not except when it is answered; and we find this by consulting a scientific text. Even computers can tell us that a question is answerable only when they find the right answer to it (by concluding the task of seeking an answer to it). Prospecting is no different: we know that it can succeed only if and when it succeeds.

There is a possible escape from all this: relativism. This is an easy way out. It always works. It is unnecessary, though, since for those who are happy with an easy way out, the easiest way out is to ignore the difficulty that relativism solves. Wittgenstein would not admit relativism, however, and certainly not a relativist theory of language. This should follow from his assertions about following a rule and about private language. This expression, "private language", is the last in the list of terms commentators misread as technical terms.[51] It appears first in Russell's

---

[48]The philosophy of Thales – all is water – was initially untestable, yet the decomposition of water has refuted it. Likewise, quite possibly the Shoah refuted the excuses for evil of Leibniz' *Theodicy*. See next note.

[49]The standard claim is that the physico-theological proof for the existence of God, the argument from design, is commonsense. The same holds for the counter-argument from the Shoah.

[50]Let me repeat: *PI,* §255: "The philosopher treats a question; like an illness." This is a pun on "treat".

[51]Commentators have observed repeatedly that the greatest and most conspicuous change in Wittgenstein's output is in style. Yet they seldom note in this context the revulsion he had against technical terminology. The reason is that they erroneously considered technical his specific terms, private language, family resemblance and language games. The revulsion is due to the philosophy that, as Russell was the first to note, Wittgenstein shared with Pascal and Tolstoy, whose philosophy expresses the wish to return to pre-critical innocence. Its inaccessibility does not prevent yearning for it. Wittgenstein had an advantage over them all (with the possible exception of Kafka), as he admitted that we have lost access to it*: PI,* §426: "We see the straight highway before us, but of

critique of the idea of the ideal language in his early *The Philosophy of Logical Atomism*. Wittgenstein used it his way: the very meaning of "language" means a tool for communication and so it cannot be private. Moreover, to speak a language one has to follow the rules of its grammar. This is impressive, but inadequate. It is an example of a paradigm case argument that Watkins (1957b) and Gellner ([1959] 2005, 63) drove out of fashion. Now, communication is a matter of degree and it is limited to a speaking community. Surely, when one invents a new idea one feels as lonely as one locked in a private language, more so when one invents a new language, whether for normal use or for computation. Moreover, learning a language is the acquisition of all sorts of aspects of it that are yet barely studied. An example is what Black called (1962, 41) an association network that operates on metaphorical discourse.

10.4 The Aftermath

The origin of Wittgenstein's anti-metaphysics was not so much his plea for clarity as his plea for intellectual honesty that he expressed very forcefully to Russell early in the day. (The two overlap: the intellectually dishonest love to muddy the waters.) Young Wittgenstein excluded metaphysics form his early system since his (Russellian) project was to put modern logic in the service of neutral monism so as to permit for verification of scientific theories. The success of this project had to turn much discourse meaningless. That discourse was not necessarily condemned: it could remain as a part of the colloquial language (*unsere Umgangssprache*) that Wittgenstein said we know too little of (*TLP*, § 4.002) and that he therefore rightly intended to leave as it is (*TLP*, §5.5563; opening), as presenting abstract problems that he left for others to study (*TLP*, §5.5563; conclusion).

The question before us now concerns the view that mature Wittgenstein inherited from young Wittgenstein. Clearly, without adjudicating on most of is early ideas, he rejected as little of them as he could. (Even then, Hintikka 1996, 47, 52 observed, he tried not to reject his early texts that expressed them but to reinterpret them. If this is true, we should ignore it as unbecoming.) Whatever he rescued of his early work had to apply to the colloquial language, since he avoided discussion of his ideal language. He therefore had to conclude that ordinary metaphysical assertions are nothing but a confusion of a certain kind: they are inherently ungrammatical. The paradigm of metaphysics surely is theology. Here is a glaring grammatical failure: "whenever someone else wanted to say something metaphysical, to demonstrate to

course we cannot use it because it is permanently closed." We can take comfort, however, in our ability approach it in small details, in life's byways, he implied, by emulating him. This explains his view of himself as the Galileo of philosophy. Kafka, incidentally, had a similar view but without self-aggrandizement.

him that he had failed to give a meaning to certain signs in his propositions" (*TLP*, §6.53). This is a twisted way to express atheism.

This looked very impressive to Wittgenstein's followers in the "Vienna Circle": it looks as if now all those who value reason should bow to the latest discovery in logic and give up theology. Wittgenstein did not like this, in deference to people whom Tolstoy took to be paragons of the truly religious, people who are sincerely religious and ignorant (of logic). How Wittgenstein reconciled atheism that rests on his strange idea of grammar with his Tolstoyan respect for the sincerely religious ignorant is perhaps the most basic problem for interpreters of Wittgenstein. No less important is the need to reconcile his atheism with the religion of some of his leading disciples. Let me skip this question, and assume that one can do this successfully: Engelmann's *Letters from Ludwig Wittgenstein* indicate fairly clearly that he solved to his own satisfaction: he suggested that sincerity overcomes all obstacles.

Now, assuming it possible to rescue religion from Wittgenstein's scalpel raises an obvious question. Is it possible to rescue other metaphysical ideas the same way? Suppose the way is to rescue religion by abandoning theology. This renders the question of God's existence irrelevant to religion. Let us suppress the question, is this achievable, namely, is the practice of religion viable that goes with indifference to theology.

We can put the question more technically. Wittgenstein applied his scalpel to mathematics and to psychology. Suppose he did so successfully as he had intended and as he claimed to have managed. Obviously, they survived the surgery and remained sufficiently free of confusion and intact. Why not do the same for metaphysics? Young Wittgenstein had a possibly satisfactory answer: metaphysics is in principle speculative and properly worded sentences are in principle verifiable. Mature Wittgenstein gave up verification in favor of existing language games.[52] Indeed, a number of religious philosophers undertook to show that Wittgenstein's theory of language games permits religious discourses as specific language games.[53] Similarly, the new vogue among philosophers of science and among philosophers of the specific sciences, is to discuss the intellectual frameworks of science and of the specific sciences, not to mention the scientific worldview; these are the traditional concerns of metaphysics, and Wittgenstein had nothing to contribute to them.[54] Does it matter whether we call

---

[52]Sextus mentioned the argument from empirical facts repeatedly as an invalid solution to the problem of induction; Hume too argued against it. To his last days Wittgenstein was painfully aware of all this.

[53]Wittgenstein's suggestion is that all language games are/should be functional. Then, Émile Durkheim's (anti-religious) functionalist reading of religion as symbolic is right too – as the religious language game.

[54]To repeat, Putnam presents Wittgenstein as a contributor to the philosophy of life. Ray Monk says Wittgenstein was a poet and a guru. The problem they were solving is, how can one praise Wittgenstein as an intellectual without ascribing to him any new idea? Russell did better: he praised Wittgenstein as an original thinker and as initially very serious.

them intellectual frameworks or metaphysical systems? Will Wittgenstein's disciples be content with the claim that he changed the name of this activity? Will they condemn students of intellectual frameworks as slum landlords or will they praise them as those who put the house in order?

Wittgenstein spoke of philosophy as the activity of clarification, not as the activity of challenging thinkers to fill gaps in given systems. He viewed all metaphysics as efforts to go beyond the limits of language. How does atomism fit this characterization, I have no idea. It is possible to declare atomism scientific rather than metaphysical. This is quite contrary to tradition. Now tradition need not bind, but when usage deviates from it, some explanation is imperative or else there will be a guarantee for failure to comprehend. Popper considered metaphysics too uninformative, and then metaphysical character is assuredly amenable to correction: add to the content of the metaphysical their and it may become scientific. This is indeed what happened around the year 1800. Still, if one insists, one can viewed the enrichment of the contents of atomism a verbal clarification of the word "atom". And then Schrödinger's equation will be one clarification of the word "electron" and Dirac's equation will be another: this will make the term "electron" mean different things for different thinkers, and the diversity of opinions on the electrons will turn a diversity of usages of a word. Except that in science crucial experiments may help render one theory more valuable than another. This is no catastrophe: we can view as competing theories the assertion that the electron is a Schrödinger electron as researcher x understand the word, or that it is as researcher y does, etc. This is not objectionable; it is, however, a farce.

As to verbal clarifications, the assumption that some of them did clear some philosophical problems away, may perhaps permit the view of Wittgenstein's program as a mere hope (Agassi and Wettersten 1980, 93). What hope? What problems was the hope that analysis would remove and how? There is a standard answer to this: the no-private-language argument removes the traditional problem of other minds. What this argument is nobody knows. Nobody. It occurs in *Philosophical Investigations* only briefly and incidentally. Commentators wrote books to say what it is. The more they discuss it the more obvious it is that nobody knows, that the proposals as to what it is are hypotheses. There is no need to go into it all, since we can follow the advice of Wittgenstein and go to the origin of that word to see how it is used. The origin is Russell's discussion against the ideal language: had it existed it would be private and so of no use. A simpler idea is Buber's response to Descartes' monumental I think, therefore I am: I exist because you exist, he said (Agassi 2010).

The distinction between the inner world and the outer world is due to Descartes and it serves psychologistic epistemology. It began with the exercise that Descartes undertook when he entered the state of utter doubt. He hoped to overcome doubt by establishing with certitude the existence of the world whose existence we normally take for granted. He failed to realize this hope to our satisfaction. This has led to the suggestion that he had better not enter the state of utter doubt, that he could not gain anything from it and that at most he could and would return to the present state with no profit. This suggestion is an error: Descartes did free us of many things that populated the world beforehand, such as goblins and angels and demons. He came up with a clear metaphysics that describes matter as extended and mind as able to think. He thus opened the road to modern science and greatly influenced its growth for centuries. Yet the experiment was impossible to begin with, as many a nineteenth century advocate of irrationalism has (rightly) argued: Descartes was in error when he allowed himself to forget all his knowledge, even his knowledge of his own body-parts, yet without forgetting his mother tongue. Hence, his experiment is a failure – yet far from fruitless all the same: he "invented a method which may still be used with profit – the method of systematic doubt", Russell said.[55]

Wittgenstein's analytic stance enabled him to fight philosophical questions, when a question, as we say today in the light of the theory of questions and answers, is a set of possible alternatives. What question did Descartes ask? How does Wittgenstein's analysis undo the question rather than his choice of an answer to it? If Descartes did ask any question, it is, how is certain knowledge possible? He answered it by saying, certain knowledge is possible if and only if one begins by doubting every idea one has and resolve to endorse all and only ideas that have passed the severest test and proven certain.[56] Wittgenstein rightly rejected this idea. He seems to have offered (1975) a different way to certainty. Whatever it is, it is an alternative to that of Descartes. Here, indeed, is the greatest difference between the early and later philosophies of Wittgenstein: he spoke in his early phase of knowledge as the choice between alternatives – between an atomic proposition and its negation – choices to be determined empirically alone (or else logically). This is logical atomism. Later on Wittgenstein repudiated it. The criticism of Descartes' method, and the declaration that other people do exist, are no refusal to entertain questions. It is the rejection of some false answers to it in favor of their true alternatives, and they do run contrary to Descartes in some sense, but not in the analytic sense, decidedly not in Wittgenstein's sense of not offering any alternative to a theory unmasked as confused.

Ayer insisted on this to the last (Hahn 1992, 644–5): Moore and Wittgenstein had different philosophies: one advocated commonsense metaphysics; the other denied the very possibility of a properly articulated metaphysics. Warnock went further than Ayer: analytic philosophy follows Moore, he adjudicated, not Wittgenstein. He invited his readers (not to ignore Wittgenstein but) to alter the ritual ancestor worship, by replacing

[55]Russell 1912, Chapter 2. Relevant here is Russell 1918, 520, as well as his 1956a, 198 on the language of the *Principia Mathematica.* "Barring the omission of a vocabulary I maintain that it is quite a nice language. It aims at being that sort of language that, if you add a vocabulary, would be a logically perfect language. Actual languages are not logically perfect in that sense, and they cannot possibly be, if they are to serve the purposes of daily life. A logically perfect language, if it could be constructed ... would be very largely private to one speaker." Compare this to *PI,* §275: "I am saying: you have not the feeling of pointing-into-yourself, which often accompanies 'naming the sensation' when one is thinking about 'private language'. Nor do you think that really you ought not to point to the color with your hand, but with your attention. (Consider what it means 'to point to something with the attention'.)" Possibly both have said the same thing about private language, but then with very different concerns.

[56]Descartes took for granted the (unproven) doctrine of prejudice of Francis Bacon (Agassi 2013, 116).

Wittgenstein with Moore on the top of the totem pole. Since he targeted for Wittgenstein the place on the totem pole as second only to Moore, even this change is minimal. One may object and suggest that it is better to ignored all ancestry in preference for the respect for the truth. This is a mistake: we do not know the truth; we know that some significant ideas are not true; we suppose that the study of great errors of the past is worthwhile. We even still teach philosophy by teaching its history. Hence, the question remains: was the error of Wittgenstein significant enough for us to remember it? What did we learn from having his philosophy and having it superseded? Much of the current analytic literature is about this issue. It is almost all too apologetic and unclear to count; since this literature will not present the mature philosophy of Wittgenstein as superseded, it cannot discuss properly the question what have we learned from it and the valid criticism of it. Rather, this literature imposes on us the question, is Wittgenstein's philosophy superseded? Only if we agree that it was, can we ask, is it an error worth remembering? What then was the philosophy of Wittgenstein and what was his contribution and is it true? What is its value on the assumption that it is true and otherwise?

This is the least biased version of the question; yet it is still biased. It is unlikely that any presentation is fully unbiased, and even if it were, we would not know that it is. (See the conclusion of Plato's *Symposium*.) For, it rests on the idea that the history of any field of study is – or rather aspires to be – the depository of the consensus about it.

Although we cannot spot our bias – or else we try to reduce it – we know that whatever we study, our very statement of our study already renders it historical, especially since Wittgenstein died in the middle of the twentieth century. This is not to deny that Wittgenstein craved recognition and assent to his philosophical views; yet he craved no less the recognition of future historians. He said (Wittgenstein 1974, 71; Monk 1990, 184), speaking of his *Tractatus* before he had found a publisher for it,

> Either my piece is a work of the highest rank, or it is not a work of the highest rank. In the latter (and more probable) case I myself am in favour of it not being printed. And in the former case it's a matter of indifference whether it's printed twenty or a hundred years sooner or later. After all, who asks whether [Kant's] Critique of Pure Reason, for example, was written in 17x or y ....

What makes a work deserve posthumous recognition? We have no general answer to this question and we do disagree about specific cases. We know of many works that experts deemed very important and that did not survive their authors – obviously not with justice, since some of these later generations rescued from oblivion. Such cases often depend on sheer fashion, which is not very interesting, and, at times, they depend on philosophy. In particular, a problem that an author addresses looks marginal until the agenda changes. We had an example here: the logical paradoxes and the problem of scope looked marginal and uninteresting until the development of the idea of formal systems and the critique of the traditional theory of meaning, both thanx to Frege. These led to Russell's idea of the meaningless that is the cornerstone of the philosophy of Wittgenstein – young and mature alike.

Chapter 11
The Message of *Philosophical Investigations*

Wittgenstein's discussions of language games and of forms of life are parts of his theory of meaning. They led him to a fragmented view of language and so he proposed a search for the rules that each of the different fragments follows. The mention of forms of life was only a hint in that direction, and the better commentators overlook mere hints. The most central question is, did Wittgenstein care about science? Young Wittgenstein instructed (*TLP*, §6.53),

> say nothing except what can be said, namely, the sentences of the natural sciences.

He meant to exclude not the vernacular but metaphysics. He changed his mind on the vernacular, which he later on took more seriously, although not seriously enough to recognize folk metaphysics. On this, he did not change his mind. Did he change his mind on science? What part of his early philosophy did he reject? This is under dispute: despite his requirement for clarity, he constantly abstained from expressing his view about science (about what theory is scientific and about induction). This is the most controversial aspect of interpreting his philosophy. It cannot easily be resolved, and it hinges on the controversy over the question, are commonsense and science in accord or in conflict? Considering staunch commonsense realism the opposite of Platonism, Russell (1953) sided with science on every detail on which it clashed with commonsense, implying that (Moore and) Wittgenstein were hostile to science (hopefully quite unintentionally).[1] It should be noted, however, that all the references to science in Wittgenstein's *Philosophical Investigations* are quite respectful but distant.[2]

Wittgenstein's response to Russell's criticism of his system of logic is not available to us. It having made him neglect formal logic and even quit philosophy for a few years indicates that he took it seriously, but not that he had no response.

[1]Presumably, Moore took for granted (as sheer commonsense) that magic is not effective.

[2]*PI*, §§79, 81, 89, 109, 392, Pt II: xii, xiv. Wittgenstein was rather critical of current psychology, anthropology and evolution – as needlessly invoking metaphysics. Kreisel 1978, 86 suggests that Wittgenstein viewed science as otherwise free of conceptual confusion.

J. Agassi, *Ludwig Wittgenstein's Philosophical Investigations*, Synthese Library 401,
https://doi.org/10.1007/978-3-030-00117-9_11

Some of his devotees have attempted to develop informal logic. It is useless. (Indeed, he never supported it.) His rejection of the distinction between the object language and the meta-language forced him to avoid reference to formal logic and thus also to neglect most of his early concerns. He then fragmented [any given] language into segments that he called language games, presumably defined by their functions as languages in use. (They are languages only in some metaphoric sense: they are not closed under the usual grammatical operations.) He thus (rightly) broke away from the traditional limitation of the study of language into its descriptive function: he stressed repeatedly that he noted a multitude of functions that a language can serve, with the proposal to study their characteristics. This is a relatively new proposal;[3] later on, the relevant literature repeatedly ascribed to John L. Austin a poorer version of it, and with much fanfare.[4]

11.1 Analysis Versus Metaphysics

Wittgenstein did not mean his analytic technique to lead to any theory akin to formal logic or to a theory of the functions of language. This lack of intention is what distinguishes mature Wittgenstein's view of the analytic method from the view of that method as Russell proposed and as young Wittgenstein and the "Vienna Circle" endorsed enthusiastically. Wittgenstein dodged Russell's criticism. Yet it made him change his mind explicitly and emphatically. He admitted, however, that he changed his mind on one point, one that Russell never discussed: the limitation on language to its descriptive role. He rejected this limitation – to logic, to the descriptive function of language. This limitation, that Aristotle has instated, is due to the problem of

[3]Wittgenstein said hardly more about the functions of language than that they are diverse; he allowed the use of nonsense as long as it is labeled poetry (*PI*, §13) without discussing its function either. The diversity of the functions of language appeared in the nineteenth-century study of the origins of human language. Out of these emerged the theory of Karl Bühler of the hierarchy of functions of language. These are the expressive, signaling and descriptive. Bartley 1974, §VI said, "Wittgenstein knew Karl and Charlotte Bühler socially and personally… Whether Wittgenstein ever made any conscious connection between Buhler's psychology and his own later thought is, however, an open question." Wittgenstein added to this his claim that the traditional concern with the descriptive is narrow. Going to the other functions of language, however, leads one to the social sciences. Gellner [1959] 2005, xi complained that analytic philosophy grants wide permission to assert flatly diverse empirical social observations, often false. This practice, he added, is justified by the Paradigm Case Argument plus the Argument from Impotence that is Wittgenstein's famous contention: philosophy leaves things as they are. Gellner followed Russell and Popper in viewing this idea as invitation to leave philosophy. Yet Wittgenstein agreed on this and lamented his disciples not leaving.

[4]Commentators have noted repeatedly that Wittgenstein had anticipated much of what his disciples propounded later. This observation, intended as a compliment, holds for any unoriginal follower of any master whose writing-style is sloppy, vague, suggestive, metaphoric or aphoristic. Gellner [1959] 2005, last sentence, emulated TLP, §7, saying, "That which one would insinuate, thereof one should speak."

knowledge. Wittgenstein argued that all aspects of language, all language games, to use his expression, deserve study. This was his kind permission, not his planned action. His aim was not to achieve progress in the social sciences but merely to exorcise metaphysics. He therefore avoided theorizing on any aspect of language except for its limits that render metaphysics not open to articulation.

Wittgenstein viewed philosophical questions as symptoms of a sickness and analysis as its cure. He could claim for this the status of an empirical psychological observation. For, indeed, philosophical questions did torment adolescent Russell[5] as well as adolescent Wittgenstein;[6] but this is not what mature Wittgenstein had in mind: his claim that philosophical questions are a sickness refers to a philosophical illness,[7] not to a psychological one; and, it seems, not of the empirical sort. It is currently unpopular, if not generally rejected. With what then do his disciples replace pain reduction? Some of them are now busy in a rather formal style of research, more formal than the style of his later years, probably more formal than he ever envisaged. As analysis is a mode of groping, its heuristic is inherently unclear; its concern is local rather than global, although it does not preclude global concerns. Moreover, traditionally, (scientific) analysis is supposed to induce desirable synthesis. This went well with Russell's hope to render philosophy scientific. Wittgenstein did not share this hope. He suggested that philosophy has no room for synthesis; its task is but the removal of the rubbish and with it the pain that it allegedly (always?) causes.

The chief use that Russell had for analysis was to remove paradoxes. Mature Wittgenstein ignored these. Otherwise, Russell was in two minds about analysis. On the supposition that the culmination of synthetic philosophy is a Hegelian system and that this system is hostile to logic and to mathematics, he took a radical anti-metaphysical stance. His comments on Wittgenstein and on his school led him to

---

[5]Russell 1929, 15:

> I continued, however, to believe devoutly in the Unitarian faith until the age of fourteen, at which period I became exceedingly religious and consequently anxious to know whether there was any good ground for supposing religion to be true. For the next four years, a great part of my time was spent in secret meditation upon this subject; I could not speak to anybody about it for fear of giving pain. I suffered acutely, both from the gradual loss of faith and from the necessity of silence.

[6]*PI*, § 133:

> "The real discovery is the one that makes me capable of stopping doing philosophy when I want to.
> – The one that gives philosophy peace, so that it is no longer tormented by questions which bring itself in question." Wittgenstein's pain deserves respect and sympathy. It is regrettably common (*Ecclesiastes*, 18:1: "In much wisdom is much grief, and one who increases knowledge increases sorrow").

[7]*PI*, §109: Wittgenstein's "Philosophy is a battle against the bewitchment of our intelligence by means of language" that he equated with traditional metaphysics. Unfortunately, he never argued about specific cases of traditional metaphysics beyond naming some central ideas such as solipsism (that no one ever endorsed). His followers differed from each other on how to reduce the pain due to metaphysical cogitations, perhaps because there are diverse kinds of it. Far from having killed all metaphysical controversies, Wittgenstein generated new ones.

suggest welcoming the option for future analysis to surprise us and show us the way. Perhaps one thing he was systematic about: he neither rejected nor endorsed Einstein's observation that he was ambivalent about metaphysics though he agreed with him that the [Wittgenstein-style] fear of metaphysics is a contemporary malady.[8]

It would be nice to say, Wittgenstein viewed metaphysics negatively and Einstein viewed it positively. More precisely, Einstein bemoaned the fear of metaphysics and Wittgenstein combatted its bewitchment. Are these opposites? I for one cannot tell, nor am I qualified to, since my position is of one openly and thoroughly bewitched by metaphysics. Still, taking the rejection of metaphysics as a foregone conclusion is not sufficiently analytic.[9] The Wittgenstein school's change on this point is agreeable; their going about it surreptitiously is regrettable: it breeds confusion.

11.2 Analysis sans Anti-Metaphysics

Analyticity *sans* hostility at its best is present in the early works of Moore and of Russell. By contrast, the style of Wittgenstein's writings is often hostile and seldom analytic. The paragraphs of his *Tractatus Logico-Philosophicus* are numbered lexicographically in a manner that rendered his text too aphoristic and puzzling (Ramsey 1923, 469). The style of the first posthumous book of Wittgenstein is free; its paragraphs are usually more comprehensible, but often not their intent (beyond a general hostility to metaphysics). He continued attacking articulated metaphysics as ungrammatical after his loss of his early apparatus for the elimination of metaphysics by analysis: he no longer discussed grammar: he claimed that some sentences follow grammar, some not, but with hardly any argument: what looked to him too metaphysical he declared ungrammatical.[10] The criterion for it, if he gave any, was generality: asking what the time is, is ordinary but asking what time is

[8]Einstein 1944, 29, on Russell's 1940 contrast between naïve and scientific realism. Russell 1944, 696 agreed.

[9]The covert way of the softening of Wittgenstein's anti-metaphysics is conspicuous in the 1996 *Cambridge Companion to Wittgenstein*. Newton Garver discusses on p. 157 Wittgenstein's "certain hostility to metaphysics combined with a continuing fascination with metaphysical problems, as well as continuing contributions to descriptive metaphysics". At the very least Garver should have noted that Wittgenstein would disapprove of this presentation. Another conspicuous example is Fogelin 2009, *Taking Wittgenstein at His Word: A Textual Study,* that ignores his words "meaningless", "meaninglessness" and "metaphysics". He used the word "metaphysical" once, ascribing to Wittgenstein a concern with some "metaphysical simples". How this tallies with Fogelin's (true) view that a major end of Wittgenstein's work was therapeutic, I do not know. Still another is Cavell 1962, where he hints that some metaphysics is legitimate by asserting the legitimacy of some confusion. See Glock 2001 for a scholarly survey of the misreading of Wittgenstein.

[10]*PI,* §90: "Our investigation is therefore a grammatical one"; the use of the word "therefore" here is questionable. "Such an investigation sheds light on our problem by clearing misunderstandings away".

metaphysical.[11] As demarcation of metaphysics, this seems impossible, since science is full of general statements. This, however, Wittgenstein dealt with in his first book (*TLP*, §6.3-4): there are no genuine generalities in science; generalities occur there as mere *façon de parler*. Yet the assertion of the existence of God is the paradigm of a metaphysical assertion. Is the assertion that a color exists (*PI*, §58) metaphysical too? Wittgenstein had no tool for discussing this question. Popper argued that it is impossible to discuss sentences in the abstract, that the same sentence can appear as a part of a metaphysical system and as a part of a scientific one. Wittgenstein held this attitude towards words (*PI*, §116): "What we do is to bring words back from their metaphysical to their everyday use." Wittgenstein did not characterize metaphysics; his claim that a general sentence is metaphysical while instances of it are commonsense is not a case to generalize with no qualification.

Other ideas notable in his later philosophy are a theory of meaning and observations on language games and on forms of life – as parts of that theory or separately, as you like it.

Any modern theory of meaning comes to replace the traditional theory of meaning as reference or denotation. That theory is both erroneous and limited – to names and descriptive phrases (*Genesis* 2:20). Its limitation hides in classical logic that was a philosophy of science, where science appeared as the (Aristotelian) theory of definitions that presents definitions as defining and classifying nouns (Bar-Am 2008, Ch. 4). This was repeatedly subject to criticism, yet with no result. As Boole expanded the scope of traditional logic, the spell broke. Frege broke away from it much further; following Boole, he kept epistemology out of his logical disquisition. Russell brought epistemology back into the discussion of logic, and Wittgenstein opened investigations into functions of language other than description.

Frege refuted the classical theory of meaning and offered a new one. Russell offered an incomplete variant of the classical theory. In his 1927 expanded om it, as he said, meaning (meaning sense) is use. Wittgenstein made it his battle cry as Russell (1940 concluding chapter) gave it up in defeat.

A simple analysis of the part of *Philosophical Investigations* that sticks together is this: the use of many words in specific contexts is natural, yet their use in general is puzzling.[12] Considered analytically, the move is from the specific to the general: an abstraction. Wittgenstein saw a profound difference between its specific and general use. The search for specific meanings he approved of and considered studies of specific language-games; the search for meaning of any word in the abstract he

---

[11] *PI*, §89 notices that "for some reason it is difficult to remind oneself". Here, presumably, he answers the question, what is his contribution if he leaves things as they are? It is his reminding us of things that we forget, thus returning to us our peace of mind (*PI*, §133). Russell's *Problems of Philosophy* ends with calling peace of mind the highest good.

[12] Wittgenstein claimed to have solved St. Augustine's puzzlement about his ability to talk about time until asked what time is (*PI*, §89–90). He viewed this as characteristic of all metaphysics and offered (*PI*, §436) it as an example of the "dead-end in philosophy".

condemned as metaphysical or dismissed as metaphorical.[13] He stressed that in a sense, sense is unanalyzable (*PI,* §§89, 108, 117–19, 246, 354). To try to analyze it is a taboo. Breaking a taboo causes loss of peace of mind.[14]

The theory that meaning is use Wittgenstein understood as the theory that the meaning of a word or of a sentence is observable in the language game to which it belongs. Did he not condemn this theory as too general? What task do we expect the theory to perform? Does it deliver the goods? The task is to explain the difference of words in their usage, said Wittgenstein. Does the theory of meaning express this difference? If Wittgenstein has described this difference well, will many such descriptions of the difference of the meaning of a given word from other words be an adequate theory of the meaning of that word? If yes, will it then render Wittgenstein's theory descriptive? Will it then be explanatory? Will it thus be scientific? If so, will it be testable? Since it has to be descriptive and not *a priori* valid, surely, it might be scientific. Did Wittgenstein agree? Would he, then, have allowed philosophical usage? If not, why not?[15] This all hinges on the readiness or refusal to identify use and usage.[16]

This raised the popularity of Wittgenstein's contrast between the kosher piecemeal use of a concept and the non-kosher metaphysical, unified use of it. Thus, the unified entity Goodness or *The* Good is metaphysical; the matrix "*x* is good" is shorthand for the kosher matrix "*x* is good for *y*," so that if *y* is not specified it has to be viewed as specified tacitly. Generally, a two-place predicate that looks like a one-place predicate is either shorthand or a pseudo-concept that generates a pseudo-statement (akin to *TLP,* §5.473 "Socrates is identical" that wittgenstein offered as an example for a psuedo-sentence).

This analysis of mine fuses ideas of young and mature Wittgenstein. It offers a way to transform the analysis of a word to an analysis of a statement – the one in which the empty word-like expression appears. Yet it is a part of a formal language in which a predicate is by its meaning a one-place or a two-place predicate, as the case may be. This proved problematic: Wittgenstein considered this technique tolerating the view of a parrot's ability to ask for food as its ability to speak (*PI,* §§345–6).

Some of Wittgenstein's followers tried to replace the general theory of truth ("Truth with a capital T") with diverse specific ones. This forced them to discuss

[13]Wittgenstein took the metaphoric understanding of magic as self-understood (*PI,* §§356, 439); he insisted that synesthesia is not (*PI,* p. 214). He was not interested in explaining metaphors but in excluding their metaphysical readings. Black developed his theory of metaphors in order to use it against the Sapir-Whorf hypothesis. An adequate theory of metaphors has to notice the frequency of the use of magical expressions as metaphors or possible metaphors and to account for the fact that the magic worldview encourages considering possible taking metaphors literally. This relates to Rousseau's theory of all language as rooted in metaphor and to Jules Michelet's profound idea of the victory of prose over poetry.

[14]Scientific research too may cause sleepless nights and the loss of peace of mind. Did Wittgenstein oppose it?

[15]Some theologians followed Wittgenstein and declared theology a language game.

[16]This is Saussure's distinction between *langue* and *parole*. Both Searle and Kripke appreciate it (Searle 1969, 17; Kripke 1972, Ch. 9) although it is an exaggeration: *langue* too is (not an ideal but) an institution.

Tarski's theory of truth, which Wittgenstein disliked as it took the meta-language seriously. The same went for methodology: Popper had said there is no scientific method; this Wittgenstein's followers admitted on condition that Popper would admit that there are many methods in many branches of science. This scarcely resembles any idea of Wittgenstein. It only looked as if it does. How should we treat it? This is hard to decide, since Wittgenstein ignored deception as an aspect of resemblance.[17]

This limits logic in a new way! Russell stressed (1912, Chapter xii: Truth and Falsity) that when one constructs a language one has to take care that it should include false statements. The theory of language games must likewise include bad games, but with the distinction between good and bad, a distinction that should be partly inbuilt (for *a priori* bad games) and partly open to the decision of the players (agreement; legislation).[18] Wittgenstein mentioned the possibility of lying to show the possibility of knowing other people's intentions. This is the same as the idea that the recognition of error is essential for the understanding of the concept of truth.

Perhaps I am in error. How would we show that? More generally, when is a theory of meaning adequate? Russell had a distinctly partial answer: a theory of meaning is objectionable if it allows for paradoxes; it is likewise objectionable if it does not allow for falsehoods. All this is incontestable. It has nothing to do with Wittgenstein's philosophy. Russell had produced some paradoxes in order to refute Frege's theory of meaning. Is there any new theory of meaning to replace the traditional theories of meaning? Frege and Russell had only partial replacements. Mature Wittgenstein gave many instances for the limitations of the traditional theories that rest on psychology: meaning is different from expectations or intents (*PI*, §577), he rightly stressed. In these examples he scored. Interestingly, he even managed to refute neutral monism – as a "misleading parallel" (*PI*, §571) – by reference to ordinary experience: physicist report their observations and psychologists observe their experiences.[19]

Wittgenstein also criticized Frege's theory of meaning: its (alleged or true) Platonism made it strictly ideal, so that he could not ascribe meaning to vague terms, which he found a very serious flaw (*PI*, §71). Whereas Russell's approach to Platonism was dismissive, Wittgenstein's was hostile.

---

[17]*PI*, §§249–50. Pseudo-science is often more intuitive than science. Presumably, this is why critical physiognomy and homeopathy interested Wittgenstein (*PI*, §178).

[18]Some commentators rejected Wittgenstein's suggestion that lying is a language game and his explanation (*PI*, §249): "(Lying is a language-game that needs to be learned like any other one.)" Their objection rests on the tacit view of language-games as utopian.

[19]This, *pace* Hintikka, is a major deviation of Wittgenstein from his early philosophy.

11.3 Anti-Essentialism Again

Here then is the place to introduce anti-essentialism, or rather Wittgenstein's version of it. He did not introduce the term. Feyerabend did – in his 1955 review of Wittgenstein's *Philosophical Investigations*. The term "essentialism" is a neologism that Popper has introduced. Initially, Popper offered the new term to replace the medieval term "realism", found unsatisfactory on two counts. First, the traditional term "realism" names a theory of meaning, the opposite of "nominalism"; today it names an ontology, the opposite of "idealism". Second, the medieval debate concerned two questions. First, do classes exist? Second, is the general name a shared proper name or the proper name of their class? Within the theory of meaning as denotation, this amounts to two options, (1) the nominalism that declares a class name a name that its members share and (2) the realist view that declared a class name a name of that class. Within ontology, however, this amounts to the realist view that classes exist (Platonism) and the view that they do not (materialism, naturalism). Popper introduced his new term in emphatic indifference to the ontological dispute: he took modern logic for granted and with it the idea that classes possess names regardless of whether they exist or not. (Logicians still tend to pretend that there are no synonyms and no homonyms. Yet in ordinary modern logic, the empty class must possess many names: a-and-not-a, b-and-not-b, *etc*..) Popper left open the question, do classes exist? His hostility to essentialism was due to its traditional confusion about definitions, a confusion that Frege, Russell and Wittgenstein never shared. Methodological nominalism, to use the jargon, is consistent with ontological Platonism. Feyerabend caused confusion in the literature about the question, was Wittgenstein an essentialist or not?

Feyerabend's intent was to close the gap between Wittgenstein and Popper. They differed on science, as Wittgenstein was an instrumentalist of sorts and Popper was neither a realist nor an instrumentalist but a putative realist, who thus changed the traditional dispute. Already Frege did that when he considered a statement meaningful if and only if it has a truth-value. Popper's putative realism is not traditional but Frege-style. As to non-realism, namely, instrumentalism, it views scientific universal statements as fictitious and thus (!) as meaningless. It is a defeatist theory, resting as it does on the recognition that there is no proof of general informative statements. Feyerabend introduced linguistic instrumentalism. It need not be defeatist, he observed, since language is indeed an instrument. It is preferable for language to be devoid of information, as its use is to make assertions and to leave their truth-values open to examination. The celebrated Sapir-Whorf hypothesis informs us that (like it or not) languages do convey information. To the extent that a language does convey a relevant item of information, that information is unreliable. On this, all linguists agree today. The attitude of Wittgenstein to this matter is unclear, but this matters little since Wittgenstein stressed that it is the application of a word or a sentence that matters, not its initial import (Kienpointner 1996, Abstract). The analysis of the Sapir-Whorf hypothesis neutralizes the metaphysics that it introduces. This is very much in agreement with Russell's view. Now most of what Wittgenstein declared meaningless, Russell declared false. In truth, thank Heavens, ideas – those that belong to magic, metaphysics, or science fiction – that language implies (Sapir-Whorf style), we consider Pickwickian: metaphorical or fictitious.

The traditional philosophical distinction between truth by nature and truth by convention is still considered the point of departure of all philosophy in the Greek mold – all of western-style philosophy. The instrumentalist approach to science presents scientific truths as truths by convention. In the wake of Einstein, the idea developed according to which the aim of science is the truth even if its theories are forever mere hypotheses. Discussing the distinction between the instrumentalist and the realist, Popper replaced the term "realism" with "essentialism". There may be a natural tendency to reject the instrumentalist theory of language in parallel to the tendency to reject instrumentalism in the theory of science – as defeatist. This, says Feyerabend, is an error: the view of language as an instrument is not in the least defeatist; it liberates: Feyerabend's view of Wittgenstein's theory of language as instrumentalist is right.

It is possible to go further, since the theory of meaning as use is a variant of the theory that language is an instrument, it invites Wittgenstein's program: to analyze any part of language through its use. It is a colossal task. It was his proposal to chop up language to fractions to characterize by their function. These fractions he called "language games". How this impinges on metaphysics is obvious: in metaphysical discussions, he declared, "language is idle". This idea is a carryover from *Tractatus Logico-Philosophicus* to *Philosophical Investigations*.[20] Assuming that metaphysics is as useless as chess, is it advisable to remove it from all human discourse? Quite possibly, Claude Shannon's communication theory that condemns the traditional hostility to redundancy thereby condemns Wittgenstein's anti-metaphysics. This depends on the damage that metaphysics causes. Assuming that it has caused pain to people like Wittgenstein, it may be humane of him to try to help them. Applying his cure to people like Russell who delight in it is the vey opposite.

This is not all that there is to it. The right attitude to metaphysics is scarcely a matter of personal taste. Bacon argued against metaphysical research as he found it the source of scholastic dogmatism. His view was dominant until William Whewell criticized him by reference to historical examples: scientific researchers may use metaphysical theories to develop exciting and fruitful research projects. To apply this against Wittgenstein and his science-minded followers, some discussion of scientific method may be necessary. Whewell assumed that research begins with the conjecturing of hypotheses and testing them.[21] In Wittgenstein's *Remarks on the Foundations of Mathematics*, in which he argued also against the claim that essences exist, he clearly echoes the classical arguments against essentialism. He asserted

[20] *PI*, §§88, 132, 291, 507. Wittgenstein borrowed this idea from Heinrich Hertz, whose effort to eliminate metaphysics from science was a part of a movement that deemed Newton's metaphysics obsolete. This movement played a significant role in the growth of physics. It became obsolete before Wittgenstein started.

[21] Whewell's claim that research begins with the search for an adequate conjecture is his version of the problem of induction.

(1978, I, 74, 99–100; II, 105; III, 38) that it is too arbitrary to count and that it is obscure. He also suggested – it is hard to ascribe to him more than a suggestion – that he accepted the classical dichotomy between naturalism and conventionalism, as he said, that one who talks of an essence is "merely noting a convention". Moreover, "to the depth that we see in the essence there corresponds the deep need for the convention". Possibly this is but an expression of hostility to essentialism and to realism at one and the same time; which is in character; Wittgenstein did not object to lumping them together this way. Yet this comes with the truth-by-nature *versus* truth-by-convention dichotomy, and Wittgenstein's endorsement of it – if he did endorse it – may prove unsatisfactory metaphysics.

Perhaps Wittgenstein rejected the dichotomy as he advocated his theory of language games that splits language into many languages, each with its own function. This idea is acceptable on the supposition that there is no reason to assume the unity of language.[22] This supposition is disappointing, since the theory of language games nullifies at least one more function. In the very last pages that Wittgenstein wrote, he examined the possibility of uncertainty with brutally honest thoroughness. Suppose everyday information turns out to be doubtful; suppose the physics textbook too is mere superstition; suppose I do not have my own name right. (After all, how did I learn it?) What then? His expression "What then?" is his *cri-de-coeur.* Moore relied on ordinary knowledge in ordinary circumstances; yet conceivable extraordinary circumstances invalidate ordinary knowledge claims. Moreover, if my experiences would alter suddenly, radically and systematically, I will naturally admit having been in error (Wittgenstein 1975, 614–17). What then is the ground for my certainty? It is my playing a language game, Wittgenstein answered his own question. This answer he found unsatisfactory. It seems he died agonizing over it. Feyerabend's ascription of instrumentalism about language to him comes to overcome this difficulty. Why did Wittgenstein not use this idea? This invites a detailed discussion.[23]

11.4 Wittgenstein's Heritage

To assess Wittgenstein's heritage properly we may want to sum them up first. What was the general idea of Wittgenstein's thought? Was it not a metaphysics?

It is hard to find a general thread going through all of Wittgenstein's writings. The study of language games chops natural language to parts that perform diverse functions, as well as to diverse dialects, and even to idiolects ("I-grammars"; Wittgenstein 1975, 614–17; Dummett 1991c, 86). These differ from chopping logic (as a research language game) into sub-systems that perform different

[22]Rhees 1998, 2, 8, 13, 15, 114–18, 247 repeatedly denied the unity of a language past the context of any language game. This is an instance of a Wittgenstein-generated philosophical dispute.

[23]Weiler 1970 on Fritz Mauthner is relevant here; he disagreed with young Wittgenstein, who deemed Mauthner's ideas irrelevant (*TLP*, §4.0031).

researches on different functions. Thus, we speak of the logic of mathematics, of science, of situations, of power, of necessity (modal logic), of questions (erotetic logic; Heijenoort 1967), of ethics (deontic logic) and of legal affairs (rhetoric; Toulmin 1958), paraconsistent logic, many-valued logic, relevance logic and more (Agassi 1978). Wittgenstein would have no trouble calling these "language games" (Peursen 1970, 111); he would have required of them all to kill metaphysics, each in the way specific to its own characteristics.[24] Other parts of his studies go nearer to classical problems of classical logic. Existential import for one. Russell disallowing naming non-existent entities still is troublesome, of course, since we do want to be able to talk about Samuel Pickwick, the brainchild of Charles Dickens who exists only in a Pickwickian sense. We do want to be able to speak of humans who may have existed and may be mere mythical figures, leaving both options open. Logic should then allow the use of names whether they designate or not, in order allow for the proper scrutiny leading to decide the question of their existence rationally. The classical functional calculus has built into it the claim that everything mentioned exists (Fy $\supset$ (Ex) Fx). Alternative logical systems (free logics) avoid this. They are problematic (they clash with Russell's theory of definite descriptions, for example). The study of this requires technical wizardry.[25]

There is more to it. As Linsky (1967, Ch. 5) has argued, Russell's theory of definite description did not solve the paradoxes concerning knowledge. This did not trouble him then, since the *Principia Mathematica* is free of opaque words like "know x" where x is a name, since PMia is extensional. This is no slight on PMia since it did perform its assigned task. Still, this does refer to the limitation of PMia as a tool for other tasks. The paradoxes that Russell has presented and did not solve, such as the paradox of the King's not having known who is the author of *Waverley*, the King's having known who is Scott, despite the identity of the author of Waverley with Scott. For overcoming this paradox, many logicians suggest, we need intensional language. The search for it is stuck. Logic seems to falter when applied to discussions concerning knowledge and necessity, perhaps also the linkage between them. These are opaque terms. Wittgenstein did not share the researches that his leading disciples conducted.[26] For a notable example, Kripke is famous for having reread Wittgenstein freely in effort to apply logic to opaque terms.[27]

[24]*PI*, §119: "The results of philosophy are the uncovering of one or another piece of evident nonsense and of bumps that the understanding has got by running its head up against the limits of language. They, the bumps, allow us to see the value of that uncovering."

[25]See Lejewski 1954 and the classical Lambert 1963. Technical wizards hardly appealed to Wittgenstein. Hintikka bravely dismissed systematically all the discussions of technicalities by mature Wittgenstein.

[26]Analytic philosophers have contributed significantly to modal logic; whether this holds also for deontic logic is questionable. All this hardly touches on their relevance to Wittgenstein's texts.

[27]Arrington and Glock 1991, 2. Kripke 1982, 2, n. 1 took credit for having made rule following a major concern. This won full recognition. "For the last twenty years, the most influential and widely discussed interpretation of the *Philosophical Investigations* has been Kripke's *Wittgenstein on Rules and Private Language*" (Stern 2004, 2). See also the first page of Gellner 1984.

Wittgenstein had a series of leading non-logician-commentators. What any of them has contributed to the understanding of Wittgenstein is unclear, except for some dubious technical contributions that they have made, ones that did not enter the consensus about him. Most of the once leading authorities on the reading of Wittgenstein's texts have meanwhile turned out to contribute merely to passing fashions. Still, they were experts; at times, they produced some details of the quasi-canonic readings of his texts and filled gaps in the quasi-canonic doctrine. Wittgenstein said repeatedly that his followers did not comprehend him, that otherwise they would have deserted philosophy, since he proved that it is an arid zone. Yet he also said the opposite: his contribution is earth shaking. The number is growing of individuals who view themselves as philosophers and as his disciples. Whereas young Wittgenstein's texts are very precise, those of mature Wittgenstein are free and literary;[28] different versions of reading them exist, satisfactory to this or that degree. There is a vast Wittgenstein industry, of an immense number of books and papers, scholarly and popular, biographical, historical and analytic. They air Wittgenstein's ideas and explore their significance. This kind of activity is generally quickly saturated, but when it comes to Wittgenstein one wonders if one can say much of his ideas. New ideas scarcely appear even in the works of the best past thinkers. One may find new texts or shed new light on the works of a leading thinker, but this kind of work eventually dries up and continuing it usually leads to rapidly diminishing returns. They then seem increasingly like instances of hero worship. How can one become a disciple without being a hero worshipper? How can one stay a philosopher and follow a Wittgenstein? What are the activities of philosophers who follow Wittgenstein? Can a disciple teach the history of philosophy, for example? How exactly?

These questions embarrass Wittgenstein devotees. Already in the heyday of his popularity the question arose, can a disciple of Wittgenstein ever teach the history of philosophy without thereby admitting philosophical doctrine as legitimate, and its study and elaboration as a legitimate pursuit?[29] Can an ungrammatically worded doctrine, condemned as inherently [?] ungrammatical, ever be recognized? Some-

[28]*PI,* Editors' Note stresses this: most of Wittgenstein's posthumous output is mere drafts, of the nature of author's notes towards proper elaboration required for their publication. The exceptions are the opening parts of *Philosophical Investigations* and *On Certainty* that display more coherence than the rest of his output, perhaps also his 1938 *Blue and Brown Books,* or at least the (revised) German edition of the latter. The style of the mature Wittgenstein is often praised; especially by Monk. Styles of philosophers signify only to the extent that they raise hermeneutic problems. There is almost no such problem in the interpretation of Wittgenstein's texts: much as they may invite interpretation, it is not that its terminology is unclear or that its sentences are convolute and allow for competing ways of unfolding them. Such problems often belong to the reading of slum landlords; but also for reading Kant the Impressive.

[29]A. J. Ayer, once a celebrated Wittgenstein promoter, became increasingly disenchanted with Wittgenstein. He then moved increasingly from post-Wittgenstein positivism towards old-style philosophy and to studying the history of philosophy proper.

how, no comfortable answer has turned up. In the citadels of philosophy Wittgenstein-style, in Oxbridge, courses in the history of philosophy went on as if nothing had changed. Yet the teaching methods in these courses were radically new. One of the most definitive documents of analytic philosophy is a book by the analytic philosopher Simon Blackburn.[30] He bravely reports that lecturers then presented every philosopher as a candidate for the status of a precursor to contemporary analytic philosophy. This, he rightly adds, distorts the history of philosophy – even when the philosophical ideas in question are indeed antecedents to those of Wittgenstein – such as those of Berkeley and of Hume, perhaps even Kant. How do they teach the history of philosophy in Oxbridge these days? There is no literature on this. It seems that the following two examples are sufficiently indicative. The 2013 *Oxford Handbook of British Philosophy in the Eighteenth Century* has no reference to Wittgenstein and no evidence of his impact, direct or indirect. The 1988 *Cambridge History of Later Medieval Philosophy: From the Rediscovery of Aristotle to the Disintegration of Scholasticism* of Norman Kretzmann *et al.* mentions him twice.[31] Its editors mention him once in their introduction: they claim there that they have learned from "insights and interests" of the analytic tradition. Oddly, they offer no elaboration and not even a hint at the kind of insight they had in mind. They refer to famous proceedings that Ernan McMullin has edited, where only the editor's input signifies here, and he was a famous staunch metaphysician in the Roman Catholic style of the day. (He was an ordained a Roman Catholic priest.)

Rorty came to the rescue: he suggested something new, not a scholarly study of some obscure text but a kind of synthesis, of Wittgenstein with the American pragmatists, which makes reasonably good sense: they were not too friendly to metaphysics. In particular, Charles Sanders Peirce attempted to construct a theory of meaning aimed to rob metaphysics of its meaning by the suggestion to treat as synonymous theories not amenable to crucial experiments between them.[32] Rorty went much further: he sought a complete list of people who did not care for foundations of knowledge or, as he called them, of non-foundationalists. He added Heidegger to his list. This makes sense since an irrationalist need not care for foundations, but it conflicts with some easily available information: though Heidegger praises Catholicism as anti-modernist, he rejected it as wanting foundations, and in his 1915 *Habilitationsschrift*, no less. Worse, Rorty rooted for democracy yet he ignored the hyper-reactionary preaching of Heidegger. He thus freed Wittgenstein's followers from the need to study meanings, or any other aspect of language, or

[30]Blackburn 1984; Hare 1964, two final paragraphs, also deserves mention here. Both, however, were far from the early mainstream, as they recognized genuine philosophical problems. To repeat, Wittgenstein's exorcism of philosophical problems is no longer in fashion. Regrettably, contrasting Wittgenstein with his followers is still not popular. Already my friendly review of Blackburn's book (Agassi 1988, 221–38) said so and he refused to publish it in *Mind* (that he was editing then).

[31]One of these two references to Wittgenstein, on page 650, says, he "has made this example the standard one", the example being my raising of my hand – an example at least as old as the writing of Nicolas Malebranche who referred to "the practical desires to move our arm, to walk, to speak."

[32]Quine's denial of synonymy is in utter opposition to Peirce's proposal.

anything whatsoever. He also freed them of the duty to fight metaphysics: by now, this fight is *passé*, anyhow. Rather, he offered them a new social function: to engage in sparkling, idle conversation. All this has little to do with Wittgenstein, to whom he paid lip service. In defense of his quaint idea, he quoted leading (conservative) political philosopher Michael Oakeshott.

Now Wittgenstein's texts readily indicate that he was trying to impart a specific technique or method (*PI*, § 464):

> My aim is: to teach you to pass from a piece of disguised nonsense to something that is patent nonsense.

And,

> What I want to teach you is not opinions but a method. In fact, a method to treat as irrelevant every question of opinion.
>
> I am not teaching you anything; I am persuading you to do something.

These are strong words (Hilmy 1985, 5). Following Wittgenstein's text and his intent should end with nothing new, he declared. Is this an achievement of which he could be proud?[33] Perhaps despite strong words he was in two minds here: he wanted to have no disciples, no cult; he also wanted others to continue his task, as he thought it was vast and worthwhile and beyond the capacities of any single individual. In his *Philosophical Investigations,* he says clearly that it is mere sketches to fill later on. There was a task to perform, he suggested repeatedly with a sense of urgency and an expression of regret at his own inadequacy for it. One can search for some guiding principles in this text (Fogelin 2009). What were Wittgenstein's problems? What were his guiding ideas? We remember that his followers in the "Vienna Circle" saw in him a scientifically minded philosopher akin to Russell only more austere about metaphysics, and that his friend Engelmann saw a great injustice in the reading of Wittgenstein as the "Vienna Circle" did. Engelmann suggested that the "Vienna Circle" had harmed Wittgenstein's reputation. How can one decide between the "Vienna Circle" and Engelmann? This may hinge on the question, did Wittgenstein accept the authority of science? We remember Bernard Williams' proof that metaphysics is impossible since, were it possible, it would be synthetic *a priori* knowledge, whereas all knowledge of the world is *a posteriori*. Whatever we may think of this attitude, we have to agree that it is pro-science. So one should ask, is this attitude suitable for Wittgenstein?[34] How does one decide such matters? No answer. Assuming that Wittgenstein lost interest

[33]An example for Wittgenstein's sensitivity to this question is his surliness to Wisdom's comment on a performance of his as boring (McGuinness 2012, 314, 322). Wisdom then proudly described one of his own books as boring (Wisdom [1934] 1963, Preface to the first paperback edition). That book in question offers not Wittgenstein-style analysis (but Moore-style): its first edition appeared before Wisdom became a Wittgenstein lieutenant and its new preface appeared well after Wittgenstein's demise. Sad.

[34]Wittgenstein 1980a, b, 56:

> It isn't absurd, e. g., to believe that the age of science and technology is the beginning of the end for humanity; that the idea of great progress is a delusion, along with the idea that the

in science, it is especially hard to decide what exactly his attitude to it was – other than that of having lost interest in it and of having carefully avoided criticizing it and of having attempted to help it free itself of its occasional descent into metaphysics/ ambiguity. It seems that Williams did attribute his proof to Wittgenstein. Was he right? Among Wittgenstein devotees, this question is under dispute. Russell declared Wittgenstein's followers careless about things scientific and he found their attitude objectionable. Some leading Wittgenstein disciples do explicitly disregard science and do claim to be doing so on Wittgenstein's authority. Possibly he saw in science a source of contemporary metaphysics and complained about the imperialism of science, to use anachronistically an expression of Feyerabend, who supported magic and divination (Agassi 1988, 424). This reading is objectionable: unlike Feyerabend and other commentators of his, Wittgenstein stuck to the positivist tradition that is distinctly pro-science, while opposing prejudices and superstitions as a matter of course. The ordinary language that he relied on he deemed free of all prejudices and superstitions, and with most of his contemporaries he took for granted the western view that magic, sorcery, witchcraft, wizardry, chiromancy, and spells and divinations are prominent among widespread prejudices and superstitions, no less than astrology and folk medicine are.[35]

Efforts to extract a worldview out of Wittgenstein's writings are frustrated. Wittgenstein himself found frustrating his "unsuccessful attempts to weld my results together into such a whole." (*PI*, p. ix) This raises the question, is the view of the world as a whole not a metaphysics? Admittedly, metaphysics is efforts to offer a picture of the world as a whole. The converse does not hold. Traditional positivists, particularly the last eminent one, Ernst Mach, insisted that the proper worldview is the sum total of all science, so that philosophy is not required for its formation. This differs from young Wittgenstein's mystical view. (On this, the "Vienna Circle" followed Mach, not Wittgenstein, as Neurath has repeatedly observed.) Is the view of the mature Wittgenstein the same as that of the young Wittgenstein? In the Preface to his *Philosophical Investigations,* Wittgenstein confesses that he had wished to present his views "as a whole", and that he should have known that his effort was bound to fail. What is this passage an expression of? Is it a positivist rejection of metaphysics? Or is it an expression of the hope to express his metaphysical opinion,

---

> truth will ultimately be known; that there is nothing good or desirable about scientific knowledge and that mankind, in seeking it, is falling into a trap. It is by no means obvious…

This opposes the scientism of young Wittgenstein in its admission that viewing science as logically contingent invites skepticism.

[35]Many anthropologists are loath to dismiss magic and all that as erroneous as they are loath to ascribe it to primitive people as too haughty. So they tend to view magic not as descriptive (and thus erroneous) but as symbolic expressions. Wittgenstein seems to have agreed. See Jarvie 1983, 118: "Applying Wittgensteinian anthropology to the Wittgenstein canon … and specifically to the contradictions and incoherence (acts without aims) of the Sacred Text, we must conclude that reading and re-statement is an art, satisfying in itself, not an opinion in the realm of mistakes. This thesis has been almost articulated" in Cavell 1962, 93. Incidentally, already Frege criticized Wittgenstein's description of his own *Tractatus Logico-Philosophicus* as meaningless: it becomes thereby as merely symbolic.

so that it is a confession that he had developed one? Or did he only express the wish to have had one? What metaphysics what views "as a whole" are we speaking of? It could be the *naïve* realist worldview implicit in western languages, and it could be the worldview implicit in science. The text I have cited from the preface of the *Philosophical Investigations* refers to Wittgenstein's views on diverse philosophical topics expressed in that book, and regret that in it he failed to present them properly: "the essential thing was that the thoughts should proceed from one subject to another in a natural order and without breaks." This is inherently do different from the view that Mach had expressed less eloquently. Russell ascribed to Wittgenstein's disciples a different view: a worldview implicit in common sense. He charged them with the ignorance of the fact that common sense contradicts science. Wittgenstein and most of his leading disciples did not wish to contradict science, and so they considered science and common sense complementary without further ado.

Thus, no matter how much the view of the world "as a whole" of Mach, Wittgenstein, or the language analysts diverged, it is not clear that they saw this as metaphysical. What then is the metaphysics that Wittgenstein abhorred?

Perhaps we may make efforts to extract out of Wittgenstein's writings a political philosophy to get some general idea about it. This will lead to the same frustration. One may wish to reconstruct any views that are implicit in any text; to expect that this will yield some profoundly enlightening results is somewhat too optimistic. We know that Wittgenstein declared himself a conservative.[36] He wanted to go to Russia soon after the Soviet Revolution, but not out of a political conviction. His notes on anthropology suggest that when he introduced his concept of forms of life he meant not much more than what is common to modes of speech. For a framework for his language games he made do with the given – very much in line with the analytic technique of not taking ordinary sentences literally that he acquired from Russell and made much less technical.[37]

Let me mention one interesting example of the current attempt to stretch Wittgenstein's ideas as far as possible and try to extract from him opinions that he did not openly express and on which he may have spent little or no intellectual effort. Knut Erik Tranøy has a significant study of Wittgenstein's views on the ethics of science (1988). He asks, what justification did Wittgenstein offer for his own philosophical activity? The standard reading of this question relates it to epistemol-

[36]Vinten 2015; Hall and Jarvie 1996, 16.

[37]Some commentators consider Wittgenstein's idea of forms of life a version of fideism. I will try to ignore this as no text supports it and as it does not solve any significant hermeneutic problem. It conflicts with Wittgenstein's repeated demand to solve problems piecemeal. In his response to Popper he rejected the problem of induction on this very ground, so that it is an error to ascribe to him fideism in the standard way – as a solution to the problem of induction. And so this ascription becomes pointless, except that it may count as an effort to unify Wittgenstein's thought. For this, however, one has to add to the ascription of fideism to Wittgenstein the axioms that he allegedly took on faith.

ogy or to the theory of rationality; clearly, Wittgenstein intended it as moral. In this reading, it should meet with no objection. Justification, he adds, is either internal, which possibly means intellectual, or public, where the standards are provided by the public at large. The author takes his cue from the fact that Wittgenstein objected to technical, professional, insider terminology only in philosophical texts, not in the texts of the diverse sciences. He tries to find what Wittgenstein's position was about the place of science in the world of morals, or *vice versa*, and he comes up with fragments at best, since there is no more than that to find in the whole of the Wittgenstein immense corpus as far as it is known.

This is not to belittle such efforts. We should center them on the question, what was the aim of Wittgenstein's efforts? Let me postpone this question to the next chapter. Let me say now that in my view, since Wittgenstein followed Tolstoy, he had no moral theory worth studying; as a severe moralist, however, he deserves a study nonetheless. That study will resemble the 1944 *The Razor's Edge of* William Somerset Maugham, written in order to record "that there lived in this age a very remarkable creature." Although neither Wittgenstein nor the hero of Maugham's tale had a special moral theory or a special moral insight, they are supposed to have contributed to our moral precepts.[38]

Appendix to Chapter 11

The meaning of the slogan "meaning is use" is under a prolonged, fierce debate. Even its attribution to Wittgenstein is under debate, as he said, (PI, § 43), "For a large class of cases – though not for all – in which we employ the word 'meaning' it can be

[38]Let me offer one incidental paragraph against the huge hagiographic literature on Wittgenstein, relying mainly on Fania Pascal who wanted to speak of him only well. Maugham's hero and Wittgenstein were both highly charismatic, the former uncommonly agreeable and the latter stern and uncommonly disagreeable. He conveyed ceaselessly the conviction that his sincerity and his struggle for perfection imposed on him his unsociability. His stern demands from himself served him as his constant excuse for his quirks regardless of their costs for others. He often used his defiance excuse for evading duties of ordinary mortals. The most important instance for this is his aggressive response to criticism of his opinions (as intolerably stupid or as insincere). As a rare alternative to this he admitted error, adding that he should commit suicide. This is the peak of insincerity. For further assessment of his character, a consideration of Bartley's studies is useful. The nearest to it is Monk 1990, which is fair but much too wary. The aspect of Wittgenstein's life that makes him so different from Maugham's hero is his sense of shame, say, about his having allowed people to mistake his Jewish ancestry. He did not lie about it but allowed people to repeat mistaken information on it (a sin of omission, Monk 1990, 372 called it). He was deeply ashamed of it, says Fania Pascal 1973, 33 (and Lurie 2012, 52, 180). He nevertheless requested his second cousin Friedrich von Hayek not to publish his paper on him in order to keep his fans ignorant of his Jewish ancestry, Hayek told me. (See Monk 1990, 279.) He published his memoir long after Wittgenstein's demise (Hayek 1977). Wittgenstein also made privately some observations uncomplimentary to Jews – putting down Jews as unoriginal (Agassi 1997, last page). Private remarks of Frege and of Heidegger (Stern 2001) are much more disconcerting.

defined thus: the meaning of a word is its use in the language." He then mentioned (*loc. cit.*) naming by pointing, possibly as an exception. He did not specify all the exceptions: mature Wittgenstein never offered a theory of meaning. Rather, he said (PI, §340), "One cannot guess how a word functions. One has to look at its use and learn from that." How then does meaning relate to function? Is to know the use of a word then the same as to know its meaning? This is under debate. He said, to know the use of a word is a necessary condition to knowing its meaning; he did not say it is sufficient (PI, § 122):

> A main source of our failure to understand is that we do not command a clear view of the use of our words. – Our grammar is lacking in this sort of perspicuity.

The meanings of sentences is even more problematic for Wittgenstein. He preferred concrete discussions to abstract ones. To refer to concrete meanings he employed diverse words; the index of his *Philosophical Investigations* is revealing. For example, Art "Application, Applying," says, "*cf.* Employment, Function, Operate, Practice, Role, Service, Use". He evidently shunned a general, abstract, definition of meaning. Hence, he could not possibly show that all metaphysical expressions are meaningless – not even had he offered a general theory of metaphysical expressions. Hence, even were all his studies successful, his general claim would remain a mere hypothesis. Hence, there is no escape from generalities. Hence mature Wittgenstein too was self-refuting. To repeat (PI, §340), "One cannot guess how a word functions. One has to look at its use and learn from that." How? To speak of use as such is to generalize. This raises the problem of induction.

(The literature approaches this situation by the discussion of the term "language" since the meaning of a word is limited to the language within which it occurs. Human language then splits into diverse languages, languages into dialects, dialects into idiolects yet there is no such thing: a private language, said (Russell and) Wittgenstein, does not exist.)

Kripke tried a way out by his take up on Frege's Morning Star: considering London beautiful and Londres ugly (not knowing that they are the same) is inconsistent. Now this is Kripke's distinction between public reference and private sense. No one will question that meaning is public, to be privately recognized. What fascinated Wittgenstein about language is that no matter how rebellious one is, one is largely a conformist when it comes to language. This is why the chief criticism of the experiment of Descartes, of his effort to start *de novo*, is by reference to his very use of French. His answer would have to be that he was using French merely as a representation of the ideal language (akin to Leibniz's *characteristica universalis*). This answer is invalidated by the dismissal of the ideal language – by Russell and by mature Wittgenstein. This is particularly so for those who present the problem of induction as if they raise it for the first time, as a great discovery. This Kripke 1982 has claimed for Wittgenstein. This makes sense: young Wittgenstein having denied the principle of induction enabled mature Wittgenstein to rediscover it. Does this allow Wittgenstein to have a theory of meaning as use based on the data he displayed in his *Investigations*?

That depends: did mature Wittgenstein give up his early views? Young Wittgenstein says (*TLP*, §4.002) that the grammar of everyday language is unknown, and yet he added that (*TLP*, §5.5563) "all the propositions of our everyday language, just as they stand, are as a matter of fact fully in logical order." He offered an obscure reason for this, particularly hard to comprehend as he also says there (*TLP*, §3.323), rightly of course, that the copula in everyday language is vague and invites clarification. Commentators take it for granted that young Wittgenstein assumed that everyday language is metaphysics-free, so that his early view includes his mature view. Quite possibly, however, young Wittgenstein found metaphysics tolerable in everyday language but not in science (Williamson 2014). If so then this change will explain his publication and rejection of his 1929 "Some Remarks on Logical Form". The contrast that he recommends to study between his *Tractatus Logico-Philosophicus* and his *Philosophical Investigations*, is then not only in scope but in a definite change of opinion about both: the ideal language of the early work came to eradicate metaphysics, failed, and gave way to *Philosophical Investigations* to do the very same job.

That places Wittgenstein deeply in the tradition of anti-metaphysics; his early work came to prove it by reference to contemporary logic. His late work is to be placed even more within tradition – the Baconian tradition that demanded the avoidance of generalizations as much as possible in the hope that the instances will sooner or later impose the right generalization on us.

Chapter 12
Analysis of Analysis

The paradigmatic case of analysis is mathematical; Newton spoke of analysis in science and chemical analysis is still paradigmatic for that; psychologists speak of analyses of all sorts, as at times social scientists do too; linguists are engaged in discourse analysis; and the critical analysis of Shakespeare's sonnets or of any other artwork is also a familiar kind of analysis. The analysis nearest to, if not identical with, Wittgenstein-style philosophy, is the most famous case of analysis in the field of foundations of mathematics, namely, Russell's 1905 study of definite descriptions. It has never left the agenda of analytic philosophy (Pelletier and Linsky 2009, conclusion). That explains Russell's having found the discovery that language includes non-descriptive items downright silly. Yet I remember vividly how exciting philosophers in England found this very discovery when they read it in the opening pages of Wittgenstein's *Philosophical Investigations* when it appeared in 1953. It is time to put all this on the agenda for analysis.

The literature of philosophical analysis takes the expressions "use of words" and "ordinary use of words" as synonyms, presumably to denote the standard use accepted by society or culture at large, to the exclusion of the above-mentioned standard uses in specific sub-societies or sub-cultures: they are exempted from the constraints of ordinarly parlance. This exempeetion should hold for metaphysicians and theologians too, yet the whole purpose of the discourse about ordinary use is allegedly to delegitimize these two (as "slum landlords"[1]). Traditionally, the interesting questions concerning the use of words often refer to their extraordinary uses. This is understandable. It was a great achievement to put on the agenda questions regarding the ordinary use of words (akin to the sociology of everyday life; Norbert

[1]Consider then real slum landlords who misuse language in order to confuse their tenants in order to exploit them. Would Wittgenstein denounce them? To the extent that he has addressed the problem of demarcation of metaphysics, his solution to it is that metaphysics is a special misuse of words: it is the bewitchment of the attempt to cross the limits of language. This invites some (metaphysical?) explanation.

J. Agassi, *Ludwig Wittgenstein's Philosophical Investigations*, Synthese Library 401,
https://doi.org/10.1007/978-3-030-00117-9_12

Elias, Erving Goffman):[2] this was the revolution in the study of ordinary grammar. The idea of proscribing the use of some words because they are metaphysical was marginal before Russell applied it rigorously to the formal language that should serve the modern study of the foundation of mathematics. The Wittgenstein literature goes much further: it takes for granted that ordinary use reflects ordinary intuitions, and that these serve as a supreme court for legitimation. We know more about grammar, namely, the rules governing the use of language, but hardly in as much detail as required. Hence, any theory of meaning, including the theory of meaning as use, is significantly metaphysical: it invites developing it to scientific versions of it (Agassi 1975, 210). There is hardly an attempt to do anything like that (Fairclough 2003, Introduction; no reference to Wittgenstein).

12.1 Meaning as Use

The Russell-Wittgenstein theory of meaning as use is still in mid-air. As Linsky (1952, 6) has observed, its vagueness is blatant. Some of its corollaries are true, yet somehow more problematic than its advocates admit. That it does not deliver the goods is obvious: as this theory applies to different translations of a given text, it leaves their differences unattended. This is the radical untranslatability thesis of Quine. Though the views of Wittgenstein and of Quine are in conflict, they are both prevalent. Hence, the received opinion on meaning is excessively confused. It is time to stop the prevalent pretense that all is well with the theory of meaning.

Frege has established to everybody's satisfaction that two expressions that have the same referent need not be synonymous; their exact difference is under dispute. Some say that they differ in sense, others that they differ in use. Meaning theorists would do the expected of them had they explained this. They do not. The choice here is between options: (a) being different in sense necessitates being different in use; (b) being different in use necessitates being different in sense; (c) being different in sense is the same as being different in use.[3] Option (a) looks suspiciously like reference to Plato's heaven; not so options (b) and more so (c). The wish (of Russell) to have no need for Plato's heaven is crucial here. Why? Why did he refuse to admit it as fictitious?

The idea that meaning dwells in Plato's heaven draws attention to the diveristy of meanings.[4] Wittgenstein rightly saw this exercise as a waste of time and he recommended to begin with the diversity. This may expose Platonism as futile and

[2]Much of the analytic literature is scarcely philosophical. It is *a priori* sociology of everyday life, contrary to Wittgenstein's hostility to the *a priori* (Hall and Jarvie 1996, 16).

[3]As both Searle 1969, 84–6 and Kripke 1972, Ch. 9 have stressed, there is a snag here, since sense may be unique whereas usage must vary, as it is context-dependent.

[4]The protest of Antisthenes that he could see horses but not the idea of the horse ("horseness") brought about a dismissive explanation: he had eyes for vision with but no mind for intuition.

baffling – at least until someone comes with a better idea as to what to do with it. Consider the case (of definite descriptions) which Russell managed to analyze with no reference to Plato's heaven; consider analysis as clarification; and consider the traditional proscription of metaphysics; you may then come up with the bright idea that perhaps analysis will show other metaphysical items as eliminable. Wittgenstein saw some immediate success: his progress showed him a way to ignore the psychology of sense, viewing language an objective entity free of psychology. The omission of psychology from meaning considerations, the (temporary) loss of the psychology of meaning, is surely welcome.[5]

Is this the whole story? It depends whom you ask. As Wittgenstein insisted on considering use and analysis as belonging to the same language (rather than to language and meta-language), he was at liberty to ask some questions about the language that he was using. What does the word "use" mean here? It is the use of the word that counts, of course. What is its use? The use that philosophers exhibit is different, since the aim of the study of meaning as use is to delegitimize their use. This is why the name of this analysis is "ordinary language analysis". What is ordinary about ordinary language? Is it the speech patterns exhibited by an ordinary speaker? If so, then, who is an ordinary speaker? How does speech possess a pattern? Obviously, some speech patterns belong to grammar, some to style. It is not always easy to distinguish between them (especially when studying a historical case): idiosyncratic traits are stylistic; jargons are not; it is hard to say where group traits stand. Analytic philosophers silently gloss over borderline cases.

What are speech patterns and, more challengingly, where do they reside? Wittgenstein had an answer: speech patterns are not visible in any specific individual's speech, be that individual a philosopher or a bricklayer. Speech patterns, said Wittgenstein repeatedly, are rules obeyed while communicating. What then are rules? Do they have references? Do they have senses? What is their status? To this, commentators agree today, Wittgenstein had a single answer: language is a form of life. This means, it is a social given; language is an institution.[6] Rules of conduct are institutions.[7] Where do institutions reside? No answer. What is the rule about rules? No answer either. In principle so.

Kripke 1982, Preface, has ascribed these questions to Wittgenstein, but he stressed that he did not insist: he declared his effort more in debt to Wittgenstein than a proposed reading of Wittgenstein's texts. This way he noted that Wittgenstein had offered no answer to these questions. This way he ascribed to Wittgenstein concern for the problem of induction concerning the description of human conduct: we can never be sure that the pattern that human behavior exhibits is the pattern it will exhibit if we study it more carefully: the assertion that any sequence we observe

---

[5]Positivists noted repeatedly that the loss of metaphysics is a loss of but an empty shell.

[6]*PI*, §380. "I could not apply any rule to a *private* transition from what is seen to words." Hence, private language is impossible.

[7]*PI*, §199. "To obey a rule . . . to play a game of chess, are *customs* (uses, institutions)." The rules of a private game, then, are different, just as a private taboo differs from a socially-enacted imposed taboo.

obeys a certain rule is forever hypothetical. For a good measure, Kripke added here the paradoxes of induction, which Wittgenstein never mentioned: he expressed indifference to paradoxes. Kripke ascribed to Wittgenstein then the question, what is the meaning of a rule? Since Wittgenstein rejected the general form of questions that have simple answers when asked in specific cases, we should present Kripke's question specifically: What does it mean to say that Ludwig plays chess by the book? What is the meaning of obeying a rule, what is the usage of the expression "to follow a rule"? What does it mean to say that person P who uses the word "rule" is thereby following the rule governing the use of the word "rule"? Since Wittgenstein rejected the meta-language, it is natural to ask, what is the rule that he followed when he asked what rule does an ordinary speaker of a language follow? What is the meaning of meaning? What is the rule governing obedience to rules?

The use of the word "meaning" – or "sense" or anything else that determines a word other than its referents – is not identical with the use of the word "use". The referents of "meaning" and of "use" may be identical; not the use of "meaning" and the use of "use": these often differ. This platitude is contested – partly at least due to the refusal to acknowledge Wittgenstein's debt to Russell. The rules governing the use of the words "meaning" and "use" differ even if we ignore the use of "use" in utterly different contexts since words and rules differ. Even the rule for the use of a word differs from the rule for the use of a rule.[8] This is too obvious. The trouble lies elsewhere: "meaning is use" differs from Wittgenstein's "meaning is revealed in use"; not seeing this, said Russell 1953, has allowed boosting the unwise view of ordinary intuition as a supreme authority. Russell's argument itself is not intuitive, and so, some Wittgenstein fans declared it unsatisfactory.[9] Though I am no side in

---

[8]The difference that troubled Wittgenstein was between the behaviorally similar expressions that describe the conduct pattern of following a given rule and the similar conduct pattern that does not: the one case is "following a rule", "behavior in accord with a rule", and such; the other is "as a rule", "habitually" and such. This difference is common and obvious. Wittgenstein declared this difference his own paradox (*PI,* §201): "Our paradox was this: a rule could not determine a course of action, as every course of action can be made out to accord with the rule. The answer was, if everything can be made out to accord with the rule, then it can also be made out to conflict with it. And so there would be here neither accord nor conflict." Indeed, we learn a few relevant ideas here: perfect rules are ideal (*PI,* §§100–3) and if we find a rule disappointing, then we may alter it (*PI,* §125). Further, we do not know what rule a series of numbers follows as we have always only a partial set of them (*PI,* §143). In brief, the games that we play or behavior-patterns that we exhibit are not clearly defined. Why did this trouble Wittgenstein? How was all that paradoxical? Why is that troubling? Suppose the game is well defined; the inability to define it clearly that troubled Wittgenstein is a version of Hume's critique. It is thus no paradox but a detour to the problem of induction. Wittgenstein's paradox echoes Duhem's proof that there can be no rule of induction by reference to under-determination (which is likewise a fact, not a paradox): all our data fit into any one of infinitely many alternative explanations. What troubled Wittgenstein then is indeterminacy. The open-endedness of language delights some and upsets others.

[9]Hallett 2008, 10: "The notion of language as a principal determinant of truth and assertability has evoked a common reaction. 'This,' writes John Mackie, 'is the basic problem for linguistic philosophy, to decide whether it is concerned with grammar or metaphysics, with language or the world. And if it is to tell us something about the world, on what evidence or on what arguments will

this dispute, I am puzzled by the concern of analytic philosophers with the limits of reason, when, obviously, the expansion of reason is so much more interesting.

12.2 Analysis Defended

Leading analytic philosophers dismiss if not ignore Russell's critique of the appeal to intuition as it is not intuitive. "Excluders", by Roland Hall, may count as a response to it.[10] It appeared in *Analysis*, the citadel of the establishment of the analytic school. There are about two dozens of references to this paper in the literature; they all refer to the term "excluder"; the paper itself is thus far unnoticed.

Hall calls "excluder" any verb that is attributive (it comes before a noun), exclusive, and vague. Hall observes that they pervade the English language unnoticed. (He offers no evidence. Let us allow his observations nevertheless.)[11] He cites Russell's discussion of the question, what properties make an object real? Russell "tries to find [such] properties, not realizing" that the word "real" is an excluder. It is negative, Hall explains, in the sense that a real animal is not stuffed. He argues that Russell misuses ordinary English. This is an extraordinary affront;[12] it gives me real pain. The noun "pain" as it is just used here is not vague, so that the adjective "real" in the last sentence is (by definition) no excluder. Hence, the adjective "real" is not always an excluder. Hence, Hall's criticism of Russell is incomplete. Russell's criticism of the ordinary-language philosophy thus still awaited an answer.

In some contexts, some verbs are excluders; this is the Wittgenstein-Kripke rule-following paradox: no amount of information about a given conduct yields the conclusion that the conduct is following a rule. (This is a variant of the problem of induction; Adler 1976, 216.) This is puzzling, since ordinary-language philosophy appeals to ordinary intuitions. Having made such song-and-dance about ousting the psychology of meaning from the discussion about meaning, how was analysis stuck

---

its conclusions rest? If we want to learn about the world, no strictly linguistic evidence will be at all conclusive'." Hallett suggests here the following idea that it seems he deems new. As the question "what we should say" in part depends on language, he asks, what part? Russell 1949 that Hallett maligns considered the inability to answer this question an obvious limitation of human thinking. Hallett's new observation appears traditionally as the observation that all observation-reports are theory-laden; it is only four centuries old.

[10]Hall 1959; Caton 1963, 66–73.

[11]Gellner [1959] 2005, 252, 265 says, members of the analytic school habitually feign observations. See also *op. cit.*, 56, 69, 76, 228, 240, 242, 302, 319, 329.

[12]Hallett 2008, 49 writes casually about "Carnap's limited awareness of everyday expressions and their everyday employment (not only in English but also in German)." He also takes it for granted that young Wittgenstein offered a metaphysics with no reference to Wittgenstein's insistence to the contrary.

with (ordinary) intuitions all over again? How does intuition enter the analytic discussion in the first place (be it ordinary or extraordinary)?

Ordinariness comes here to exclude technical uses as permissible and yet metaphysics as impermissible. The very same terms, even the very same sentences, which, when used by philosophers incur Wittgenstein's wrath, he emphatically declared quite innocuous when used ordinarily — as the commendable return of words to their ordinary use! This includes their everyday use and their use in science; [13] it excludes metaphysics.[14] How then does intuition enter this discussion? Cavell (1979, Introduction) has come up with a brilliant answer: Russell's discussion of the views of the language analysts about scientific issues, he said, is irrelevant, since analytic philosophy applies intuition only to grammar, not to content. Now in Wittgenstein texts the function of grammar is to disqualify philosophical utterances as ungrammatical. This is far from being intuitive, or else Moore would not reject it. Moreover, since the same expression whose use Wittgenstein proscribes when used philosophically he allows when used otherwise, grammar alone does not suffice as a tool for locating the culprit expression. Hence, Cavell is in error. How then is metaphysics recognizable? The answer is not in Wittgenstein's writings; it stands out distinctly for anyone who in the 'fifties moved in analytic circles, especially in Oxbridge: whatever sounds odd is suspect and is probably unacceptable. Odd.

Obviously, that solution did not satisfy everyone. The search for an adequate answer to Russell's challenge continued. The latest attempt is Hallett's 2008 *Linguistic Philosophy: The Central Story*. It answers many charges against the cult of common usage. Hallett's answer falls back on Hall's mode of argument, one that he uses repeatedly but with no reference to Hall. Russell notes that the speech of Oxbridge fellows is extraordinary; Hallett observes (2008, 93) that the opposition of Oxbridge philosophers is not to street English ("Russell confused their stance with reverence for common sense") but to specific [metaphysical] expressions of Russell. Like Hall before him, Hallett finds the expression of intent of Oxbridge philosophers sufficient to rebut Russell. He too does not trouble himself to ask, was the intent met successfully? This is odd, since one task of critics is to point at the failure to achieve the expected outcome. Ordinary-language philosophers do not aim at doing useless work; Russell's criticism is that their output is useless despite their good will. Hall suggests that reasserting the intent of ordinary-language philosophers to be useful

[13] A large portion of *Philosophical Investigations* is devoted to examples of terms that are specifically applicable comfortably and that are inapplicable or problematic as universal terms. Consider the first example there. It is of St Augustine *à la* Wittgenstein, who knew what is the time but not what is time. This is fine; it does not do the job Wittgenstein assigned it, though: the admittedly metaphysical question, what is time, is legitimate by his own rules, as we all use freely the metaphysical concept: "Time and tide wait for no man".

[14] Analytic philosophers have observed repeatedly that theology is neither scientific nor commonsensical. This claim is an appeal to the demarcation of science as empirically verified. Thus, the ordinary-language school was in error as it disassociated itself from the "logical" positivist school (the "Vienna Circle"), since it shared its verification theory of meaning. See Wisdom 1958.

suffices as a response to the claim that their output is useless. Someone should tell him that the better response is to point at the useful output of ordinary-language philosophers and at its usefulness.

Hallett has a new idea about commonsense (*op. cit.*, 94). When people wish to have a debate about some things, he observes, they have to agree that they talk about the same things. Ostensive definitions do not suffice to determine this: the parties to the debate have to agree about some characteristics of the things they have the debate about, and these characteristics serve as additional pointers. They need not agree about all of the characteristics of the things that they wish to discuss in order to know that they discuss the same things; otherwise, they will not be able to disagree about them.[15] This is a complex interplay between medium and message: some information serve as linguistic markers. Discussants typically fix their gaze on messages, Hallett suggests, since the medium, the mother tongue, is taken for granted as a second nature. It is not surprising, then, that critics of commonsense beliefs should frequently fail to discern which aspect of the beliefs they are targeting – the factual (to be debated) or the linguistic (used as markers). Ordinary language philosophy clarifies such confusions, Hallett says.

This is an observation that confusion is abundant and an explanation for that observation. Now, since some speech is systematically clearer than other (for example, the writings of Russell as compared with those of Wittgenstein), the question that Hallett raises is, how can we improve our communications? In some situations, ambiguities can cost lives. In such situations, speech patterns are set carefully in accord with simple commonsensical procedures. Hallett's explanation does not fit these procedures, as they show concern not with markers but with ambiguities.

What helps smoothen communication is shared concern rather than shared opinions, the wish to solve a problem is the best marker and Hallet ignores it altogether. Nevertheless, he is right: all authors on language admit that language is limited and hence so is thinking. Yet we keep pushing this limit; Wittgenstein and his crowd have not yet noticed this fact: analytic philosophy is inherently a-historical. Hallett stands out here. He deserves congratulations for his having opened his book by citing Russell's introduction to Gellner's book, as well as Gellner himself, to say this. He answers them (rather inadequately, but never mind that). Language progresses by developing new ideas and new avenues for their communication (Pole 1958). In the process, some odd ideas appear; some of them stay. The most bizarre idea that appeared in England in the wake of the publication of Wittgenstein's *Philosophical Investigations* was the idea that calling an expression odd was the way to dismiss it, worse than calling it stupid (Joad 1950, 10). There was no rule for distinguishing oddity: if a bunch of Oxbridge philosophers agreed that an expression is odd, then odd it was and so it was condemned with no right of appeal.

[15]Carnap made this mistake in his *Aufbau*: he rendered all sentences tautologies by his very rule of identifying every object by the full sets of its characteristics: It turns all identifications into definitions.

12.3 The Limits of Language

The idea that what sounds odd is suspect prevailed in analytic circles for decades. It exited as stealthily as it entered. It seems to be associated less with Wittgenstein than with Moore. He objected, for example, to the inference whose conclusion is identical with its premise (a; therefore, a.), since it is odd sounding. It is, indeed. So is the innovation that zero is a number, observed Popper; so is the empty class. Gellner asked, does all this deserve the investment of much intellectual effort? He described the activities of Oxbridge philosophers at the time in detail and he deemed it by and large a waste of time. He had little patience with the concern shown in Oxbridge at the time around the oddity of the truth of any conditional with a false antecedent (a contrary-to-fact conditional or a counter-factual conditional). It is, indeed, odd that "If today is Tuesday, then I am a king" is true six days of every week. Yet the oddity is a matter of a situation, not of grammar, observed Popper in his logic lectures at the time: where the throne is rotational, some individual may utter this sentence all week long, as a matter of course, with no oddity, and in truth. The oddity is not always due to violation of the presuppositions that ordinary language embeds, whatever these are. What sounds odd may depend on circumstances more than on expression, observes Paul Ziff: it sounds odd indeed when upon a casual street encounter of friends, one of them says to the other, "pass me the salt".[16] This observation of Ziff is odd, except as a very strong critique of the way many a philosophy professor practiced their craft then, especially in Oxbridge. Incidentally, it hardly applies to Wittgenstein. Indeed, in the last days of his life, writing his *On Certainty*, he performed exactly such exercises as Ziff did, and even more bizarre ones. Moore ended his famous "The Refutation of Idealism" by expressing certainty about the identity of his hands as parts of the external world. This would be natural, were one's hands cut off and mingled with other people's severed hands, comments Wittgenstein, and were one asked to identify one's own hands and make sure of one's identification.[17] I do not know why Wittgenstein took recourse to such a gruesome case: speaking of photos of hands would be no less convincing. Moreover, he could imagine all sorts of situations that would make it reasonable to utter the expression in question, such as the one depicted in the folk song "*Alouette, gentille Alouette*" which is supposed to teach children the names of parts of the body. Why did Wittgenstein proscribe proving the truth of realism (*PI*, §402) like young Moore? Because he wanted realism to be taken for granted, not as a conclusion of a debate!

The writings of mature Wittgenstein include many instances of the rejection of odd assertions as odd. He was not very consistent in his effort to distance himself from Moore's arguments from commonsense and stick to the claim that

[16]Ziff 1960, 28. Anscombe 1963, 293 review of Ziff's book considers it a challenge to do more analysis. This response applies to any criticism, of course.

[17]This echoes a real incident in Congo Free State that served Joseph Conrad for his "Heart of Darkness". Wittgenstein's failure to see Moore's point is not due to his anti-metaphysics alone. The need to explain Moore is more general, as Morris and Preti 2015 and Preti 2017 have discussed it in great detail.

philosophical utterances are all ungrammatical: he endorsed Moore's advocacy of healthy common sense. What however is common sense? Both Moore and Wittgenstein considered sensationalism commonsense; obviously, it is not. Another example is so famous it has a name: Moore's paradox. It is not. It is an absurdity, namely, an incredible assertion. Whatever assertion is too incredible, Moore declared false. The example in question is "p and I do not believe that p" where "p" is shorthand for a sentence. Now it matters little whether Moore was right here or not. Yet, in principle, whatever sentence is strange sounding, observed Popper, in some appropriate strange situation will sound quite apt. In the case of Moore's paradox, consider "p" to be a description of an utterly unbelievable statement, such as that the prophet Elisha brought a dead boy back to life. A religious person may say in truth, "The Bible describes it, and so it must be true, but I cannot possibly believe it". From this statement, the statement called Moore's paradox follows.

Another example is the sentence "I am asleep now". Moore objected to it because of the inability to say it in truth. In some cases, it is true: somnambulists learn to say this in their sleep. Even the sentence "I am dead" or "I am dead now" or "now I am dead" is not as contrary to commonsense as it sounds. Even if in this discussion we exclude ghost stories (as belonging to another universe of discourse), this is no license for excluding it from last wills and testaments; and these can, and at times do, include this very sentence.

Consider a paragraph in Wittgenstein's *Philosophical Investigations* of which he was particularly proud. In it, he said (*PI,* Part II, xi, page 222, lines 6–9), a "whole cloud of philosophy condenses into a drop of grammar". It is his claim that he had refuted the assertion that I know only what I think and experience. He said, "I know what you think" is grammatical (and at times true), but "I know what I think" is not. I beg to differ. "I know what I think", is grammatical. Substitutions between first and second person singular is always grammatically permissible. It is also easy to conjure a situation in which it is required. Upon hearing "I know what you think, you think that ...", one might rightly (and angrily) respond with, "Do not tell me what I think; I *know* what I think."

The trouble with this discussion is not that it refutes Moore and Wittgenstein; this means little, considering that Moore clung to commonsense in the sense that takes care of my criticism, and that Wittgenstein's *Philosophical Investigations* is an unfinished posthumous publication. This does not hold for Malcolm, however. He has praised this passage, claiming that this passage is a clear example of the book's marvelous combination of poetry and precision.[18] I doubt Wittgenstein would have approved of this sad display of slavish defensiveness.

Did Wittgenstein consider opinions that common usage insinuates? Did he deem them authoritative? Did he consider the ordinary intuitions behind them a supreme

---

[18]Fann 1967, 47–8, 72. See also there, page 6 for more references. Monk 2005, 124 admired Wittgenstein's style. Remarkably, Wittgenstein could view his *Tractatus Logico-Philosophicus* as poetry; Frege viewed this (not as a compliment but) as a defect.

court?[19] If he did this, then he did it as his adamant refusal to recognize metaphysical expressions as legitimate: he repeatedly denied that he had a theory, and, in his defiance, he did not even consider the possibility that he held a metaphysical theory. Against this Gellner stated ([1959] 2005, last sentence), "That which one would insinuate, thereof one should speak."

One question hovers over all of Wittgenstein's writings: what was his theory of rationality? Did he assume that we have to justify what we say or is Rorty right in having denied this? Rorty denied that Wittgenstein's concern with certainty had to do with justifications of opinions. He offered an alternative explanation for Wittgenstein's interest in certainty or in my ability to tell that you are in pain. Haller disagreed with Rorty and declared (1988, 108) that in Wittgenstein's opinion "the justification of knowledge does not ... lose its foundation" despite his (Wittgenstein's) anti-metaphysics: a metaphysics-free foundation is conceivable. To seek a Wittgenstein text to decide between Rorty and Haller is futile. It is better to ask, why did Wittgenstein wish to avoid theorizing? He often agreed with Moore, and common sense was for Moore a fund of generalizations.[20] Taking meaning to be ordinary use and ordinary use to be subject to ordinary intuitions already suggests that Wittgenstein did endorse the judgment of common sense as authoritative, as foundational – at least as preventing unacceptable deviations from grammar – and so as a fund of generalizations. Does that mean that he took commonsense (only about grammar?) as the required foundation? For an answer to this, evidence requires examination – both of Wittgenstein's texts and of texts that display his influence.[21]

On this, evidence goes both ways. Wittgenstein said he only excluded violation of grammar and of usage, not corrections of factual errors, as these belong to science. He repeatedly exhorted his disciples not to take the opposite side when arguing against metaphysicians: to disagree with them is to declare their views false and so meaningful! Will analysis destroy all metaphysics?[22] Russell said, no. Wittgenstein said, yes. On what ground? Is it possible to produce a debate on this disagreement? Will it not be metaphysical?

What made Wittgenstein hostile to metaphysics? This is an irrelevant question: we should ignore psychological conjectures, especially if they refer to the possibility of dogmatism. Nor is there any need for such a conjecture: the hostility is traditional (ever since Bacon's methodology became popular) and Wittgenstein tried to support

---

[19]Kripke 1972, 138 said we have no choice, even though we know that some marginal commonsense opinions are false (to use his example, "the whale is a fish"). Maxwell said, today's science is tomorrow's commonsense. Russell was more cautions, saying, when commonsense heeds science, it improves.

[20]Duncan 2007, 54: "Although heavily influenced by Moore, Wittgenstein ... was generally contemptuous of Moore's version of 'philosophy of commonsense'".

[21]Haller declared Rorty's view absurd, as it makes Wittgenstein a precursor of Popper or Feyerabend. This is a witty dismissal of Rorty's view as frivolous.

[22]Ayer 1985 conclusion declares Wittgenstein the greatest after Russell, since he was hostile to some metaphysics; Wittgenstein's hostility to all metaphysics Ayer dismisses. (Hallett 2008, 48, faults Carnap for confusing these two options.)

it by the use of the new logic. Two conjectures explain why Wittgenstein clung to his hostility to metaphysics, each sufficient and each in accord with all available information. One is that he feared that solipsist metaphysics has the upper hand, while (rightly) judging it intolerable. (So he interpreted it – *TLP, §§5.62, 5.64* – as other than ontological.) The second conjecture is Russell's: Wittgenstein was a mystic and lost interest in logic. Both may be true.

Perhaps here Engelmann's *Letters from Ludwig Wittgenstein* is helpful. Engelmann cited the following assertion from a letter of Wittgenstein:

> My work consists of two parts: the one presented here plus all that I have not written. And it is precisely this second one that is the important one.[23]

This is the standard claim of mystic teaches that they cannot say their message, but only point the way to it. Russell (1914, last three paragraphs of Chapter 1) dismisses mysticism as both trite and useless. Carnap allowed it as it is (traditionally) outside language, in accord with Wittgenstein's edict. His views still demand critical assessment.

12.4 Wittgenstein's Analytic Approach

The analytic approach requires a destructive analysis of all metaphysics. Perhaps this requires a similar constructive analysis of some science too – in order to purge it of metaphysics.[24] Perhaps this requires a similar analysis of analysis, in order to find whether the program of removing confusion is better performed wholesale (Wittgenstein) or piecemeal upon demand (Popper). Before that, the confusion over the authority of commonsense requires an analysis and a demolition job.

The Sapir-Whorf hypothesis is, languages are theory-laden in different ways; hence, commonsense is different in different societies. Now some Wittgenstein devotees tried to argue against this. Black argued (1962, 41) against it in his interesting study of metaphors. His study has justly become a classic, yet the critique of the Sapir-Whorf hypothesis that is its core is weak. It did not change its being the received opinion among linguists. The demand for justification has rendered the Sapir-Whorf hypothesis relativist: different societies use different frameworks, and justifying them all makes them all true despite the difference between them. Some philosophers of the social sciences, notably Peter Winch (1958), suggested that Wittgenstein was a relativist. Not all Wittgenstein's fans agree, possibly due to severe criticism of Winch by I. C. Jarvie (1972, 44). Alternatively, ordinary language

[23]Janik and Toulmin 1973, 192 also quote this assertion of Wittgenstein.

[24]This was the program of Heinrich Hertz and of Ernst Mach. Popper noted that any testable theory entails untestable ones. Instead, Popper 1959 §20 demanded (like Peirce before him) to choose the most testable theory first: this renders it preferable to omit from a theory under examination all metaphysical ballast.

philosophy supports current commonsense – in ethnocentric neglect of other tongues (Markus 2012, 22) and in a conservative mood (Pole 1958, 80–2, 96).

Wittgenstein knew of course that ordinary language reeks with metaphysics, theology and myth. He rightly ignored it (*PI*, §426):

> Here ... we get the same thing as in set theory:[25] the form of expression we use seems to have been designed for a god, who knows what we cannot know; he sees the whole of each of those infinite series and he sees into human consciousness. For us, of course, these forms of expressions are like ornaments which ... we cannot do much with ...... we go by side-roads; though we see the straight highway before us, of course we cannot use it as it is permanently closed.

This is an allusion to Kafka. It throws a strange light on mature Wittgenstein: the wish to transcend the limits of language is strong and painful. This imposes a Kafkaesque reading on what Wittgenstein says elsewhere in it as it deviates from his critique of essentialism and of theology (*PI,* §371):

> essence is expressed by grammar

and thus (*PI,* §373)

> Grammar tells us what kind of object anything is. (Theology as Grammar.)[26]

Scholars have failed to make sense of these observations. It is sensible to ignore the exegetic literature that fails to clarify this passage but conceals its failure. Still, let me notice one obvious example: *Wittgenstein Bibliographic Guide* (Frongia and McGuinness 1990, 6):

> the apparently unsystematic character of his arguments often end up by giving the impression of impenetrable obscurity.

Do not worry; the "impenetrable obscurity" is a mere impression. Now texts are difficult to read as the ideas in them are hard to comprehend or as they are opaque. Opacity may find elucidation or not. If not, then it raises suspicion: the situation may involve a foul play. In that case, only elucidation may alleviate this suspicion. The art of rendering opaque texts lucid differs from the art of facilitating reading difficult texts (Zemach 1989, 423). Mature Wittgenstein texts are at times lucid and at times opaque, seldom difficult to comprehend,[27] no matter what the editors of the *Wittgenstein Bibliographic Guide* say.

Consider texts that are difficult to comprehend. If they matter, then some commentators try to elucidate them. They do so by unpacking the content of the difficult passages. An example of unpacking is the witty example of Ayer (1936, Introduction) for a meaningless pseudo-sentence: "the Absolute is lazy." The Absolute is

[25]*PI,* § 426. Wittgenstein refers here to the most basic idea of standard abstract set theory, the one-to-one correlation between infinite sets, where a mere rule for it plays the role of its execution. This is astute.

[26]This passage, they say, alludes to a (vague) remark by Martin Luther. This is of no useful consequence.

[27]The ordinary use of the attribution of difficulty to comprehend is often the expression of inability to consider false a seemingly false statement of a bigwig plus the inability to reinterpret it as true.

Hegel's name for God; Aristotle said, as God is perfect, He does not move: He is the unmoved mover. Ayer used for his joke two allusions to unpopular famous philosophers. The joke is serious, since the meaning of the word "perfect" is highly problematic and traditionally much unpacked to no avail: in the wake of Wittgenstein we may say, it is unproblematic in context and paradoxical otherwise. It follows that it has either no need or no possibility of unpacking. Hence, efforts to unpack it are useless. This is Wittgenstein's standard message.

The way Wittgenstein treated religious expressions embedded in ordinary language is an example of analysis. In an Introduction to an intended book he said (1975, Preface), he wanted to say, "This book is written to the glory of God", but, fearing misunderstanding, he settled for its cognate, "This book is written in good will". This is moving; nevertheless, most people will deny his suggestion that these two expressions are synonymous. Russell (1905, 489) had stressed that his analysis did not transcend ordinary understanding. An assertion like "I thought your yacht was larger than it is", he said, is figurative: people analyze it without notice when they hear it! This is a humble sort of analysis, and Russell presents it as obviously kosher. It should be all the more so when forms of life enter the picture. The researches of Wittgenstein's disciples are (at least are supposed to be) in the wake of Wittgenstein.[28] They may ascribe him a tacit metaphysics; as long as the metaphysics they ascribe to him is there by mere implication, as not explicitly assertable, he might have allowed it. He would have approved even of some explicit version of it as long as the one who asserts it withdraws it at once. A few of his fans, Austin in particular, did so habitually. Naturally, little of this remains in print. This absence of printed records repeatedly led to expression of surprise and disappointment. This Warnock's obituary of Austin (2013, 16–17) asserts – also indirectly, of course. This agrees with the possibly true piece of gossip according to which Austin was supposed to act as the rightful heir of Wittgenstein and Warnock as the rightful heir of Austin. [29] No objection.

---

[28]Hintikka appeared as a follower of Wittgenstein. He never specified. He assumed that meanings imply possibilities. The sense of a word is then a class of all of its possible references. Models and modalities thus equal meanings and theories. Sense and reference of the morning star differ, as the two do not coincide in all possible worlds. The set of all possible empirical tests serve as filters that aim at narrowing down the set of possible worlds to one. This is not far from critical rationalism; his view of science as games against nature is more so. It is still unfeasible, as it tacitly equates meaning with science. Carnap and Hempel proved this long ago, as they tried to refute Popper's theory: to that end, they assumed tacitly that Popper had endorsed this equation. They never recognized his protest. This is cheap.

[29]The opinion reflected in Warnock's obituary is quite the contrary: it is a disclaimer: Austin was in no way under Wittgenstein's influence and admiration for ordinary language did not sway him. Warnock's evidence goes the other way, especially in his discussion of the problems that Austin was facing. Austin's 1956 Presidential Address to the Aristotelian Society ("A Plea for Excuses") excited the London philosophical community. It presented a task (that he called a problem): to offer an alternative to Russell's theory of definite description closer to ordinary language. This refutes Warnock's characterization of Austin's attitude.

The initiator of the theory of meaning as use was Norwegian Arne Naess, at the time an authoritative Wittgenstein follower. He investigated in the early period ordinary parlance by means of questionnaires. Unfortunately, he took the criticism that he met not as incentive for a retry but as a proof of futility. He left the Wittgenstein circle. The tenacity of the Wittgenstein ethos contrasts sharply with the cavalier abandonment of empirical projects related to it. Later on, a critic of the Wittgenstein methods, Benson Mates, demanded that scientific research replace the appeal to intuition. Rather than claim that the Wittgenstein-style method is not designed to extract metaphysical views out of the use of language, but, on the contrary, to eliminate them, the analytic establishment defended intuition. The dispute fizzled out with the demise of the verification theory of meaning. Wittgenstein is supposed to have rejected it soon after he returned to philosophy after his self-imposed exile, before he moved from his early to his later views. Haller (1988, 16) reports that

> in a conversation of ... 1929 ... Wittgenstein makes some comments regarding his own view concerning verification, saying, 'no sentence is completely verifiable'.

We learn that a mere hypothesis may have value even if it is meaningless: we may consider it a rule.[30] This may become a way to admit metaphysics, at least as rules or as a given framework; Wittgenstein's forms of life then become such frameworks.[31] Naess left the Wittgenstein school of thought.

The central question for anyone who seriously wishes to make further sense of Wittgenstein's writings concerns the possibility of keeping his position as a significant thinker while ignoring his anti-metaphysics stance despite its centrality in all of his posthumous writings. Consulting his output is of little help. Commentators are of less help, as they are often eager to read their own ideas into his writings rather than make sense of his – except for Russell, who declared Wittgenstein a mystic and left it at that and except for Popper who saw Wittgenstein's output as a vain elaboration on valuable ideas in Russell's style. Whatever the right idea is, however, clearly, Wittgenstein did suggest – already in his first book – that he did not advocate the total neglect of metaphysics. He kept his rejection of all metaphysical assertions; yet he dropped his rejection of rules (as a part of his admission that the logic of propositions is too narrow) and he reinserted metaphysics not as assertions (explicit or unassertable) but as rules where some rules generate language games. This looks

[30] *Ibid.* 17; see also *ibid.* 103, where Wittgenstein's last work seems to tolerate hypotheses – on the ground of a "*Weltbild*", a world-image, to wit, a metaphysics, no less. See also *ibid.*, 109. Stegmüller 2012, 10, 106 legitimized unverified hypotheses by reading them as expressions of hope. See Agassi and Wettersten 1980.

[31] Ironically the idea that the meaningless may be rescued as a rule appeared in efforts to save from the proverbial dump-heap not metaphysics but science. This was a move initiated by Mill in the pre-history of analytic philosophy, revived by Schlick and adopted by Toulmin. Russell had presented the metaphysical theory of simplicity as a rule: seek the simplest hypothesis. Both Wittgenstein and Popper followed suit.

daring, but it amounts to no more than the observation that prescriptions are not descriptions, though their rationale is describable.

What then could language analysts use to demarcate the rules that replace the metaphysics that stems from the rules that are moral or logical? Take the easier problem first: what demarcates science from metaphysics? Bernard Williams, we remember, had an answer to this question: were there any metaphysical statements, he said, they would constitute synthetic *a priori* knowledge, which is impossible. This is, again, the theory of meaning as [complete] verification smuggled through the backdoor, or else, since "no sentence is completely verifiable", it would be possible to reinstate metaphysics as sets of hypothetical statements, not as knowledge, and so not as *a priori* knowledge.

Except for Haller, no language analyst discusses explicitly the problem of demarcation of science from metaphysics.[32] It is therefore scarcely reasonable to find an answer to it in the secondary Wittgenstein literature.[33] Yet in a true analytic spirit, there is a partial answer here. Some metaphysical assertions are not kosher, and there is no evidence with which to spot them and only grammar helps do this. This is a reasonable suggestion. It has run its course.

Wittgenstein's *On Certainty* is his last piece of writing. It is a comment on Moore's defense of common sense. In Moore's view, commonsense knowledge largely consists of certain knowledge – especially of the existence of material things, of the mind, and of this knowledge. Moore did not have the right to assert these views as commonsense. By commonsense it is silly to assume that solipsism is true, but commonsense does not declare solipsism refuted for sure: "In this world nothing can be said to be certain, except death and taxes." Wittgenstein sought a better foundation for commonsense than Moore. His first move was in line with his general strategy: whereas Moore disagreed with skepticism, Wittgenstein said, when countering that view one should not assert the opposite. That is to say, skepticism is not false but meaningless. Why? What grammatical error does the skeptic commit? Wittgenstein's answer is thin: the skeptics support their view, he said, by keeping some options open that ordinary use precludes. Wittgenstein supplemented this with the idea that for discussing all this we need language and with it society and social institutions. This is a sufficiently strong argument against the kind of skepticism that invites solipsism, and the very existence of language suffices to refute solipsism. Fine. Is it sufficient reason to claim that the endorsement of skepticism is a violation of grammar? How do we understand it if it violates grammar? Perhaps since it is an effort to transcend language our understanding of it is partial. Even so, transcending

---

[32]Only Rorty 2010, 135 noted that Wittgenstein had shunned the problem of demarcation.

[33]Remarkably, Warnock 2013, 5–7 attributed to Austin a demarcation of philosophical problems; in a true Austin style he took it back (2013, 5, note 1). What is the metaphysics to exorcise is unclear. Yet the difference between complete and partial verification is clear enough to avoid with ease (McGuinness 1985).

language is not necessarily violating grammar. Notoriously, Wittgenstein used the words "grammar" and "logic" loosely.[34]

Suppose the rules of analysis are the rules of logic alone. (The alternative is the view that grammar and logic overlap.) Did Wittgenstein suggest that the grammar of ordinary discourse, the grammar of natural languages, includes general rules in addition to logic? The grammars of natural languages obviously differ, and so they are not merely logic. Frege, Russell and young Wittgenstein assumed that logic is the rules that all languages share, that logic is the essence of grammar. This assumption did not survive scrutiny.[35] Still, there is a lacuna here: without the idea of the ideal language, we may claim this. As logic includes rules that all language share, a language that includes no other rules is quite possible. It is still not the ideal language, since there is also the meta-language. Wittgenstein's refusal to entertain this and his refusal to admit the meta-language leaves a lacuna in his philosophy, but this is no criticism. Still, with this lacuna he had no hope to develop a theory that bans all metaphysics in all possible languages, except by showing that each metaphysics is illegitimate by the rules of logic strictly understood. This renders objectionable the liberty that Wittgenstein took with the word "logic".[36]

To comprehend Wittgenstein better, a framework was invited, preferably his worldview. *Wittgenstein's Vienna by* Janik and Toulmin is a famous example. The concern of this book is regrettably only Wittgenstein's ambience. An example (Wittgenstein 1980a, b, 6–7) is an oft-quoted remark he wrote in 1930 in an Introduction to a book he never published:

> I have no sympathy for the current European civilization and do not understand its goals, if it has any.

This quote is enigmatic. Many expressions of Wittgenstein are present in the literature about his moral stature, about his self-doubts, even about his enthusiasm for Schopenhauer, Spengler and even Otto Weininger. I hope that all this is insignificant. More insightful is the remark that he rarely spoke of politics and that when he did, he presented *naïve* conservative views. It scarcely tells us of his attitudes to science. One such expression is "You can fight, hope and even believe without believing scientifically". It may be in spirit anti-scientific but it may just as well describe uncontroversial facts. In 1930, he spoke against the idea of progress; by contrast, the "Vienna Circle" took their faith in progress for granted. Since then, the pro-progress Popper exposed progressivism as metaphysical and rejected it. Young

---

[34]David Pole 1958, 68–9 rightly dismissed Wittgenstein's assertion that privileged access to sensations is a matter of grammar.

[35]Chomsky postulated a universal grammar: generative grammar generates all specific natural grammars. He never explained this and he never elaborated beyond offering some dazzling instances that possibly refute some possible alternative theories.

[36]Jorge Luis Borges describes a language (Tlon) that is inherently idealistic; he could not say how. Can such a language exist? Wittgenstein had to answer in the negative. He did not explain.

Wittgenstein's anti-metaphysics was obviously traditional pro-science *cum* anti-metaphysics. As to his mature philosophy, some see it as anti-metaphysics pro-science; others see it as rather indifferent to science. This is far-fetched: the attacks on prejudices that fill Wittgenstein's writings to the end bespeak decidedly traditional anti-metaphysics that is pro-science (*op. cit.*, 7). He seems obviously inconsistent on this; for example,

> the place I really have to get to is a place I must already be at now. Anything that I might reach by climbing a ladder does not interest me.

As the evidence in this dispute, however, it is weak: it may be a passing thought and it may present a personal predilection that philosophically explains his concentration on therapy and no more. It resembles some Zen Buddhist idea that is obviously independent of any view about science. The manuscripts that circulated among close followers of Wittgenstein soon after he returned to Cambridge (1958, 18) seem too exploratory to qualify as firm evidence. As their target may very well have been Russell who had just brought him back to Cambridge, it looks not too pleasant:[37]

> Our craving for generality has [as one] source ... our preoccupation with the method of science. I mean the method of reducing the explanation of natural phenomena to the smallest possible number of primitive natural laws; and, in mathematics, of unifying the treatment of different topics by using a generalization. Philosophers [read: Russell[38]] constantly see the method of science before their eyes, and are irresistibly tempted to ask and answer in the way science does. This tendency is the real source of metaphysics, and leads the philosopher into complete darkness.

Finally, Wittgenstein showed great concern for resemblances, even for the vague ones that he viewed reasonably as resembling family resemblances. This concept has acquired much commentary, despite the fact that using it he meant to draw attention to the fact that we often observe resemblances without being able to put our fingers on them. This appears also, and frequently, in pseudo-science; it may display either an anti-science attitude or a *naïve* view of science. In this, his display of interest in resemblances resembles his display of interest in physiognomy (*PI*, §§285, 536–7, 539, 568, 583; pp. 178, 181, 210–11, 218) that is decidedly a pseudo-science. His

---

[37]*Blue Book*, 17. Dummett 1960 (his review of Gellner's critique of analytic philosophy) opens with observing that both Oxford and Wittgenstein owe Russell a grudge. This is the insinuation that Russell's support of Gellner is a retaliation of sorts. This is the worst insult to Russell. Dummett 1978 ix–xiii, in which he discussed his own review, seems an effort to make amends. For, it displays marked respect for Russell. As implicit withdrawals are confusing, it is hard for me to ascribe this conduct to Dummett. Gellner's book, he concludes his initial review, "does not even have the smell of honest or seriously intentioned work" as he does not distinguish between Wittgenstein's two schools. He does not say there why this is so serious an error. In the 1978 addendum to it, he adds some agreement with some of Gellner's criticism of the analytic school. This exposes his initial review as a partisan hatchet job: as not very candid.

[38]Kenny 2006, xiii says, this text expresses anti-scientism. The project that Russell was working on with Wittgenstein's help was the attempt to render philosophy scientific; this is how Kenny presents scientism.

view of science was perhaps a jot too *naïve*, but it was staunchly favorable. In his youth, he went much further, as he forcefully expressed scientism: all that is sayable is sayable scientifically; about the rest, one should be silent. In his later works, he expressed a distaste for scientism that some of its devotees (erroneously but understandably) took to be hostility to science. All this said, it might be fair to admit that Wittgenstein's conservative political attitude prevented him from sharing with others (say, Russell and Popper) their enthusiasm for scientific and technological progress.

12.5 The Limits of Language and of Thought

A passage of Wittgenstein that expresses an idea that divides admirers from others deserves citing again:

> My work consists of two parts: the one presented here plus all that I have not written. And it is precisely this second one that is the important one.

It is this kind of writing that made Russell distance himself from his beloved Wittgenstein. Taking it seriously may tempt one to try to articulate the ideas to which Wittgenstein alludes here. This would be a serious error, since Wittgenstein declared not articulable what he found important and yet did not try to articulate. This is his idea about the limits of language that he was proud of as his discovery (PI, §119).[39] Clearly, you cannot state a limit here (*PI,* §145). To describe limits, one needs to describe its two sides (*TLP,* §6.4311). Now, obviously, everyone says the unsayable equally; what Wittgenstein accused slum landlords (such as Plato and Aristotle!) is that they pretended to have managed to say the unsayable. How then could Wittgenstein call his own this common characteristic of not saying the unsayable? "My work" consists of both the sayable and the unsayable, he said, ignoring the view of metaphysics as unsayable being ubiquitous. How are we to understand his claim? The diverse mystic teacher hint at the unsayable that they want to say but cannot. This, said Russell, exempts them from critical scrutiny. Wittgenstein did not wish to escape scrutiny; his aim was to avoid and prevent the kind of self-delusion that engagement in metaphysics is (*TLP,* §5.5563; *PI,* §97).

Here, then, is my basic presentation and my basic criticism of the doctrine of the mature Wittgenstein.

As we know, both our language and our thinking are limited. Admittedly, some thinkers were optimists enough to deny this. Today we agree: we are intellectually inherently limited. What the study of language has shown is that our intellectual limitation is not removable because we use human language. This is common to all mystic thinkers; it is perhaps the single philosophical idea that is present in every

---

[39]It is hard to express the sense of puzzlement that this claim should provoke. Even on the assumption that Wittgenstein was indeed as ignorant as he systematically pretended to be, even then it is very hard to suggest that he had never bumped into the ubiquitous mystical claim that language is severely limited.

advanced culture, ease, west, north and south. This idea came up in a new context, in the philosophy of mathematics, when the idea of a formal language gelled and made it impossible to dismiss the paradoxes – semantical, set-theoretical or any other. It is very important, therefore, to notice that the idea of the formalization of language was not arbitrary. Diverse thinkers, Frege, Hilbert, Peano and Russell were interested in it because of problems within mathematics that required increased precision of wording to solve. Moreover, as Mach has asserted, as Wittgenstein did, and as Bar-Hillel discussed in detail, it is in formal systems that the paradoxes are troublesome. It was the urgent need of mathematics to get rid of paradoxes, as opponents Brouwer and Fraenkel agreed. This fact has raised a new kind of problems, philosophical problems that touched upon the limits of language. Wittgenstein stressed that language is limited, and declared that this shows the barrenness of metaphysics. We need not argue that throughout the history of western thought, metaphysical discourses have proven fruitful. Suffice it to argue that Wittgenstein could not know that metaphysics is barren, that he declared it as a dogma. Likewise, he was right in declaring that it robs some people of their peace of mind, but he was in serious error in denying that it brings some people blissful peace of mind.

We know more: we know that the limits of language and the limits of thought are not fixed. Wittgenstein erroneously said that it is: "we see the straight highway before us, but of course we cannot use it, because it is permanently closed", he wrote (*PI*, §426). When we compare our metaphysical views with those of our ancestors, we can see progress, however limited it is. Wittgenstein ignored all this as prior to the development of the idea of formal languages that puts it all in a new light. What the new light is precisely he could not say, much less why this new light makes even the small progress in metaphysics impossible. It is no accident that the friendly critics of Wittgenstein, such as Hintikka and Horwich consented to be, wanted to reduce the rank of metaphysicians that he condemned as slum landlords – from all to most.

Doubtless, the mystics of all countries are right: the limits of thinking is in part at least due to the limits of language. Even rationalist thinkers like Maimonides and like Russell agree on this. Yet we also know, *pace* Wittgenstein, that these two limits are different. Intellectual advancements that came within given verbal frameworks testify to that as much as the advancements of the verbal frameworks testify, even when these advancements led to intellectual ones. The recent advancements of the theories of language may be in part due to insights of Wittgenstein, and to the extent that this is so we are indebted to him, but they also render his views obsolete. Indeed, their indebtedness to him and their having rendered his doctrine obsolete are one and the same move.[40]

Admittedly, Wittgenstein was right when he observed (as Kant did before him) that knowledge of a limit includes knowledge of something about both sides of it. This does not mean that we never know anything about the limit of language. If

[40]This is a general methodological hypothesis that is derivative of Popper's solution to the problem of induction (Agassi 1975, Ch. 3).

we look at this limit historically and we discover that at some point the limit has moved forward, then we may be able to speak of the limit as it was before its move forward. We can then make a conjecture about today's limit. The ancient distinction between opinion and knowledge, between *doxa* and *episteme* is at the heart of the observation of Kant and of Wittgenstein. It is obsolete: the most advanced sciences extant make no claim to be *episteme* any more. This does not mean that all conjectures have the same scientific status. The question, which ones do, is the problem of the demarcation of scientific theories. We need not discuss it here beyond the observation that we no longer deem science *episteme*, so that we do not dismiss all conjectures with equal ease. Quite possibly, then, we will have diverse conjectures, good and not so good, about the limits of language and of science.

This may turn out to be an exciting project. It will be indebted to the inventors of mathematical logic but it will be free of some of their suppositions, including the supposition of Wittgenstein that "we see the straight highway before us, but of course we cannot use it, because it is permanently closed." Access to it need not be permanently shut, even though it will never be as free as the road to common sense. We should realize that the impression is misleading that to speculate and come up with all sorts of metaphysical opinions is all too easy: the adequacy *criteria* for metaphysical ideas should make this a serious venture. They will dismiss of course idealism of all sorts, especially the solipsism that troubled Russell and Wittgenstein so. Thus, the very distaste of (Russell and) Wittgenstein for metaphysics, poor as a rationale for the opposition to metaphysics that it is, signifies as a motive power for the removal of the conflict between metaphysics and common sense.[41] Once we hope to have metaphysics help research (Einstein), it will be more attractive and less accessible.

[41]We may remember that Berkeley declared his idealism commonsense. His evidence was from the claim that the alternative, materialism, is contrary to commonsense. Of course, he was in error as two wrongs to not make a right, but he presents the ideal as a commonsense metaphysics. This ideal is too good to be true: as Maxwell has noted, in time commonsense adjusts to scientific novelties.

Chapter 13
Conclusion: The Place of Wittgenstein

> *Philosophical Investigations* is thought by some to be the greatest philosophical work of the twentieth century. It shatters certain images of man that are both deeply embedded within the Western mind and utterly familiar in it. It profoundly alters one's conception of thought, consciousness, sensation, linguistic understanding and the self. It is probably the most powerfully disturbing work of philosophy to have been written since Hume's *Treatise*. It is among the intellectual monuments of the age (Ahmed 2010, 1).

This is the usual high praise for *Philosophical Investigations*. The praise rests on a fuzzy claim: the book "returns to the themes that had dominated Wittgenstein's *Tractatus Logico-Philosophicus*". It is the opening of an introduction to a book that serves as elaborations on this fuzzy claim. The elaboration is mostly further pretense that some trite comments on a trite or a puzzling quotation from Wittgenstein is profound.[1] (The exceptions go the other way all the way; Child 2011, 75.)

13.1 Things to Avoid

Reasons for the popularity of a thinker are often cheap. This is a burden. As poor reasoning is ubiquitous,[2] we may ignore it. Serious reasons for popularity should live up to some reasonable standards. Even the reasonably admissible works often invite

[1]Russell (1959a, 217) observed that Wittgenstein fans cite his platitudes as if they were discoveries. The view of (Moore and) Wittgenstein of self-identity as meaningless has generated many pointless publications, as did Wittgenstein's view of some obvious tautologies (like "two is a number") as meaningless. Unfortunately, the proliferation of Wittgenstein-style platitudes has spilled over to other fields, including neuroscience. See Bennett et al. 2007, for example the flat quote from Wittgenstein on page 19.

[2]Russell, "Faith and the Mountains". As Shaw has observed (1912, end of Preface), every successful movement suffers joiners from among the rabble and the mixed multitude. Their vulgar explanations for their having joined the bandwagon gives them away.

J. Agassi, *Ludwig Wittgenstein's Philosophical Investigations*, Synthese Library 401,
https://doi.org/10.1007/978-3-030-00117-9_13

reinterpretation (especially when they are seemingly inconsistent); their multitude prevents this. (See Arrington and Glock 1991, 1).

This is achievable with reasonable ease once we follow supreme rule of interpretation: avoid partisanship. The ideal is to achieve understanding of a text that does not divide followers from opponents: they should both agree that the text is significant and why. To this end, the preliminary rule is, do not waste too much time on worthless texts. In order to abide by this, interpreters should agree that we could agree on ideas that we find of too little value to spend much time to interpret, and that some texts we do value even though we do not endorse them. Thus, all modern economists value Adam Smith's *Wealth of Nations* highly, but no modern economist fully agree with him, not even his most ardent fans.

As to the reading of Wittgenstein, the first thing to avoid doing that is the choice of the reading that makes Wittgenstein right (Conant and Diamond 2004, 46, 81).[3] For example, we should not allow our reading of his texts depend on whether we take his therapeutic method to be applicable or not. Nor should we confuse our own views on clarity with those of Wittgenstein, young or mature. We should recognize the differences between his early and late philosophy as between the demand for utter clarity and the demand for reasonable clarity (*PI*, §107). We may disagree about the place of this distinction. Consider the view that the [utter] clarity of comprehension of a [scientific] thesis requires also knowledge of what are the [empirical] tests that may apply to it. This requirement amounts to what the members of the "Vienna Circle" called the verification principle. Since verification is impossible, utter clarity is also impossible. Yet the demand in question reappears in the analytic literature, for example in the writing of doyen analytic philosopher Dummett (Weiss 2014, 10, 96). Early Wittgenstein's assertion that whatever can be said can be said [utterly] clearly is patently false. Too many analytic philosophers falter here. Worse. Young Wittgenstein conflated meaning and comprehensibility; Dummett confuses them. Whereas comprehensibility improves, as he observes, when testability is apparent, clarity does not depend on experience so simply. Quine (1988, 117) says, "I see metaphysics, good and bad, as a continuation of science, good and bad, and grading off into meaninglessness." Whether he is right or not, he clearly speaks of meaning in the abstract, not of the comprehension that varies from one person to another.

It is easy to overlook the confusion that Dummett has displayed in his defensive attitude towards the ideas behind analytic philosophy, since the matter is highly sophisticated. Yet this sort of confusion deserves discussion, since it is very common. Confusing clarity with absolute clarity is a case of confusing a quality with its extreme version. Repeatedly analytic philosophers showed that the very good is possibly a sub-category of the good and possibly a different category; both senses of goodness are permissible, but confusing them may cause trouble. Carnap's theory of

[3]It would be easier to cope with the Wittgenstein commentaries had there been a rule for their authors to say outright whether they agree with Wittgenstein about the non-existence of philosophical problems.

explication is a tool for overcoming this kind of confusion. This theory handles qualities sufficiently well understood to require no analysis but still its diverse kind of representation are; so he split a quality (or a relation etc.) to three versions called classificatory, graded and quantitative. The classificatory quality will be positive or negative, say, hot or not-hot, namely cold, graded, say, warmer or colder, and quantitative, namely, temperature. He did not discuss the question whether hot and cold are different qualities or the one is the negative of the other, since this was a question for science to resolve, not for philosophy that only puts clearly what science has decided upon, and the modern theory says that cold is the negative of hot. Now Carnap offered what he called adequacy rules. Hot or cold are more or less than some determined temperature. Hotter is having higher temperature. Carnap forgot one more adequacy rule, the rule of *a fortiori*: what is hotter than hot is hot too. Unfortunately, his own major application of this tool was in his 1950 solution of the problem of induction by equating confirmation with probability *à la* Keynes. It was a dreadful mess, just because he confused confirmed with more confirmed, since the most probable is the tautology that is not open to any empirical confirmation by definition. (As a tautology is true in all possible worlds, its probability is highest and impossible to raise, so that the idea of confirmation as probability clashes with the idea that confirmation is evidence that raises probability.)

Analysis began with clarifying much more difficult confusions. This raised the question, is analysis limited to difficult questions? No, said Russell: people solve the easy ones, not noticing that they do so. His first, path-breaking example of language analysis, we remember, was, "I thought that your yacht is bigger than it is." This example is not obscure and not suggestive, but it requires analysis, easy though it is. Presumably, the same holds for the expression, "it is impossible to say that . . ." or "it is impossible to write that . . ." Reading such an expression, invites learning from its context what it means. At times it is, "it is impossible to say in truth that . . .", and at times it is, "it is impossible to answer such a question by saying something like . . ." In Wittgenstein's texts it means, "it is impossible to say meaningfully that . . ." where examples are seemingly meaningful statements like "there exist at least three churches".[4]

This shows that Wittgenstein used the word "meaning" not in the sense that Frege or Russell did, yet as alternatives to them. This is not quite kosher. The efforts to analyze puzzling texts to make them reasonable is not advisable unless one expects them to be enlightening in particular ways, and then if one publishes then one should show that this is worthwhile. To assume that it is because its author is Wittgenstein or some other great light is an error.

Gellner's 1959 *Words and Things* is the first serious, severe critique of the analytic school of thought. Yet it is partial: its condemnation is total. I hope that this concluding chapter will add to its judgment some praise of Wittgenstein and of his school, but only after paying Gellner the high tribute that he deserves. The first addition is the repeated recognition of Frege, Moore and Russell as members of that

[4]McGuinness, 1988, 291; Monk 2012, 182.

school and of their analytic contribution as serious. The second addition is the repeated recognition that the current analytic literature is mostly less serious than its predecessors. Yet recent contributions to it comprise a distinct improvement over the studies that Gellner has bulldozed. More significantly, the analytic literature is superior to the very best of the Continental or phenomenological literature (as it displays methodological essentialism with almost no exception). Only philosophers who face the metaphysical problems that engage some phenomenologists in the clear analytic style may count, and even in this trend, leading philosophers like Putnam are too confused. Contrary to the opinion received in the current analytic literature, neither Wittgenstein nor analytic philosophy in general offers much by the way of improving philosophy or the philosophy of life. Berlin said, mature Wittgenstein was more humane than young Wittgenstein was. Let Wittgenstein's fans argue in favour of Berlin's observation if they can; otherwise let them add it to Wittgenstein's philosophy. One way or another there remains the allegation that the attraction of Wittgenstein's philosophy is in his having turned a blind eye to both Auschwitz and Hiroshima. They tacitly use the pretext that these terrifying experiences are irrelevant to the lives of concerned citizens.

The view of Gellner's criticism as decisive invites the normal follow-up: what is the gain due to its presentation and criticism? Gellner said, in the present case little of value is accrued. This is a slight exaggeration.

There is the virtue and the narrowness of any philosophy that takes common sense for its guide. The virtue of common sense is obvious. First, it is problem oriented. The very objection of Wittgenstein to metaphysical problems forced his opponents to discuss their agenda. Although he raised too many new questions to take seriously, he thereby made philosophes rethink their agenda. Stroll (1994, 4) presents him as a great philosopher just because he has raised many new questions. He also had always new and fresh *aperçus*. Let me offer one instance. He said, logicism is suspect *ab initio*, since logic and mathematics look so different: logic is a theory of valid inferences and mathematics is the outcome of the use of logic. Yet Wittgenstein was in error here: the view that mathematics is a part of logic is the view that its statements are analytic. It is easy to remedy this: read Wittgenstein to say, the analytic statements of logic, such as the laws identity or of contradiction, differ from mathematical statements, such as the axioms of Euclidian geometry. Indeed, we can take theorems in a mathematical text and view some of them as logical, such as that of the reflexivity and transitivity of the laws of equality, and other theorems as mathematical, such as Pythagoras' theorem. More generally, the logical side of any study and its mathematical side are different from the body of that study and we can present them separately. We thus have mathematical physics and mathematical economics and so on. Remarkably, Wittgenstein held a seminar on mathematics with Alan Turing in the audience, yet when the book based on his lecture notes appeared it received reviews that expressed extreme disappointment. Wittgenstein himself had severe doubts as to the value of that seminar: the book ends with great doubt as to its value.

All this has aspects that are to the good. We should admit, however, that these are sparks rather than a big light. Moreover, one must heed the proposal (Bunge 1974a and b) to remove from the agenda all versions of subjectivism [including some Wittgenstein texts], since it was traditionally imposed on philosophers by the futile quest for certainty.[5] In addition, what is needed is to state what criterion one uses for determining the order of priority of questions to discuss in order to avoid sinking into trivialities.

The seriousness with which Russell and Wittgenstein took the idea that rationality is proof is remarkable: together with the new physics their studies that this very seriousness had invited has helped the removal of the search after certainty or certainty-surrogate and put the problem of rationality in center stage instead (Bartley). This is indebted to Wittgenstein's pioneering effort to use the new logic in the development of a new, clear philosophy.

Let me explain this. Although no one could meet the demand for certitude, efforts to approach it were standard fare ever since the great Newton discussed certainty as a goal worthy of efforts to approach. The overthrow of this idea was necessary in order to allow a fallibilist idea of rationality to supersede it: *humans are fallible yet rational and able to improve their rationality.* What is still missing is the popularization of the logical fact that these are conflicting research programs, one seeking to establish the authority of reason and the other seeking rationality as a *desideratum.* The unbridgeable disagreement between Moore and Wittgenstein was about the status of commonsense metaphysics: Moore (1953, 76) found it "largely correct" or "generally correct" whereas Wittgenstein declared it not given to articulation.[6] On this, Russell won: he had to dismiss their views since he took for granted that commonsense is improvable by admitting corrections due to advancements of science. One may of course choose to ignore what commonsense owes to science, but then it is advisable to make that choice openly. Those who ignore this force others to ignore them.

If one wishes to discuss what Wittgenstein presented as commonsense certainty, one has to explain why, and take account of the following. Skeptic Robert Boyle called commonsense certainty moral certainty. He also offered a working criterion for that, referring to knowledge as recognizable in courts. (Wittgenstein mentioned courts too, but as an example, not as a criterion.) Whereas young Wittgenstein declared meaning absolutely clear and thus context-free, mature Wittgenstein discussed meaning in natural languages, in context and with some vagueness that he compared with the friction without which controlled movement is impossible (*PI*, §§107, 130). Yet in his very last work (§59) he grossly misread the ordinary use of "I know" as the claim for infallibility. A court of law may recognize that some specific piece of information that experts had claimed to be knowledge has turns out to be

[5]Wittgenstein never gave up his quest for certainty. In his *On Certainty,* he tried to present it as ordinary. Yet its roots are obviously in the idea that an unrefuted falsehood is a menace. This idea is not ordinary and it is rooted in the traditional theory of rationality that is far from commonsense since it is commonsense to admit fallibility. Ziff, Paul 1984, 100 admits fallibilism yet meets it with the claim for having an infallible method!

[6]Moore and Wittgenstein disagreed about commonsense in a few ways, but all their disagreements seems marginal (Stroll, 1994) in comparison with their shared view of it as authoritative.

mistaken. This may lead to public investigation as to whether the claim of experts was a reasonable error. Outside courts, people may leave the status of such a claim undecided and they may find it necessary to decide for themselves which of the alternatives to act on. In or out of courts, proofs occur and experts admit them iff they follow some given standards. The standard vary in degrees of severity for journalists, the police, civil courts, criminal courts, applied science and mathematics. They are all improvable, and so they are all not proven perfect (Agassi 1975, Chap. 16, Agassi and Meidan 2008, Introduction.).

Wittgenstein ignored all this, common knowledge though it is. For, his aim was to debunk metaphysics as meaningless. His examples of meaninglessness thus naturally tend to come from metaphysics, but not all of them. He even noted that meaningless expressions that are not metaphysical are easier to generate than metaphysical ones (*PI,* §169). Metaphysics is all and only nonsense produced in efforts to say the unsayable. My criticism of his analysis comprises a counter-example to his sweeping claims. He claimed that metaphysical doubt is nonsense and that commonsense offers examples of absolute certitude.

The popularity of Wittgenstein as the greatest philosopher has the best justifications possible. First, he killed all metaphysics. Second, he invented a new philosophical technique. As to the first justification, those who take it seriously have to respond to criticisms of it. The second needs elaboration: what is the benefit from the use of Wittgenstein's new technique? Naturally, this question invites a detailed discussion.

13.2 Wittgenstein's Self-Appraisal

Wittgenstein's leading disciple Georg Henrik von Wright said,[7] "The later Wittgenstein, in my view, has no ancestors in the history of thought." Now the program of purging science of its metaphysical component characterizes the philosophy of Mach most. (Indeed, Neurath considered the Vienna Circle followers of Mach rather than of Wittgenstein.) The idea considered specific characteristic of young Wittgenstein, the idea that metaphysical assertions are ungrammatical, even that was not new with him. What was his own is the idea that he had proven this. Even were this true, mature Wittgenstein admitted, it would be insufficient because it relates only to science. Wittgenstein did not quite admit that he had no general proof of his anti-metaphysics. Rather he offered a program: watch language in action and see that metaphysics is idle: it plays no role in productive thinking. This is the idea characteristic of mature Wittgenstein as it transpires mainly from his *Philosophical Investigations*. It is the idea that pertains to the new technique of which he was proud. It is inapplicable to the history of science, since there the productivity of metaphysics is most transparent. This raises the question, is common discourse or is science a better

[7]Wright in Malcolm 2001, 14; see also Ayer 1971, Cloeren 1975, 513 and Sluga 1998.

field for analysis. This is how Popper considered the situation as the background for his critique of ordinary language analysis (1959, Preface to the English Translation). Nor is this all that there is to it. Politics too shows the limitation of ordinary language analysis: few ideas are politically as central as the siblinghood of humanity and as liberalism, and these are metaphysical ideas implemented with success that would have astonished their originators. Politically speaking, here the proposal makes excellent sense to add comparative studies to the ordinary practices of ordinary language analysis: the ordinary language of any democratic society is considerably superior to the ordinary language of any traditional society.[8]

The doctrine of Wittgenstein is straightforward: modern logic has shown that articulated metaphysics is dead: metaphysical doctrine does not exist: "Philosophy is not a body of doctrine but an activity" (*TLP*, §4.112). Mature Wittgenstein added this: the purpose of the elucidation is curing people of the metaphysically caused mental anguish. Analytic philosophers repeatedly declared this idea as Wittgenstein's, and as one that right-minded philosophers endorse. This was declared the revolution in philosophy, celebrated in the fifties as described in *The Linguistic Turn*, a volume of essays (1967, 1972) by the most prestigious philosophers in England then, based on broadcasts on the educational program of the national radio, the Third Programme of the British Broadcasting Corporation. Embarrassing as history is, it is advisable not to ignore it and not to beautify it.

Wittgenstein said, metaphysics is alive and active, though permanently out of touch (*PI*, §426):

> although we see the straight highway before us, of course we cannot use it, as it is permanently closed.

Kant had said, metaphysics is a mistress with whom we quarrel but to whom we regularly return. Wittgenstein said, metaphysics is beyond reach – beyond verbal reach, that is: metaphysics is unsayable. As Russell has critically observed, Wittgenstein has said a lot of what he declares unsayable; it is wiser to consign all that to another language, he suggested: to the meta-language. This would render his message sound: philosophy is unsayable in the object language; its assertion belongs to the meta-language; all philosophy is within the meta-language (Hooker 1975).[9] The version of anti-metaphysics that Carnap advocated allows philosophers to perform only analysis.[10] It turned out that the relations between the object-language and the

[8]Even Israeli Hebrew differs from traditional Hebrew in representations of political wisdom. This kind of progress English commentators hardly noticed – due to the tremendous liberalism of Shakespeare as expressed, say, in *Henry V* and in the final scene of *Romeo and Juliette*. See Bloom and Jaffa 1964, 1–22.

[9]Carnap had said, the right philosophy is the analysis of science and the wrong one is the mix between the meta-language and the object language. Hooker improved upon him.

[10]Carnap [1934] 2014 tried to translate the meta-language to the object language and was stopped when he learned of Gödel's proof that it is impossible. Tarski demanded that the meta-language include translation of the object language; this surprised Carnap greatly (Coffa 1991, 281, 299, 368, 372).

meta-language deviate surprisingly from commonsense.[11] Only the need of formal logic for the meta-language is obvious: paradoxes are deadly for formal mathematics (Bar-Hillel 1966). In response Ayer (1985, 65–66) said, mature Wittgenstein's dismissal of the meta-language is right for ordinary discourse, as its speakers need not worry about paradoxes; it is mathematicians who must do something about them.

Wittgenstein praised his output (1967 §41):

> Much of what we are doing is a question of changing the style of thinking.

This expressed his ambition (Wittgenstein 1974, 71; Monk 1990, 184):

> Either my piece is a work of the highest rank, or it is not a work of the highest rank. In the latter (and more probable) case I myself am in favor of it not being printed. And in the former case it's a matter of indifference whether it's printed twenty or a hundred years sooner or later. After all, who asks whether the Critique of Pure Reason, for example, was written 17x or y.

Moore (1953, 26) has reported:

> He said that what he was doing was a 'new subject', and not merely a stage in a 'continuous development'; that there was now, in philosophy, a 'kink' in the 'development of human thought', comparable to that which occurred when Galileo and his contemporaries invented dynamics; that a 'new method' had been discovered, as had happened when 'chemistry was developed out of alchemy'; and that it was now possible for the first time that there should be 'skillful' philosophers, though of course there had in the past been 'great' philosophers.

Here (in the wake of Russell's expression of hope to render philosophy scientific?) Wittgenstein compared himself to Galileo![12]

> The real discovery is the one that makes me capable of stopping doing philosophy when I want to.—The one that gives philosophy peace, so that it is no longer tormented by questions which bring itself in question.—Instead, we now demonstrate a method, by examples; and the series of examples can be broken off.—Problems are solved (difficulties eliminated), not a single problem. (*PI*, 133).

We see here a patient obsessed with philosophical doubt cured by losing interest in philosophy.[13] Since even among philosophers few are so obsessed, the problem is not very urgent. Yet after World War II, most English philosophers agreed with his self-appraisal as the greatest philosopher around. Perhaps there were external reasons for that. For, admittedly, when he became popular his works helped greatly to ease the publication pressure that just then began to become a great burden.[14] The

---

[11]Russell declared all his philosophical works other than logic and epistemology as not philosophy but contributions of a concerned citizen. Wittgenstein disapproved of them.

[12]Already Russell did that: in his 1914 book, he compared the liberation that the new logic has brought about to the one that Galileo's work did.

[13]Hermine Wittgenstein 1984, 2 contains a report that her brother was obsessed with philosophy and that he suffered terribly, in an almost pathological state of agitation. There is no reason to take this report lightly.

[14]The public image of Wittgenstein was naturally complex. Bartley 1982–83, 170 said, it was a carefully crafted artefact. (See also notes 20 and 33 there.) He presented Wittgenstein as tormented by his sexual preference. The torment by philosophy that he discussed and that he had suffered from

idea that he brought a new hope conflicts with the popular image of him as melancholic in constant mental anguish.[15] His hope for a new assessment, of the new philosophy as a new skill that one can acquire, is largely justified, but only as a hope. It never evolved as a therapy, not even after philosophical counseling became available as a therapy of sorts since the early 1990s. Wittgenstein's contribution rests on his defunct thesis that the end of academic metaphysics has arrived and perhaps on his aborted view of analytic philosophy as therapy of sorts. Can the reputation of the new philosophy survive this disappointment? Will it survive as the encouragement of trivial pursuits under the guise of serious philosophy?

What did Wittgenstein consider the great value of his philosophy? Locke considered his clearing rubbish no more than preparatory to constructions that others performed; did Wittgenstein consider the significance of his contribution the same way? His declared intention was to avoid offering any view on mathematics or on science or even on commonsense, but merely to remove puzzles. Yet he did express there a view of mathematics distinct enough to be identifiable with the minority group among the philosophers of mathematics (intuitionist or even constructivist), that he rightly feared is insignificant. Commentators praise his honesty as exceptional. Praising a person this way, says the Talmud, is a slight on that person's peers: honesty is best self-understood. Moreover, comments on a thinker's contribution have to relate to ideas, not to conduct.

Wittgenstein said little about ethics or politics. It does not concern the choice of a code but of humility and sincerity about it. He did choose a code: it was Tolstoyan. He found distasteful talking about it. As to the philosophy of life (*Weltanschauung* or worldview), Mach discussed it before him, saying, the right worldview is the sum total of all science. Wittgenstein endorsed this in his early days. Hayek said (1957, 20), the imprint of Mach made Wittgenstein's first book popular. Later on, Wittgenstein (1980a, 7) took a worldview to rest on a metaphysics the he considered (1968, §97) pretentious super-concept. This raises the question already raised here a few times: what was mature Wittgenstein's attitude to science? As attitudes belong to the philosophy of life, understandably, he left us in the dark about them all. In 1930, he said (1980a, 7) he valued clarity for its own sake. Commentators often see this as his dissent from his Viennese followers, who embraced the progressivism that he decidedly repudiated. He was moderately conservative, and so decidedly rather indifferent to science. At times, he looked unusually withdrawn from anything practical, yet he saw himself as a practical man (like his father)[16] and this trait as

as an adolescent, may very well have been a displacement – which is in the nature of obsession. In any case, Bartley 1082–3, 174, considered obvious that this preference was not the source of his philosophy but it was "of central importance in understanding the man and his influence" nonetheless.

[15]The most indicative of this seems to me the lovely letter of Keynes to Wittgenstein of 27.12.24, inviting him for a visit to his country home: "Then perhaps you could work too, — and at any rate be as morose as you might feel inclined, without upsetting anybody."

[16]Drury 1984, 110. Quoted in Schroeder 2006, 2. Hayek 1977, 22 describes Wittgenstein as impractical.

the root of his conservative politics and its reflection as a very significant motive behind the whole of his philosophy of language.[17]

The best description of the attitude to science of mature Wittgenstein seem to be that of Bouveresse (2013, 43). He describes it as "general lack of enthusiasm for science, and for a culture dominated like ours by scientific thinking"; "his concept of what a science is, broadly speaking, might have been much looser or more accommodating than that of members of the Vienna Circle or Karl Popper". Still, he was more intolerant of pseudo-science than they were. He examined Freud's texts in order to purge it of metaphysics, not of pseudo-science.[18] He discussed the idea of the unconscious, introducing the possibility of unconscious pain. In the context of this study he ignored his having suffered the anguish that he declared he had found a method to cure.

> Fania Pascal suggests that 'we can understand his cavalier attitude towards Freud once we realize that he himself felt he had no need of Freud' (Bouveresse 2013, 8).

Wittgenstein felt so as he could not conceive of the philosophical anguish, especially the fear of solipsism, being true, as possibly a displaced Freudian anxiety. This is what John O. Wisdom has suggested.[19]

What then was Wittgenstein's explanation of his comparison of his contribution to those of Galileo and of Kant? The answer we have is but that it is his having invented a new skill for professional philosophers.

13.3 Wittgenstein as a Moralist

The choice of a philosophy of life, need one say, should rest (not on authority but) on a rational decision that should follow some discussion of it. It is not clear whether Wittgenstein agreed. Perhaps he suggested that one does not choose one's philosophy of life; and perhaps he suggested that one chooses a philosophy of life intuitively, perhaps even unknowingly (*On Certainty*, §94). Are these two responses

---

[17]Schroeder 2006, 4, 9, 129, 154, 227; Wittgenstein 2012, 427, 785, 805, 1076, 1099, 1102, 1175, 1361, 1432, 1478. Notice that on page 1076 Wittgenstein refers to language as an institution. When in 1958 Pole presented Wittgenstein's philosophy of language as conservative, he met with unfair derision. See my angry letters on it to the editor of *Times Literary Supplement*, 22 and 29 May 1959.

[18]Bouveresse 2013 contrasts Wittgenstein with Popper repeatedly. Early Popper used the word "metaphysical" as synonym with "non-scientific", the way members of the "Vienna Circle" used it, but definitely in refusal to entertain any theory of meaning, much less a theory of the meaningless.

[19]J. O. Wisdom's metaphysical position, 1958, 161, is unwavering realism, and as a matter of course. This Wittgenstein would have appreciated, even envied. What Fania Pascal 1973, 38 said about his attitude to Freud and his mental anguish that she describes with no reference to his attitude to Freud (Pascal 1973, 32, 35) appear as one story. It is very much in line with Wisdom's astute observations.

one? In any case, he ignored the choice and did not discuss the place of rational debates in the life of reason.[20]

Wittgenstein's Tolstoyan moral code is evidently politically irresponsible. It is the dogmatic endorsement of a political attitude that is either extreme anarchist or extreme conservative, but decidedly while being politically utterly passive (as illustrated in his grand novel *War and Peace*). Tolstoy preaches conversion to anarchism as a personal moral revival and as a political ideal attainable only by the restoration of public morality, decidedly not by political means and decidedly not through political public debates. Like the Quakers, he preaches the advancement of a political cause — pacifism — by living a clean spiritual life. This is anti-political. Anti-political attitudes of all sorts are prevalent and the political (scarcely the intellectual) need to criticize them publicly is urgent. At best Wittgenstein's philosophy is irrelevant to all this.

All sorts of moral doctrine accompany all sorts of moral conduct. This renders Wittgenstein's personal conduct irrelevant to the assessment of his views and of their impact.[21] In particular, that Wittgenstein was a difficult person makes no difference to the present discussion, except that it explains his withdrawal from the people with whom he could argue. As Ramsey said of him in a letter to Keynes (Wittgenstein 1974, 20, 29),

> The people in England he wants to see are few; Russell he can no longer talk to, Moore he has some misunderstandings with ... yet he did speak with them on and off.

Since Keynes described Wittgenstein as "a very old and intimate friend of mine", we may view Wittgenstein's relations with intimates as Keynes has described their relations (*op. cit.*, 130):

> I alternate between loving and enjoying you and your conversation, and having my nerves worn to death by it ...

Not all aspects of one's personal characteristics, however, are utterly irrelevant to the understanding of one's ideas. Wittgenstein's attitude to Russell was confused, and is thus confusing. He was most respectful to him in his presence; not quite so in his absence. He both encouraged and discouraged this kind of folly. This matters not: in any case, we should ignore folly as much as possible. This is nowadays happening, now that the personal acquaintances of Wittgenstein are all dead.[22]

---

[20]The view that it is possible to avoid debates rationally is pervasive. Those who share it naturally view the ascription of it (to Wittgenstein or to Carnap or to anyone else) as a complement; others view it as expression of excessive *naïveté*. Wittgenstein spent most of his intellectual life arguing.

[21]The impact of Wittgenstein may very well be in his offer of a tacit excuse for silence over the most urgent new political problems that the world faced after World War II. This study is not the proper place to discuss this thesis.

[22]Probably the last philosopher to have reported from personal memory on an encounter with Wittgenstein is Hintikka, who died in 2015. The personal impact of Wittgenstein is still very much alive. See Fann 1987, 615 for the review of J. N. Findlay, *Wittgenstein: A Critique*: "After reading the book one wonders why the author bothered to write a book to criticize such a 'simplistic' (72), 'arbitrary' (17), 'elementaristic' (12) and 'mechanistic' (39) philosophy. The answer may lie in an unexpected dimension: Wittgenstein's 'intellectual charisma' (20) and his 'unbelievable personal

So much for Wittgenstein as a moralist thinker. This aside, there still is the fact that he considered his own influence unhealthy. This is not a moral matter but simple commonsense: he considered himself a practical person and was proud of it. Yet he viewed his philosophy as having no positive practical implication. He therefore suggested to his students that they should draw the consequences and do something other than philosophy. Few of his close disciples did: most of them became professional philosophers. What did they do? Remove intellectual rubbish, of course. How much intellectual rubbish is there to remove? Will the operation ever end? Is the expectation to remove the same item repeatedly? Are there replicas of an item that appear endlessly? Does intellectual rubbish grow regularly like physical rubbish? Does it threaten to overflow like modern industrial refuse? Wittgenstein spoke regularly of metaphysics as a serious temptation (Cavell 1962, 71). Is this why he had the self-image as a new Kant or a new Galileo? Did he consider the world with constant weeding out of the ever-regenerating weed of metaphysical assertions inherently better than the world we live in today? As he was politically a conservative who had distaste for the progressivism of his peers, perhaps he was in two minds about all contributions to progress, including his own.

The intellectual rubbish that Locke wished to remove was a dominant view of the world that the educational system was imparting and that tradition kept despite its refutations. The rubbish that Wittgenstein wished to remove was different: he stressed repeatedly that he had no wish to correct errors. Many people who are not scientifically adept are prone to make empirical assertions that science has refuted. Worse: democratic societies often keep obsolete ideas circulating by the use of popular votes that endorse them. This exasperates and makes a considerable part of the population express the wish to replace democratic control by expert control over their political systems. Some consider the popular errors that persist despite scientific corrections superstitious, prejudicial or dogmatic. Wittgenstein tried to avoid discussion of all error, unless he thought it interfered with his performance of his self-appointed task as a philosopher. This he did only incidentally. At times, he spoke of grammatical illusions, as a special kind of error, but as no one has discussed the question, what is this kind of error, he did not elaborate:

> 'Language (or thought) is somewhat unique' – this proves to be a superstition (not a mistake!), itself evoked by grammatical illusions.

And on these illusions, on the problems, the pathos falls back."[23]

This is one of the most obscure passages in the *Philosophical Invesstigations*. My attempts to find a context for it has failed. (This criticism is not of Wittgenstein but at his editors. As they stress, the manuscript that they have published is unfinished.)[24]

---

beauty.' 'As the extreme beauty of Wittgenstein is not often spoken of, it seemed fit to mention it here: certainly it contributed, even if unconsciously, to his immense influence at Cambridge' (19)."

[23]*PI*, §110. My literal translation. The definite article in this text indicates that it is sketchy as it refers to an intended but unwritten previous sentence that would require the definite article.

[24]To expect an author to avoid all malapropism is excessive. It is also hardly problematic: usually readers can use contexts to retrieve the intended texts from their distorted versions, at times even

Wittgenstein and Moore both defended of commonsense knowledge as true. Some of their most celebrated disciples argued in this vein. I find all of it embarrassing: what they consider commonsense is a basket full of superstition. It is irrelevant to any Wittgenstein-style treatment of any metaphysical doctrine, since the treatment comprises dismissal of the meaningless, not the erroneous: an attack on a claim as erroneous is an assertion of its negation; an attack on a claim as meaningless, however, is equally an attack on its negation – considering them both equally confused. (This is what makes the attack linguistic.) Unlike Wittgenstein-style dissolution of a metaphysical doctrine, Moore-style criticism of a metaphysical doctrine is the establishment of its negation. Thus, as Moore attacked McTaggart's view of time as unreal, he claimed to have proven that time is real. (Otherwise, he explained, he could not have had breakfast and be going to have lunch.) Wittgenstein found the destruction of the claim that time is unreal achievable without establishing that time is real.

Too few analytic works repudiate the willful mix of the two strategies, those of Moore and of Wittgenstein. The reason that the method of criticism of metaphysics Wittgenstein-style is confused with its Moore-style relative is that the latter conveniently stands in for the former when it fails. The use of a combination of the Moore and the Wittgenstein techniques has received critical scrutiny, but it remains popular: it is very comfortable.

13.4 The Task of Philosophers

Whatever the task of philosophers may be, from its very beginning this task included some analysis, some clarification, some concern with language. With no discussion of detail, the conclusion is imposing that the way this was done has immensely advanced – due to Boole, Frege, Russell and Wittgenstein. Their impact will be manifest in diverse situations, from the examination of legal documents to generative grammar.

This acknowledgement to Wittgenstein would not have satisfied him. He did think he was great; yet he deemed this thought sinful. Seeking the truth on such matters is not easy, since the value of a philosophy much depends on the criterion of value, and, naturally, a philosophy recommends itself. This is not always so. Thus, analytic philosopher Geoffrey Warnock judged Wittgenstein less great than other analytic contemporaries: his assessment appeared in 1958, during the peak of the Wittgenstein fashion, declaring that the father of analytic philosophy is Moore, not Wittgenstein. As the present study is not about Moore, we need not address here the question, how much of Moore's philosophy was linguistic. Obviously, there is a

without notice. This is what redundancy is for, says Claude Shannon. In analytic philosophy Davidson 1986 and commentators (see Reimer 2004 for references and for comments) studied this phenomenon without benefit from Shannon.

great affinity between the two. This means little, since Moore made many obviously true assertions that Wittgenstein had to agree with regardless of his anti-metaphysics. Of course, when Moore made commonsense metaphysical assertions (especially realism), Wittgenstein could not express agreement with him, even in cases in which he did agree with him, and even wholeheartedly. Here we see the significance of the difference between the philosophy of Wittgenstein and the one ascribed to him by the vulgar: he agreed with Moore on realism but not with his assertion of it. Why not? Because there is no occasion for saying it in everyday life. Moore ended his most remarkable lecture by declaring that his hand was a real object in the external world. Wittgenstein said, we remember, he could imagine a situation in which one would say, his is my hand, but the end of Moore's lecture is not one of them. Why? Because it is metaphysical.

Things are worse when the metaphysical assertions in question appear in scientific contexts. Thus, Newton and Leibniz agreed that time is real; this is agreeable to Moore; not their disagreement about it: the one asserted that it is an independent co-ordinate, and the other viewed it as a property of matter. (Einstein [1954] 2012, xvi expressed a preference for the view of Leibniz over that of Newton.) This dispute is obviously metaphysical. Is it confused? Moore had no opinion on this. Wittgenstein said, yes. Thus in Wittgenstein's view the dispute wants a clarification. Would the clarification delete the disagreement? Yes, said analytic philosophers: it may undergo translation into a scientific dispute or into a commonsense dispute: as such, a dispute can come into conclusion; an endless dispute is sham. Suppose this is so. It follows that a metaphysical dispute, even if confused, may be worthwhile. This is far from the Wittgenstein-style clarification as that will be the rejection as ungrammatical of metaphysical assertions. It will be the (commonsensical) recognition that new and valuable ideas may first appear in a wording that can be improved upon, that the limits of language are not a given but a challenge.

Neither Wittgenstein nor Moore could oppose Einstein's views on time. Those who doubt this can take ancient atomism instead. It and nineteenth-century atomism share a metaphysics. Popper ([1935] 1959, xxiii, 16, 277) took this as a refutation of Wittgenstein's anti-metaphysics. Wittgenstein's paradigm case is St. Augustine on time: he knew what it is until asked. This question is meaningless, Wittgenstein said. Unintentionally this is anti-science.

The thesis of Moore, [current] commonsense-metaphysics is true, conflicts with that of Wittgenstein, metaphysicians attempt in vain to articulate the not given to articulation. Both are false, even though each has interesting true instances. Perhaps the techniques of the two can complement each other. Ayer discussed the displeasure about Moore that accompanies Wittgenstein's text *On Certainty* (Vesey 1974, 227). Possibly, his discussion is an effort at reconciliation: Wittgenstein had not expressed disagreement with Moore, he observed; he did express wonderment at Moore's having made this or that obvious point. Yet he found it impossible to assert or deny the metaphysics that Moore defended. Inspired by Russell's analysis, Wittgenstein declared both idealism and materialism ungrammatical; Moore expressed realism and rejected idealism emphatically, contrasting them in the traditional manner.[25]

In a letter to Moore (October 1944) Wittgenstein said that Moore's term "nonsense" is "highly misleading". He admitted the examples of Moore are examples of "absurdity", but only in the ordinary sense of the word.[26] Strictly speaking, he suggested, it is "something similar to a contradiction", whatever this may mean. Russell's examples of science in conflict with commonsense seriously undermines much of the output of analytic philosophers. Some efforts to argue against it appear from time to time; thus-far they are pathetic. This need not be so: analytic philosophers can discuss conflicts of commonsense with science and the way this improves commonsense. They can then discuss pre-scientific commonsense – magic – too. This is a reasonable venue, now that no significant philosopher advocates Wittgenstein's view of metaphysics as meaningless or Moore's view of commonsense: it is sheer commonsense that commonsense opinions are merely the default option and no more than that.[27] This is how devoted followers of Wittgenstein deeply disagreed about matters religious. Religious analytic philosophers tend to consider science as an instrument devoid of metaphysical import, whereas others take it literally Frege's way.

Hence, when we view the practical corollaries of Wittgenstein's philosophy as frustrating, we should notice not only the question, what are philosophers to do, but also, and even more so, what are theologians to do, especially since this category includes most of the religious leaders of the many religious communities that flourish in the liberal world. Wittgenstein and his Viennese followers considered theology the central item in religion that is unable to deliver the goods, as it comprises pseudo-statements. The members of the Vienna Circle took it for granted that this way they were combatting religion and its evils; Wittgenstein did not share their views. What then did he think he was doing combatting theology? What did it matter that some religious people spouted meaningless utterances? Why did he take it to be his task to stop them?[28]

[25]Moore distanced himself from Wittgenstein (Schilpp 1952, 675–6): "It is not true that I have either said or thought or implied that analysis is the only business of philosophy. . . . In fact analysis is by no means the only thing I have tried to do". His early lecture course, published after Wittgenstein died (Moore 1953), opens with the characterization of metaphysics (Moore 1953, 1) as a "general description of the universe". Warnock continued to distort Moore's opinions (Klemke 1962.)

[26]The term "absurd" appears in existentialist philosophy in a quite different sense, yet it is still not a technical word: readers have no trouble comprehending it. (This is an empirical observation.)

[27]This is admittedly not universal commonsense. The use of default options as a default option is the contribution of computers to our civilization. It comes from traditional legal practice as reflected in Talmudic law and English property law, contrary to traditional legal philosophy prior to the fallibilist contribution of H. L. A. hart.

[28]Gellner protested, saying, some liturgy is intentionally meaningless. Similarly, when the Vatican ordered to say the mass in the vernacular, Feyerabend complained.

13.5 The Public Impact of Wittgenstein's *Investigations*

The program of purging science of metaphysics is traditional ever since the scientific revolution. Its task was to help researchers perform better research. Wittgenstein efforts in this direction inspired some scientists – in a manner may have been beneficial or detrimental to the progress of science: this hardly interested him.

We are still too ignorant to be able to assess Wittgenstein's impact on religion or on science. The same goes for aesthetic, not to mention metaphysics and the impact of Wittgenstein's proscriptions. Possibly he effected a kind of censorship and possibly this has raised standards of rational discourses.[29]

The thrust of Wittgenstein's method is the intent to apply it in order to eliminate all disagreements or at least the ones that are in principle undecidable. Already Charles Sanders Peirce contemplated the idea that any theories that have no crucial experiment to decide between them are synonymous. This is contrary to the practice of trying to enrich the contents of theories in order to generate crucial experiments. To limit the right to articulate and discuss views on the sacred or the beautiful is to cause harm. Doing this Wittgenstein encouraged obscurantism (Gellner [1959] 2005). In addition, no assessment was ever made of the number of individuals whom Wittgenstein drove away from philosophy – for good or ill.

Clarification, the removal of confusions, can be excessive waste of time (as it is easier to muddle than to clarify). Otherwise, it is eminently rational, as it helps the removal of error and the improvement of thinking. Any ambiguity can lead to a confusion and any effort to find the confusion a deep idea is a standard example that invites a clarification Wittgenstein-style. For example, dividing the scale of grey between black and white to two differs from discussing the two extremes in disregard for the middle. Each option is useful in some circumstances. Confusing them is as common as it is disastrous.[30] Now this kind of analysis conforms to Wittgenstein's characterization,[31] yet usually it has nothing to do with metaphysics. Wittgenstein's hostility to metaphysics boomerangs: he hated something beyond its confusion and pretense that he said it was. Worse: if his allegation is true, then his hostility to metaphysics is too little to spend a lifetime on, and his comparison of himself with Kant or with Galileo is then somewhat exaggerated.

Russell had found Hegelianism attractive but harmful; Wittgenstein considered solipsism this way. This situation made Russell ambivalent about metaphysics; it made Wittgenstein determined. He took for granted the intuition that solipsism is / may be imposing, and that it is the source of mental anguish. This conflicts with

[29]Possibly this made Wittgenstein attractive to philosophy professors of the kind he disliked most.

[30]Reichenbach suggested to replace the two extreme option of truth and falsity – 0 and 1 – with intermediate values – all numbers between 0 and 1. He evidently confused different options, as Carnap's theory of explication indicates.

[31]This explains how it came to be that a number of papers in the philosophical press clarified the difference between permissible classifications, such as that of qualities, one in which the very good (or the excellent) is good and the other in which it is not.

Freud's theory of the source of all mental pain being due to early childhood traumas, as it allows for intellectual illness that leads to mental illness. Consider the fear of solipsism. Wittgenstein's treatment of it aims at reestablishing a natural peace of mind. Let this be so. The not tormented can still ask, is normal intuition peaceful? Is it authoritative? There is no clear answer to this question in the works of Freud or of Wittgenstein: they promised cures, but they did not offer theories of health. What is the status of their therapeutic methods? This may depend on their technique. On this, Freud has offered much more than Wittgenstein has. The latter is language analysis, an improvement on old-style analysis. What exactly this is still raises dispute among aficionados. This seems obvious: philosophical analysis leads to skepticism with cruel logic. This is why Wittgenstein repeatedly studied certainty: he tried to disprove this; he argued that everyday experience comprises certain knowledge. He admitted that this certainty is qualified. Hence, it is simpler to claim that such hypotheses qualify it as the hypothesis that the sun will rise tomorrow (*TLP*, §6.36311). Both this hypothesis and its negation disparage the threat of solipsism. (This seeming refutation of solipsism is question-begging as it rests on the choice of a universe of discourse; by contrast, the choice that neutral monism favors permits solipsism.)

Following Hume and other eighteenth-century commonsense philosophers, Wittgenstein pitched commonsense against both ruthless[32] skepticism and solipsism. Yet commonsense is alterable. Wittgenstein never said commonsense (or commonsense intuition) is authoritative (the way Frege did) though he did not say the opposite either (the way Russell did). He did side with Russell against Frege, however. This, as usual, causes confusion among his supporters.[33] This is so particularly since it counters Wittgenstein's aim to abolish solipsism by boosting common sense.

What then was the specific contribution of Wittgenstein and his disciples? Many Wittgenstein passages strongly suggest that he had no intent to contribute any novel idea. They mislead. Other Wittgenstein passages express tremendous ambition. Clearing the rubbish in the sense of Locke or in some other sense is not the contribution that Wittgenstein took pride in. What is the contribution he claimed to have made?

This question hardly pertains to the enormous Frege-Russell revolution in logic, to which he contributed as a student of Russell who collaborated with him as a young genius whom Russell designated his heir and successor. The revolution aimed specifically at problems in the foundations of mathematics. Even for mathematicians who had no interest in foundations had to notice this, as it changed all mathematics and established the idea of formal languages that is in the foundation of many a mathematical system and comprises the bread-and-butter of all computer science and

---

[32]The Platonic equation of rationality with certitude imposes ruthless skepticism. To tame it one has to reject Plato' theory of rationality.

[33]Iglesias 1984, 285: "Russell development up to 1913 may thus be seen as a movement from 'constituents' to 'constituents and form'. Wittgenstein's, on the other hand, begins with 'constituents and form' and moves on to 'constituents'." "At the time of 'Notes on Logic' [1913] both Russell and Wittgenstein share the belief that in order to understand a proposition one has to understand its 'indefinable' elements, and these are constituents and forms." Constituents and forms are concepts and statement-forms respectively; or, analysis old-style and new-style respectively. Wittgenstein's theory of forms of life may read as a reversal to old-style analysis.

artificial intelligence and all that. The greater importance of this revolution, however, is due to its having opened the road to a new philosophy of science. Under the influence of the revolution in physics, and under the influence of Einstein and of Popper, there is great opposition to Bacon's proposal to delete all errors of the past. Such opposition is wise, since deleting the past amounts to deleting the present. This is the greatest divergence from our predecessors of the Age of Reason: Newton denied that his theory superseded those of Kepler and Galileo. Most of us (all philosophers except those philosophers of science who cling to the study of the problem of induction in the old mold) know how to distinguish between silly errors, perhaps ones that may pass as superstitions or prejudices, and reasonable ones, as scientific errors are.[34] A metaphysical error may likewise win appreciation — especially if it is pro-science. Carnap was willing to admit the pro-science metaphysics: hence, he opposed metaphysics only because he saw it as an obstacle for positive scientific progress. (This is why opponents of metaphysics call their opposition "positivism".) Carnap was in error: even if a theory is not useful for science, perhaps it should be preserved and taught. (Example: logical atomism.) Moreover, in the opinion of Einstein, the hostility to metaphysics impedes the growth of science.

Now whatever deserves support, in science, art, politics, morality, or anything else human, studied critically, in the light of Wittgenstein's philosophy or not, invites examination as well as comparison with existing alternatives. Wittgenstein's ideas, especially his theory of language, evolved this way: he said, his idea that philosophical words have no specific meanings came to restore these words to their natural home, to ordinary parlance. His theory of meaning as use came to replace the views that Frege supported, of sense as a separate domain. It came to replace Russell's theory of denoting too, but for different reasons. This is still engaging analytic philosophers in arcane debates.

Members of Wittgenstein's school refuse to discuss sharp criticism of his ideas. This is both regrettable and unbecoming, considering that he impressed Russell so much just because in his youth his criticism of Russell was sharp and at times deadly. The intent of Wittgenstein to push all the way his idea of the ineffability of some ideas that he held, and his doing so with the aid of his theory of sense as use, caught wide interest. It is therefore worthwhile to see both what is its allure and what is left of it after it underwent some valid criticism. Were Wittgenstein alive and willing to discuss objections to the view that he had advocated, we would wish to hear him. Possibly, he would respond to these the way he responded to Russell, saying he did not comprehend the objections, so that he could not respond to them. We may nonetheless try to imagine a debate with Wittgenstein. Russell said he found Wittgenstein's *Philosophical Investigations* boring. Did Wittgenstein or his followers respond to his response with any measure of adequacy?

Wittgenstein's disciples have removed some traditional confusions that may impede some interesting discussions, thereby performing a public service even

[34]All modern liberal legal systems distinguish between negligent and reasonable error. English common law allows as reasonable even superstitions received in the pertinent background community.

when they are boring (Wisdom 1963, Preface to the first paperback edition). Now the sophists performed such tasks already of old. The specific in the teaching of Wittgenstein is his insistence that philosophers can do no more than that, since there are no real philosophical problems.[35] Wittgenstein found debunking significant. He was in the habit of deriding with much uproar some books that he saw in quarters of peers, friends and acquaintances. This is of questionable value.[36] Private libraries harm no one even though they normally contain much rubbish. Recommending the burning of worthless books is offensive. When Hume said that a book containing no empirical and no mathematical assertion deserves the flames, he was only expressing figuratively his view of their being non-books. He would obviously shudder at the suggestion to read his recommendation literally. What then was Wittgenstein's attitude to the ideas of Plato, for example? There is little discussion on this, such as that of R. M. Hare (1964, 62). Nor is it necessary: suffice it to agree that civilized people will not sanction the burning of books, let alone Plato's, regardless of one's opinions about them.[37]

Hence, Wittgenstein's comparison of metaphysicians to slum-landlords is objectionable. Admittedly, it is no more objectionable than the disrespect that some art critics flaunt towards some insignificant works of art or that some researchers show towards some silly theories. Possibly, Wittgenstein showed towards metaphysics the consideration that psychotherapists show towards the delusions of their patients — with no disrespect for them. Wittgenstein discussed the treatment of highly sophisticated confusion, one that "the understanding got by bumping into the limits of language" (*PI*, §119). Readers whom Wittgenstein's texts disturb may find in them challenges that others overlook.[38] A number of leading Wittgenstein apologists have noticed that. They say, commentators who find Wittgenstein's works unchallenging approach them the wrong way: they have at least to imagine themselves disturbed by the ills he attempted to treat. I admit: I find this too exacting, perhaps because I allow people their confusions and help them with clarifications only at their requests.

Now the psychotherapeutic literature is informative on the agonies of patients, on the methods of treating them, and on the theories behind these methods. The parallel philosophical literature is almost non-existent. In particular, following Freud,

---

[35]*TLP*, §6.53: "to say nothing except . . ." *PI*, §38: "philosophical problems arise when language idles"; §109: "Philosophy is a battle against the bewitchment of our intelligence by means of language"; §255: "The philosopher treats a question: like an illness." This last remark is a pun on "treat" that survives translation.

[36]As gossip has it, Wittgenstein borrowed books from the university library only at night: he displayed contempt for all sorts of books and great interest in pulp fiction; he did this quite expressively. In brief, gossip describes him as pretending to be an anti-intellectual. It also describes him as a super-candid intellectual.

[37]David Pinset, diary of May 1912, reports young Wittgenstein surprised that philosophers he had admired were "stupid and dishonest" and sources of "disgusting mistakes". I do not know what to make of this.

[38]This is where Wittgenstein idiosyncrasies and eccentricities enter the picture.

psychotherapists would avoid treating non-patients.[39] Now psychological practice raises moral issues: therapists should respect the wish of patients to resist therapy. This is particularly so, Buber observed, when the patient is religious and the therapist is anti-religious (like Freud). Wittgenstein would have agreed. What then is Wittgenstein's therapy to say on this? Possibly, Wittgenstein aimed at curing only willing metaphysicians.

Wittgenstein's position regarding philosophic therapy is open for critical debate. There is yet no job-description for philosophers as therapist who cure philosophical agonies. At least those who say that they do not suffer but find philosophy and metaphysics delightful have the right to stay in the positions of their choice, be these positions in accord with Wittgenstein's teaching or not. Some say that Wittgenstein the therapist first tried to cure himself (in parallel with Freud's self-analysis); others say he tried to cure not any individual patient but philosophy in the abstract. There can be no objection to either view as long as we respect the expressed wish to stay confused. Wittgenstein was clearly aware of this. He echoed it in a very strong passage:

> The philosopher is one who must cure many diseases of one's understanding before one can come to the notions of common sense. If in the midst of life we are in death, so in sanity we are surrounded by madness.[40]

The subject of this passage is one who cures one's own understanding: Wittgenstein, for example. This passage also says what the aim of the cure is: to acquire common sense. This sharply contrasts with the cavalier attitude of some of the rather famous followers of Wittgenstein. Pedantic Austin was reputed to have ridiculed as ungrammatical the answer that the Good Lord gave to Moses from the Burning Bush when asked about His identity.[41] Though Austin knew no ancient Hebrew, much less its grammar, he dared criticize the grammar of the Good Lord's answer to the query of Moses.[42] Wittgenstein had said his intention was to leave things as they are, and prominent among these are the religious beliefs and rites that he respected when sincere. His position is superior to Austin's even on the understanding that he was inconsistent. (Russell always expressed preference for humane inconsistency over cruel consistency for advocates of cruel philosophies.)

Wittgenstein suggested that with the arrival of clarity, commonsense prevails and then peace of mind reigns supreme. His biographers say, he never gained peace of mind: he always struggled against the skepticism that he felt was but a step away from the solipsism that he identified with metaphysics in general and that he (rightly)

---

[39]This is questionable, as preventive medicine prevents pain. As Wittgenstein-style preventive medicine does not exist, we may overlook this criticism, valid and significant though it is.

[40]Wittgenstein 1977, V, 53. See also Crary and Read 2000, Introduction. Their claim that their reading of Wittgenstein is new is not very convincing. It was the rage in England when Wittgenstein's *Philosophical Investigations* appeared (1953).

[41]Munitz 1965, Sweeney 1997, 506.

[42]Oracles use vague language carefully crafted. At times, these are language games: intentionally ungrammatical: some religious rituals use meaningless expressions or expressions with forgotten mysterious meanings (Gellner [1959] 2005, 60; Jarvie 1972, 60).

considered an intolerable sickness. Did he find a cure for it? As a report by his sister has it, doubts tormented him from the first time he had encountered philosophy. As the manuscript that he wrote on his deathbed, *On Certainty*, shows, he did not achieve it. He said (to Joan Bevan, the owner of the house where he lied on his lonely deathbed), "tell them I had a wonderful life". Is this true (Klagge 2001, x)? Does it matter? I does if you think that genuine healers should be able to heal themselves. Notoriously, this is not always right.

Unlike Wittgenstein, Russell considered idealism and solipsism errors, not confusions. Unlike Russell's view, Wittgenstein's view invites proof. Early in the day, he gave a proof. It is invalid. He never admitted this and never suggested an alternative. He only suggested the technique of looking at the use of expressions. Severe criticism of this suggestion appeared in a widely discussed paper of Chihara and Fodor (1964).

Wittgenstein suggested that the allure of metaphysics is strong and evil. As he went on discussing this idea, his philosophy required increasing entanglement. As long as we identify rationality with certitude and certitude eludes us, we are in trouble. As long as we cannot refute idealism or solipsism, we cannot show that our resistance to it is rational. That is why Denis Diderot (and other eminently sane people) found the task of refuting it so very urgent: he had no urge to engage in metaphysics; idealism and solipsism had no appeal to him and caused him no pain; yet he considered them a philosophical hurdle to overcome in order to reestablish rationality. The way Wittgenstein offered to treat the situation would not have spoken to Diderot. Popper's lowering the standard of rationality would have appealed to him, as he would have understood it and he could deliberate on it.

Wittgenstein did not discuss rationality. He said about it one and only one thesis: the understanding is limited to the limits of language. This he argued in his early work, yet the argument holds only within his artificial language that is too limited, as he later admitted. He then asserted that language is limited so that it has no room for metaphysics. He illustrated it; he said, observing language in use will exhibit no metaphysics and no philosophical problems as these "arise when language *idles*". This is hardly an explanation, much less a proof. The strongest rebuttal of it is the claim (Einstein's) that science can scarcely progress without metaphysics. A few leading followers of his tried to offer proofs. They all fall short.

13.6 Conclusion

Wittgenstein's glory is his *Tractatus Logico-Philosophicus* viewed as neutral monism – that philosophers Hume, Mach and Russell admired – presented as logic Russell's way. Wittgenstein left out of it both the vernacular (*TLP*, §5.5563) and the scientific mode of speech (*TLP*, §§6.341*ff.*), leaving everything as it previously was, provided that metaphysics has no place in his sysstem.

All this, especially the anti-metaphysics, is *passé*; the best way to view the anti-metaphysics of Wittgenstein is as a protest against chitchat ("slum landlords"). Alas, anti-chitchat is a kind of chitchat, and even a popular one. He felt strongly puzzled about skepticism: it goes with commonsense a long way, but falls short. Beyond its commonsense limit, it is odd, ludicrous, and nevertheless disturbing. The story that Russell narrates is telling: when young Wittgenstein doubted that there are no rhinos in the room, Russell sarcastically advised him to look under the furniture. Not that Russell trusted commonsense more than young Wittgenstein, but that young Wittgenstein sought a criterion for certainty. The view of certainty as limited to immediate experiences (Quine 1988) messed things up. It is an error: you cannot possibly doubt your being "born of a woman, short of days and sated of vexation" (*Job* 14:1). In his last days, Wittgenstein improved his view: street names on signs here are certainly not in Chinese (Wittgenstein [1969] 1975, §70).

Why does skepticism trouble philosophers? Wittgenstein said, they are confused; their confusion is due to the deceptive allure of metaphysics. Popper disagreed; skepticism trouble those who wish to be rational, identify rationality with proof, and learn that proof is impossible (skepticism). This is indeed upsetting. The rejection of the identification of rationality with proof legitimizes the commonsense-certainty that Robert Boyle called moral certainty, even though it comes with no criterion. Mature Wittgenstein gave up the search for a criterion.

Following Russell, young Wittgenstein sought certitude more doggedly than anyone before him did. This is just admirable. Unless this is recognized, his first book will keep puzzling us. Its failure is comparable to Maxwell's failure to rescue the hypothesis of the aether and to Russell's failure to rescue logicism. Our intellectual traditions that does honor significant, enlightening failures, but only sporadically, not systematically. Pity.

Those who persist in the search for certainty often seek help from psychology. We should welcome all psychological attempts to explain certainty: there is much too little scientific study of it and of the alternatives to it. (The few extant efforts in this direction are mere expressions of hostility to theology or to metaphysics.) Yet psychology cannot help us sort out the philosophical puzzle at hand: all psychological theory is doubtful and is better put to test than trusted (Popper). Psychology is seldom as compelling as the assumptions we live by, and so it cannot alleviate our sense of puzzlement. It nevertheless enters the discussion, invited or not. The ancient skeptics (the Pyrrhonists) conjectured that undermining all sense of certitude would cause peace of mind. They never doubted this conjecture. A part of their observation is still true: often the refusal to doubt is an expression of fear, and whenever possible with little cost, it is advisable to overcome fear. We may doubt this: it is a matter for empirical psychology to investigate. It is thus irrelevant here.

The analytic literature repeatedly recognizes much of all this, usually not in the clear manner required by the standards that it advocates. This indicates that possibly there is an obstacle here. Since this study exhibits a bias towards critical rationalism, it should not be surprising that it suggests the hypothesis that the obstacle still is the shameful attitude of the fathers of the analytic school towards Popper. Let me propose again that readers consult the authoritative *Stanford Online Encyclopedia of Philosophy*, on articles such as Carnap, Rudolf and Reichenbach, Hans, whose defensiveness is obvious, say, in comparison with the article Wittgenstein, Ludwig.

This is not to blame the analytic literature; rather, it is to observe its failings and to explain them: that literature is the tail end of a venerable tradition of the Enlightenment Movement that began with Francis Bacon: we are prone to idleness and intellectual idleness leads to self-delusion and to the neglect of learning. The preventive medicine for this ailment is the suspension of belief. This way philosophers ignored the suspension of disbelief; they hardly ever discussed it, except for those who were obviously biased. Obviously, art appreciation depends on both (similarly to what Buber called proximity and distance): the flickering between belief – absorption into the artwork – and disbelief – the recognition of the artificiality of the artefact – is essential for the enjoyment of art. The literature about science asserts repeatedly the demand for the suspension of belief, in disregard for the demand for the suspension of disbelief that is essential for the ability to read radical novelties. (Russell complained that Wittgenstein could not postpone his criticism of a new idea to when Russell finished developing it.)

What else characterizes analytic philosophy? This is still under dispute. More so is the question, how much of it is Wittgenstein's? Yet this discussion often takes place *sub rosa* (Agassi 1987, 1988, 102). Wittgenstein fans often present the concern with language as his new and invigorating source of philosophical pursuits. Critics find it a mere expression of a regrettable indifference to traditional philosophical problems. It is, however, the unearthing of the metaphysical systems that stand behind common parlance – at least according to the Sapir-Whorf hypothesis. As language hides traces of diverse defunct systems, chiefly magical, the Sapir-Whorf hypothesis dooms to failure any attempt to learn metaphysic from the analysis of common parlance. Analysts decide what expressions in their mother tongue are metaphorical and what are literal. This reflects nothing but their own metaphysical preconceived notions – especially their anti-magic. It is advisable to state it openly.

Important failures deserve respectful examination; other failures it is better to ignore. Important failures comprise milestones in the progress of the human spirit. The paradigm case here is the failure of Maxwell to reconcile Newtonian mechanics with his own electro-dynamics – a failure that Einstein viewed as the significant background to some of his own early achievements. The failure of Russell's logicism is another impressive example. Quine (1960, 19n.) admired Carnap's *Der logische Aufbau der Welt* (1928) as a grand failure, akin to Russell's *Our Knowledge of the External World* (1914). He could just as well appreciate that way Wittgenstein's *Tractatus Logico-Philosophicus* (1921–1922) that he sadly dismissed as too cavalier about mathematics.

This raises a general question: by what criterion do we judge some failures contributions to human knowledge? It is doubtful that one adequate answer to this question is possible. This is my Wittgenstein-style response to the question; unlike Wittgenstein, however, let me suggest that this question deserves exploration nevertheless.

Sextus Empiricus dismissed the argument against skepticism from the certainty met in daily experience. He found this argument exasperating. This made him view doubt irrelevant to daily experience (except that it offers peace of mind). He was in error. Doubt is significant as it keeps openness to criticism and to possible alternatives to our convictions (regardless of peace of mind). The certainty that we may experience is not wholly psychological. It also depends on some very general assertions that we make. It is easy to see the superiority of current commonsense over the commonsense that prevailed in the modern world a century or two ago, not to mention the commonsense that prevails in any magically-oriented society. We once believed in tales that our parents told us; we lost our faith in the tales but (usually) not in our parents. This selectivity of loss of faith deserves study. Yet this much we already know: some general assumptions change, and then what we take for granted alters too. Take any field of inquiry that you are familiar with, and ask yourself if you can share the certainty of your predecessors of a century or two ago and if not why not. This exercise solves our puzzle (Michael Polanyi, Popper): we cannot doubt all background beliefs, or even know them all; nevertheless, with effort we can find some of them, try to criticize them, and at times also open the road to improvement. Such improvement is bliss.

The recognition of this is skepticism new-style (Einstein, Russell, Shaw, Popper, Polanyi): the skepticism that may be consistent with all sorts of hypotheses as well as with many alternatives to them – a skepticism identical with active fallibilism. Try the following exercise: take a mature Wittgenstein text, say *Philosophical Investigations* or *On Certainty*, and try to modify minimally any assertion in it that looks to you inconsistent with fallibilism. See if it still makes sense. I suspect you will agree that for the former this makes less sense than for the latter, since Wittgenstein identified fallibilism with Pyrrhonist skepticism. Fallibilist philosophy has as its patron saints Robert Boyle, Solomon Maimon, Heinrich Heine, Russell (1940, 128), Einstein (1944), Shaw, Popper and Polanyi. As to the literature that qualifies as current language analysts, non-Pyrrhonist fallibilism leaves behind most of it. The question is, was the exercise worthwhile, and if so, what did we learn from it? My answer is, yes, we have learned from it, though too little, and at an unnecessarily high cost; it is all of the ideas that Wittgenstein and his followers have articulated by minimal modifications that some unanswerable criticism has demanded. Most of these ideas have merged fruitfully into the current corpus of logical knowledge and into the current wordings of diverse post-Russellian philosophical problems, solutions and controversies.

"I thought your yacht was larger than it is."

"Tell them I had a wonderful life."

References

Adler, J. E. (1976). Evaluating global and local theories of induction. *PSA: Proceedings of the Biennial Meeting of the Philosophy of Science Association, 1976*(1), 212–223.

Agassi, J. (1963). Empiricism vs. inductivism. *Philosophical Studies, 14*, 85–86.

Agassi, J. (1964). The nature of scientific problems and their roots in metaphysics. In Bunge, 1962, pp. 189–211; Agassi 1975, Chapter 9, pp. 208–239.

Agassi, J. (1970). Philosophy as literature: The case of Borges, discussion of Jorge Luis Borges. *Other Inquisitions, Mind, 39*, 287–294.

Agassi, J. (1971). What is a natural law? *Studium Generale, 24*, 1051–1066 Reprinted in Agassi 1975.

Agassi, J. (1975). *Science in flux*. Dordrecht/Boston: D. Reidel Publishing Company.

Agassi, J. (1978). Logic and logic of. *Poznań Studies in the Philosophy of the Sciences and the Humanities, 4*, 1–11.

Agassi, J. (1981a). The secret of Carnap. Review of Jaakko Hintikka. *Rudolf Carnap, Logical Empiricist, Materials and Perspectives, Philosophia, 10*, 57–88.

Agassi, J. (1981b). *Science and society*. Dordrecht/Boston: D. Reidel Publishing Company.

Agassi, J. (1985). Two-valued logic in ordinary circumstances. *International Logic Review, 32*, 83–86.

Agassi, J. (1987). Whatever happened to the positivistic theory of meaning? *Zeitschrift für allgemeine Wissenschaftstheorie, 18*, 22–29.

Agassi, J. (1988). *The gentle art of philosophical polemics: Selected reviews and comments*. La Salle: Open court.

Agassi, J. (1993). Conditions for interpersonal communication. *Methodology and Science, 26*, 8–17.

Agassi, J. (1994). Review of Wayne A. Patterson. *Bertrand Russell's Philosophy of Logical Atomism, Canadian Philosophy Review, 14*, 44–45.

Agassi, J. (1995a). Contemporary philosophy of science as thinly masked antidemocratic apologetics. In Gavroglu, et al., pp. 153–170.

Agassi, J. (1995b). Naming and necessity: A second look. *Iyyun, 44*, 243–272.

Agassi, J. (1997). To salvage Neurath (Cartwright et al., *Otto Neurath: Philosophy between science and politics*). *Philosophy of the Social Sciences, 28*, 83–101.

Agassi, J. (1998). In retrospect: Wittgenstein – The end of a myth (Review of Brian McGuinness, *Young Wittgenstein*). *Protosociology I, 10*, 238–241.

Agassi, J. (2000). To close the Wittgenstein project. *Iyyun, 49*, 313–329.

Agassi, J. (2005). To renew a rational debate. *Iyyun, 54*, 317–323.

Agassi, J. (2006). On proof theory. In Rahnfeld 2006, pp. 264–282.

J. Agassi, *Ludwig Wittgenstein's Philosophical Investigations*, Synthese Library 401,
https://doi.org/10.1007/978-3-030-00117-9

Agassi, J. (2008). *A philosopher's apprentice: In Karl Popper's workshop*. Amsterdam/New York: Rodopi.
Agassi, J. (2009). The advantage of theft over honest toil. (Review of Landini, 2007.). *Philosophy of the Social Sciences, 29*, 507–546.
Agassi, J. (2010). Martin Buber: His place in twentieth-century philosophy, In H. L. Israeli (Ed.), *The fortitudes of creativity: In honor of Shlomo Giora Shoham*. (Part 2, pp. 71–82). Cambridge Scholars Publishing; New edition edition.
Agassi, J. (2013). *The very idea of modern science: Francis Bacon and Robert Boyle, Boston studies in the philosophy and history of science*. Dordrecht: Springer.
Agassi, J. (2014). *Popper and his popular critics: Thomas Kuhn, Paul Feyerabend, Imre Lakatos*. Cham/New York: Springer.
Agassi, J. (2016). *Beg to differ: The logic of disputes and argumentation*. Cham: Copernicus Books.
Agassi, J., & Laor, N. (2000). How ignoring repeatability leads to magic. *Philosophy of the Social Sciences, 30*, 528–586.
Agassi, J., & Meidan, A. (2008). *Philosophy from a skeptical perspective*. Cambridge/New York: Cambridge University Press.
Agassi, J., & Wettersten, J. R. (1980). Stegmüller Squared. *Zeitschrift für allgemeine Wissenschaftstheorie, 11*, 86–94.
Ahmed, A. (Ed.). (2010). *Wittgenstein's philosophical investigations: A critical guide*. Cambridge: Cambridge University Press.
Alam, R. (1990). *Bertrand Russell's logical atomism*. New Delhi: Mittal.
Ambrose, A., & Lazerowitz, M. ([1972] 2014). *Ludwig wittgenstein: Philosophy and language*. Hoboken: Taylor and Francis.
Ammereler, E., & Fischer, E. (2004). *Wittgenstein at work: Method in philosophical investigations*. London: Routledge.
Anscombe, G. E. M. ([1959, 1963] 1971). *An introduction to Wittgenstein's Tractatus*. London: Hutchinson University Library.
Anscombe, G. E. M. (1963). Semantic analysis by Paul Ziff. *Mind, 72*, 288–293.
Anscombe, G. E. M. (1990). Wittgenstein: Whose philosopher? *Royal Institute of Philosophy Supplements, 28*, 1–10.
Anscombe, G. E. M. (1995). Cambridge philosophers II: Ludwig Wittgenstein. *Philosophy, 70*, 395–407.
Anscombe, G. E. M. (2011). *From Plato to Wittgenstein: Essays* (M. Geach & L. Gormally, Eds.). Exeter: Academic.
Arneson, P. (2007). *Perspective on philosophy of communication*. West Lafayette: Purdue University Press.
Arrington, R. L., & Addis, M. (2001). *Wittgenstein and philosophy of religion*. London: Routledge.
Arrington, R. L., & Glock, H.-J. (1991). *Wittgenstein's philosophical investigations: Text and context*. London/New York: Routledge.
Austin, J. L. (1956–1957). A plea for excuses. *Proceedings of the Aristotelian Society, 57*, 1–30.
Auxier, R. E., & Hahn, L. E. (2006). *The philosophy of Jaakko Hintikka*. Chicago: Open Court.
Ayer, A. J. (1936). *Language, truth, and logic*. Burwood: Royal Blind Society of New South Wales.
Ayer, A. J. (Ed.). (1959). *Logical positivism*. Glencoe: The Free Press.
Ayer, A. J. (1969). *Metaphysics and common sense*. London: Macmillan Education.
Ayer, A. J. (1971). *Russell and Moore: The analytical heritage*. London: Macmillan.
Ayer, A. J. (1982). *Philosophy in the twentieth century*. London: Phoenix.
Ayer, A. J. (1985). *Wittgenstein: The man and his philosophy*. New York: Random House.
Ayer, A. J. (1991). *A. J. Ayer: Memorial essays* (A. Phillips Griffiths, Ed.). Cambridge: Cambridge University Press.
Bacon, F. (1620). *Novum Organum scientiarum*. London: John Bill.
Baker, L. R. (1984). On the very idea of a form of life. *Inquiry, 27*, 277–289.

Baker, G. P., & Hacker, P. M. S. (2009). *Wittgenstein: Rules, grammar and necessity: Volume 2 of an analytical commentary on the philosophical investigations. Wittgenstein: Rules, grammar and necessity. Essays and exegesis of §§185–242*. New York: John Wiley & Sons.
Bar-Am, N. (2008). *Extensionalism: The revolution in logic*. New York: Springer.
Bar-Am, N., & Agassi, J. (2014). Meaning: From parmenides to Wittgenstein: Philosophy as 'footnotes to parmenides'. *Conceptus, 41*, 188–208.
Bar-Elli, G. (2002). Analysis without Elimination: On the philosophical significance of Russell's 'On denoting'. In Matar & Biletzki, 2002, pp. 167–81.
Bar-Elli, G. (2005). Meaning and use in Wittgenstein's *Tractatus*. *Iyyun, 54*, 375–402.
Bar-Hillel, Y. (1950). On syntactic categories. *Journal of Symbolic Logic, 15*, 1–16 reprinted in his 1964, 19–37.
Bar-Hillel, Y. (1962). A prerequisite for rational philosophical discussion. In Kazemier 1962, pp. 1–5.
Bar-Hillel, Y. (1964). *Language and information: Selected essays on their theory and application*. Reading: Addison-Wesley.
Bar-Hillel, Y. (1966). Do natural languages contain paradoxes? *Studium Generale, 19*, 391–397.
Bartley, W. W., III. (1973). *Wittgenstein*. New York: J.B. Lippincott.
Bartley, W. W. III (1974). Theory of language and philosophy of science as instruments of educational reform: Wittgenstein and Popper as Austrian schoolteachers. In Cohen and Wartofsky, 1974, pp. 307–337.
Bartley, W. W., III. (1982–1983). Wittgenstein and homosexuality. *Salmagundi, 58–59*, 166–196.
Beaney, M. (1998). What is analytic philosophy? Recent work on the history of analytic philosophy. *British Journal for the History of Philosophy, 6*, 463–472.
Bell, D. (1979). *frege's theory of judgement*. Oxford/New York: Clarendon Press/Oxford University Press.
Benacerraf, P. (1973). Mathematical truth. *The Journal of Philosophy, 70*, 661–679.
Bennett, J. F. (1989). *Rationality: An essay towards an analysis*. Indianapolis: Hackett.
Bennett, M., Dennett, D., Hacker, P., Searle, J., & with Introduction and Conclusion by Daniel Robinson. (2007). *Neuroscience and philosophy: Brain, mind, and language*. New York: Columbia University Press.
Ben-Yami, H. (2017). Vagueness and family resemblance. In Glock, 2017, pp. 407–419.
Berlin, I. (1955). *Historical inevitability*. London/New York: Oxford University Press.
Biletzki, A. (2003). *(Over) interpreting Wittgenstein*. Dordrecht/Boston: Kluwer Academic Publishers.
Biletzki, A. (2014). Ludwig Wittgenstein. In *Stanford encyclopedia of philosophy*. Cambridge University Press: Cambridge.
Biletzki, A., & Matar, A. (Eds.). (1998). *The story of analytic philosophy: Plot and heroes*. London/New York: Routledge.
Black, M. (1934). The principle of verifiability. *Analysis, 2*, 1–6.
Black, M. (1944). The "paradox of analysis.". *Mind, 53*, 263–267.
Black, M. (1962). *Models and metaphors: Studies in language and philosophy*. Ithaca: Cornell University Press.
Black, M. (1971). *A companion to Wittgenstein's tractatus*. Ithaca: Cornell University Press.
Blackburn, S. (1984). *Spreading the word: Groundings in the philosophy of language*. Oxford: Clarendon Press.
Bloom, A., & Jaffa, H. (1964). *Shakespeare's politics*. New York: Basic Books.
Boole, G, (1847). *The mathematical analysis of logic, being an essay towards a calculus of deductive reasoning*. Cambridge: MacMillan and London: John Bell.
Boole, G. (1854). *An investigation of the laws of thought. Cambridge: Macmillan, Barclay and Macmillan, London: George Bell.*
Bostock, D. (2012). *Russell's logical atomism*. Oxford: Oxford University Press.
Bouveresse, J. (2013). *Wittgenstein reads freud: The myth of the unconscious*. Princeton: Princeton University Press.

Bradley, R. D. (1987). Wittgenstein's essentialism. *Australasian Journal of Philosophy, 65*, 43–55.

Broekman, J. M. (2016). *Meaning, narrativity, and the real*. Cham: Springer.

Bunge, M. (Ed.). (1964). *Critical Approach to Science and Philosophy*. In Honor of Karl R. Popper.

Bunge, M. (1974a). *Treatise on basic philosophy, 1. Semantics I: Sense and reference*. Dordrecht: Springer.

Bunge, M. (1974b). *Treatise on basic philosophy, 2. Semantics II: Interpretation and truth*. Dordrecht: Reidel.

Bunge, M. (1979a). *Treatise on basic philosophy, 3: Ontology I: The furniture of the world*. Dordrecht: Springer.

Bunge, M. (1979b). *Treatise on basic philosophy, 4: Ontology II: A world of systems*. Dordrecht: Springer.

Bunge, M. (1980). *The mind-body problem: A psychobiological approach*. Oxford: Pergamon.

Burge, T. (1973). Reference and proper names. *The Journal of Philosophy, 70*, 425–439.

Canfield, J. V. (1986). *The philosophy of Wittgenstein: The later philosophy*. New York: Garland Publication.

Carnap, R. (1932). *Überwindung der Metaphysik durch Logische Analyse der Sprache, Erkenntnis* (The Elimination of Metaphysics through Logical Analysis of Language, pp. 219–241. http://www.ditext.com/carnap/elimination.html).

Carnap, R. ([1934] 2014). *Logical syntax of language*. Routledge: London.

Carnap, R. (1937). *The logical syntax of language* (A. Smeaton, Trans.).

Carnap, R. (1943). *The formalization of logic*. Cambridge, MA: Harvard University Press.

Carnap, R. ([1947] 1956). *Meaning and necessity: A study in semantics and modal logic*. Chicago/London: University of Chicago Press.

Carnap, R. (1956). The methodological character of theoretical concepts. *Minnesota Studies in the Philosophy of Science, 1*, 33–76.

Carnap, R (1963). Intellectual autobiography. In Schilpp 1963.

Carnap, R. (1966). Probability and content measure. In Feyerabend & Maxwell 1966, pp. 248–260.

Carnap, R. ([1928] 1967). *The logical structure of the world* and *pseudo-problems in philosophy* (*Der Logische Aufbau der Welt* und *Scheinprobleme in der Philosophie*). London: Routledge & Kegan Paul

Carter, W. R. (1989). *The elements of metaphysics*. Philadelphia: Temple University Press.

Caton, C. E. (1963). *Philosophy and ordinary language*. Urbana: University of Illinois Press.

Cavell, S. (1962). The availability of Wittgenstein's later philosophy. *The Philosophical Review, 71*, 67–93.

Cavell, S. (1976). *Must we mean what we say*. London/New York/Cambridge: Cambridge University Press.

Cavell, S. (1979). *The claim of reason*. Oxford: Clarendon Press.

Chapman, S. (2009). *Key ideas in linguistics and the philosophy of language*. Edinburgh: Edinburgh University Press.

Chappell, V. C. (Ed.). (1964). *Ordinary language: Essays in philosophical method* (pp. 64–74). New York: Dover Publications.

Chihara, C. S., & Fodor, J. A. (1964). Operationalism and ordinary language: A critique of Wittgenstein. *American Philosophical Quarterly, 2*, 281–295.

Child, W. (2011). *Wittgenstein*. London: Routledge.

Church, A. (1943). Reviews. Willard V. Quine. *Notes on Existence and Necessity. The Journal of Philosophy, 40*, 113–127 and *the Journal of Symbolic Logic, 8*, 45–47.

Cloeren, H. J. (1975). The neglected analytical heritage. *Journal of the History of Ideas, 36*, 513–529.

Coffa, A. (1991). *To the Vienna station: The semantic tradition from Kant to Carnap* (L. Wessels, Ed.). Cambridge: Cambridge University Press.

Cohen, R. S., & Agassi, J. (Eds.). (1982). *Scientific philosophy today: Essays in honor of Mario Bunge*. Dordrecht: Reidel.

Cohen, R. S., & Laudan, R. (Eds.). (1983). *Physics, philosophy and psychoanalysis: Essays in honor of Adolf Grünbaum*. Dordrecht: Reidel.
Cohen, R. S., & Wartofsky, M. W. (1974). *Boston studies in the philosophy of science* (p. 14). Dordrecht: Springer.
Collingwood, R. G. (1924). *Speculum mentis*. Oxford: Clarendon Press.
Conant, J. (2001). *Philosophy and biography*. In Klagge, 2001, pp. 16–50.
Conant, J. (2005). Wittgenstein's late criticism of the *Tractatus*. In Pichler & Säätelä, (Eds.), 2005, pp. 172–204.
Conant, J., & Diamond, C. (2004). On reading the tractatus resolutely. In Kölbel & Weiss 2004, pp. 46–99.
Cook, J. W. (1994). *Wittgenstein's metaphysics*. Cambridge: Cambridge University Press.
Cook, J. W. (2000). *Wittgenstein, empiricism, and language*. New York/Oxford: Oxford University Press.
Cook, J. W. (2005). *The undiscovered Wittgenstein: The twentieth century's most misunderstood philosopher*. Amherst: Humanity Books.
Copleston, F. C. ([1956] 2002). *Contemporary philosophy: Studies of logical positivism and existentialism*. London: Burns & Oates
Craig, E. (1998). *Routledge encyclopedia of philosophy*. London: Routledge.
Crary, A. (Ed.). (2007). *Wittgenstein and the moral life: Essays in honor of cora diamond*. Cambridge, MA: MIT Press.
Crary, A., & Read, R. (Eds.). (2000). *The new Wittgenstein*. London: Routledge.
Darwin, C. (1989). *The works of Charles Darwin, volume 21: The descent of man, and selection in relation to sex*. Washington: New York University Press.
Davidson, D. (1986). A nice derangement of Epitaphs. In Lepore 1986, p. 157.
Davidson, D. (1999). Intellectual autobiography. In Lewis Edwin Hahn 1999, pp. 1–70.
de Rijke, M. (Ed.). (1997). *Advances in intensional logic*. Dordrecht: Kluwer.
Demeter, T. (Ed.). (2004). *Essays on Wittgenstein and Austrian philosophy: In honour of J. C. Nyiri*. Amsterdam: Rodopi.
Dewey, J. (1929). *Experience and nature*. New York: Norton.
Diamond, C. (2000). Does Bismarck have a beetle in his box? In Crarey & Read 2000, pp. 262–292.
Diamond, C. (2004). Criss-cross philosophy. In Ammereler & Fischer, 2004, pp. 201–220.
Drury, M. O'C. (1984). Conversations with Wittgenstein. In Rhees, 1984, pp. 97–171.
Duhem, P. (1905). Physics of a believer. In *The aim and structure of physical theories*. Princeton: Princeton University Press.
Dummett, M. (1960). Oxford philosophy. *Blackfriars, 41*, 74–80. Reprinted in Dummett 1978 as Ch. 24, 431–6.
Dummett, M. (1973). *Frege: Philosophy of language*. Cambridge, MA: Harvard University Press.
Dummett, M. (1975). *Frege*. Oxford/New York: Oxford University Press.
Dummett, M. (1978). *Truth and other enigmas*. London: Duckworth.
Dummett, M. (1981). *Frege: Philosophy of language*. London: Duckworth.
Dummett, M. (1991a). *Frege: Philosophy of mathematics*. Cambridge, MA: Harvard University Press.
Dummett, M. (1991b). *Frege and other philosophers*. Oxford: Oxford Clarendon Press.
Dummett, M. (1991c). *The logical basis of metaphysics*. London: Duckworth.
Duncan, S. M. (2007). *Analytic philosophy of religion: Its history since 1955*. Tirril/Penrith: Humanities-Ebooks.
Ebert, P. A., & Rossberg, M. (2009). Cantor on Frege's foundations of arithmetic. *History and Philosophy of Logic, 30*, 341–348.
Eddington, A. S. (1928). *The nature of the physical world: Gifford lectures (1927)*. New York: Macmillan Co.
Edmonds, D., & Eidinow, J. (2001). *Wittgenstein's poker: The story of a ten-minute argument between two great philosophers*. New York: Ecco.
Edwards, P. (1967). *The encyclopedia of philosophy* (Vol. 4). London: Free Press.

Einstein, A. (1944). Remarks on Bertrand Russell's theory of knowledge. In Schilpp, 1944, pp. 278–291.
Einstein, A. (1945). Appendix II: A testimonial from professor Einstein. In Hadamard 1945, pp. 141–142.
Einstein, A. (1949). Autobiographical notes. In Schilpp, 1949, pp. 1–95.
Einstein, A. ([1954] 2012). Foreword to concepts of space: The history of theories of space in physics: Third, enlarged edition, Cambridge MA: Harvard University Press.
Engel, S. M. (1975). *Wittgenstein's doctrine of the Tyranny of language: An historical and critical examination of his blue book*. The Hague: Martinus Nijhoff With an Introduction by Stephen Toulmin.
Engelmann, P. (1967). *Letters from Ludwig Wittgenstein. With a memoir*. (L. Furtmüller, Trans., B. F. McGuinness, Ed.)
Erneling, C. E. (1993). *Understanding language acquisition: The framework of learning*. Albany: State University of New York Press.
Fairclough, N. (2003). *Analysing discourse: Textual analysis for social research*. London: Routledge.
Fann, K. T. (1967). *Ludwig Wittgenstein: The man and his philosophy*. New York: Dell Pub. Co.
Fann, K. T. (1987). Wittgenstein: A critique. *Journal of the History of Philosophy, 25*, 614–615.
Fann, K. T. (2013). *Symposium on J. L. Austin*. London: Routledge.
Feigl, H. (1960). Mind-Body, not a pseudo-Problem. In his *Inquiries and provocations*, Volume 14 of the series *Vienna circle collection* (pp. 342–50).
Feyerabend, P. K. (1955). Wittgenstein's philosophical investigations. *Philosophical Review, 64*, 449–483.
Feyerabend, P. K. (1975). *Against method*. London: NLB.
Feyerabend, P. K. (1987). *Farewell to reason*. London: Verso.
Feyerabend, P. K. (1996). Herbert Feigl: A biographical sketch. In Feyerabend & Maxwell 1966, pp. 3–13.
Feyerabend, P. K., & Maxwell, G. (Eds.). (1966). *Mind, matter, and method: Essays in philosophy and science in honor of Herbert Feigl*. Minneapolis: University of Minnesota Press.
Findlay, J. N. (1953). Wittgenstein's philosophical investigations. *Revue Internationale de Philosophie, 25*, 203–216.
Findlay, J. N. (1984). *Wittgenstein: A critique*. London/Boston/Melbourne: Routledge and K. Paul.
Fischer, E. (2011). How to Practise philosophy as therapy: Philosophical therapy and therapeutic philosophy. *Metaphilosophy, 42*, 1026–1068.
Flew, A., & MacIntyre, A. C. (1955). *New essays in philosophical theology*. London: SCM Press.
Floyd, J. (1995). On saying what you really want to say: Wittgenstein, Gödel, and the trisection of the angle. In Hintikka 1995.
Floyd, J. (2007). Wittgenstein and the inexpressible. In Crary, 2007, pp. 177–234.
Fogelin, R. J. (2009). *Taking Wittgenstein at his word: A textual study*. Princeton: Princeton University Press.
Føllesdal, D. (1996). Analytic philosophy: What is it and why should one engage in it? *Ratio, 9*(3), 193–208.
Forster, M. N. (2004). *Wittgenstein on the arbitrariness of grammar*. Princeton: Princeton University Press.
Frege, G. (1984). *Collected Papers* (P. Geach & R. H. Stoothoff, Trans.).
French, P. A. (1993). Why did Wittgenstein read Tagore to the Vienna circle? *ProtoSociology, 5*, xxx.
French, P. A., Uehling, T. E., & Wettstein, H. K. (1979). *Contemporary perspectives in the philosophy of language*. Minneapolis: University of Minnesota Press.
Freud, S. (1923). *The ego and the Id*. New York: Dover Publications.
Friedman, M. (1999). *Reconsidering logical positivism*. Cambridge: Cambridge University Press.
Frongia, G., & McGuinness, B. (1990). *Wittgenstein: A bibliographical guide*. Oxford: Basil Blackwell Scientific Publications.

Gabriel, G. (2009). Carnap's 'elimination of metaphysics through logical analysis of language'. A retrospective consideration of the relationship between continental and analytic philosophy. *Linguistic and Philosophical Investigations, 8*, 53–70.
Gandon, S. (2012). *Russell's unknown Logicism: A study in the history and philosophy of mathematics*. Basingstoke/New York: Palgrave Macmillan Ltd, cop.
Garver, N. (1996). Philosophy as grammar. In Sluga and Stern 1996, pp. 139–170.
Gattei, S. (2008). *Thomas Kuhn's linguistic turn and the legacy of logical empiricism*. Farnham: Ashgate Publishing, Ltd.
Gavroglu, K., Stachel J., & Wartofsky M. (Eds.). (1995). *Physics, philosophy and the scientific community,* In Honor of Robert S. Cohen. Boston Studies Philosophy Science. Springer: Boston Studies, 163.
Gellner, E. (1957). Ideal language and kinship structure. *Philosophy of Science, 24*, 235–242.
Gellner, E. ([1959] 2005). *Words and things: An examination of, and an attack on, linguistic philosophy*. London/New York: Routledge Classics. Foreword by Bertrand Russell, with a New Introduction by Ian Jarvie.
Gellner, E. (1973). *Cause and meaning in the social sciences* (I. C. Jarvie & J. Agassi, Eds.). London/Boston: Routledge & Kegan Paul
Gellner, E. (1984). The gospel according to Ludwig. Review of *Wittgenstein on Rules and Private Language* by Saul A. Kripke: *The American Scholar*, 53, 243–4, 250–1, 254–6, 258–60, 262–3.
Gellner, E. (1991). *Spectacles and predicaments: Essays in social theory* (I. C. Jarvie & J. Agassi, Eds.). Cambridge/New York/Port Chester: Cambridge University Press.
Gellner, E. (1992). *Postmodernism, reason and religion*. Abingdon: Taylor & Francis.
Gibson, J., & Huemer, W. (Eds.). (2004). *The literary Wittgenstein*. London: Taylor and Francis.
Glock, H.-J. (2001). Wittgenstein and reason. In Klagge 2001, pp. 196–220.
Glock, H.-J. (2004). Was Wittgenstein an analytic philosopher? *Metaphilosophy, 35*, 419–444.
Glock, H.-J. (2008). *What is analytic philosophy?* Cambridge: Cambridge University Press.
Glock, H.-J. (Ed.). (2017). *A companion to Wittgenstein*. Hoboken: Wiley.
Glock, H.-J., & Hyman, J. (2017). *A companion to Wittgenstein*. Hoboken: Wiley.
Gödel, K. (1944). Russell's mathematical logic. In Schilpp, 1944, pp. 123–153.
Goldstein, L. (1999). *Clear and queer thinking: Wittgenstein's development and his relevance to modern thought*. London: Duckworth.
Goldstein, L. (2004). Wittgenstein as soil. In Kölbel & Weiss 2004, pp. 148–178.
Gomperz, H. (1941). The meanings of 'meaning'. *Philosophy of Science, 8*, 157–183.
Grant, H. (1990). Geometry and politics: Mathematics in the thought of Thomas Hobbes. *Mathematics Magazine, 63*, 147–154.
Grattan-Guinness, I. (1992). Russell and Karl popper: Their personal contacts. *Russell: the Journal of the Bertrand Russell Archives, 12*, 3–18.
Grice, P. (2001). *Aspects of reason* (R. Warner, Ed.). Oxford: Oxford University Press.
Griffin, N., & Jacquette, D. (Eds.). (2009). *Russell vs. Meinong: The legacy of on Denoting*. New York: Routledge.
Hacker, P. M. S. (1975). *Insight and illusion: Wittgenstein on philosophy and the metaphysics of experience*. London: Oxford University Press.
Hacker, P. M. S. (1996). *Wittgenstein's place in twentieth-century analytic philosophy*. Oxford: Blackwell.
Hacker, P. M. S. (1997). *Wittgenstein on human nature*. New York: Routledge.
Hacker, P. M. S. (2000). Was he trying to whistle it? In A. Crary & R. J. Read (Eds.), *The New Wittgenstein* (pp. 353–388). London: Routledge.
Hacker, P. M. S. (2001). *Wittgenstein: Connections and controversies*. Oxford: Clarendon Press.
Hacohen, M. H. (2002). *Karl Popper: The formative years, 1902–1945: Politics and philosophy in interwar Vienna*. Cambridge: Cambridge University Press.
Hadamard, J. (1954). *An essay on the psychology of invention in the mathematical field*. New York: Dover.

Hager, P. J. (2012). *Continuity and change in the development of Russell's philosophy*. Dordrecht: Kluwer.
Hahn, L. E. (1992). *The philosophy of A. J. Ayer*. La Salle: Open Court.
Hahn, H. (1995). *Collected works*. In H. Schmetterer, & K. Sigmund, (Eds.), Vol. I.
Hahn, L. E. (1999). *The philosophy of Donald Davidson*. Chicago: Open Court.
Hailperin, T. (1981). Boole's algebra isn't Boolean Algebra. *Mathematics Magazine, 54*, 172–184.
Hall, R. (1959). Excluders. *Analysis, 20*, 1–7 Reprint, Caton 1963, 67–73.
Hall, J. A., & Jarvie, I. C. (Eds.). (1996). *The social philosophy of ernest Gellner*. Amsterdam: Rodopi.
Hallen, B., & Sodipo J. O. ([1986] 1997). *Knowledge, belief, and witchcraft: Analytic experiments in african philosophy*. Stanford: Stanford University Press
Haller, R. (1988). *Questions on Wittgenstein*. Amsterdam: Rodopi.
Hallett, G. L. (1991). *Essentialism: A Wittgensteinian critique*. Albany: State University of New York Press.
Hallett, G. L. (2008). *Linguistic philosophy: The central story*. Albany: State University of New York Press.
Hanfling, O. (2013). *Philosophy and ordinary language: The Bent and genius of our tongue*. London: Routledge.
Hare, R. M. (1964). A question about Plato's theory of ideas. In Bunge 1964, pp. 61–81.
Harman, P. M. (2001). *The natural philosophy of James Clerk Maxwell*. Cambridge/New York: Cambridge University Press.
Harvey, H. (1961). The problem of the model language-game in Wittgenstein's later philosophy. *Philosophy, 36*, 333–351.
Harvey, H. (1963). A reply to Dr Carney's challenge. *Philosophy, 38*, 170–175.
Hayes, C. J. (2003). *The grammar of politics: Wittgenstein and political philosophy*. Ithaca: Cornell University Press.
Hazewinkel, M. (2013). Intermediate logics. *Encyclopaedia of Mathematics, 5*, 135–137.
Heidegger, M. (2006). *Towards the definition of philosophy*. New York: Lang.
Hempel, C. G. (1935). On the logical positivists' theory of truth. *Analysis, 2*, 49–59.
Hempel, C. G. (1950). Problems and changes in the empiricist criterion of meaning. *Revue Internationale de Philosophie, 4*, 41–63.
Hempel, C. G. (1965). *Aspects of scientific explanation and other essays in the philosophy of science*. New York: Free Press.
Hermann, J. (2015). *On moral certainty, justification and practice: A Wittgensteinian perspective*. Basingstoke: Palgrave Macmillan.
Hilbert, D., & Ackermann, W. (1928). *Grundzüge der theoretischen Logik*. Berlin: Springer.
Hilmy, S. (1985). *The later Wittgenstein*. Oxford: Blackwell.
Hintikka, J. (1958). On Wittgenstein's 'solipsism'. *Mind, 67*, 88–91.
Hintikka, J. (1967). Individuals, possible worlds, and epistemic logic. *Nous*, 1, 22–62.
Hintikka, J. (Ed.). (1975). *Rudolf Carnap, logical empiricist: Materials and perspectives*. Dordrecht: Springer.
Hintikka, J. (1981). On denoting what? *Synthese, 46*, 167–183.
Hintikka, J. (1991). An impatient man and His papers. *Synthese, 87*, 183–201 (A shorter version appeared in *The Times Literary Supplement* 1990, p. 1030.).
Hintikka, J. (1993). On proper (Popper?) and improper uses of information in epistemology. *Theoria, 59*, 158–165.
Hintikka, J. (Ed.). (1995). *From Dedekind to Gödel*. Dordrecht: Springer.
Hintikka, J. (1996). *Ludwig Wittgenstein: Half-truths and one-and-a-half-truths*. Dordrecht: Kluwer.
Hintikka, J. (1997). No scope for scope? *Linguistics and Philosophy, 20*, 515–544.
Hintikka, J. (2003). What does the Wittgensteinian inexpressible express? *The Harvard Review of Philosophy, 11*, 9–17.
Hintikka, J. (2004). Wittgenstein's demon and his theory of mathematics. *Demeter, 2004*, 89–108.

Hintikka, J. (2006). Intellectual autobiography. In Auxier & Hahn, 2006, pp. 1–84.
Hintikka, J., & Remes U. (1974). *The method of analysis: Its geometrical origin and its general significance* (Boston Studies in the Philosophy of Science, 25). Dordrecht: Springer
Hooker, C. A. (1975). Philosophy and meta-philosophy of science: Empiricism, Popperianism and realism. *Synthese, 32*, 177–231.
Horwich, P. (2012). *Wittgenstein's metaphilosophy*. Oxford: Oxford University Press.
Howes, B. (2007). 'Rethinking' the preface of the *Tractatus*. *Philosophical Investigations, 30*, 3–24.
Hylton, P. (1989). The significance of *on denoting*. In A. Savage & H. Feigl (Eds.), *Rereading Russell* (pp. 88–107). Minneapolis: University of Minnesota Press.
Hylton, P. (1992). *Russell, idealism, and the emergence of analytic philosophy*. Oxford: Oxford University Press.
Iglesias, T. (1984). Russell's 'theory of knowledge' and Wittgenstein's earliest writings. *Synthese, 60*, 285–332.
Ignatieff, M. (2011). *Isaiah berlin: A life*. London: Vintage.
Jacquette, D. (1997). Haller on Wittgenstein and Kant. In Lehrer & Marek, 1997, pp. 29–44.
Jammer, M. (1954). *Concepts of space. The history of theories of space in physics*. Cambridge, MA: Harvard University Press With a Foreword by Albert Einstein.
Janik, A., & Toulmin, S. (1973). *Wittgenstein's Vienna*. New York: Simon and Schuster.
Jarvie, I. C. (1972). *Concepts and society*. London: Routledge.
Jarvie, I. C. (1983). Book review: remarks on Frazer's Golden Bough. *Philosophy of the Social Sciences*, 13, 117–118.
Jarvie, I. C., & Agassi, J. (1967). The rationality of magic. *British Journal of Sociology, 18*, 55–74.
Jarvie, I. C., & Laor, N. (Eds.). (1995). *Critical rationalism, the social sciences and the humanities*. Dordrecht: Kluwer Academic Publishers.
Joad, C. E. M. (1950). *A critique of logical positivism*. London: Victor Gollancz.
Jolley, K. D. (Ed.). (2010). *Wittgenstein: Key concepts*. Durham: Acumen.
Katz, J. (1990). *The metaphysics of meaning*. Cambridge, MA/London: MIT Press.
Kazemier, B. H. (1962). *Logic and language: Studies dedicated to professor Rudolf Carnap on the occasion of his seventieth birthday*. Dordrecht: Reidel.
Kenny, A. (1993). *Aquinas on mind*. New York: Routledge.
Kenny, A. (2006). *Wittgenstein* (Rev ed.). Malden: Blackwell.
Keynes, J. M. (1920). *A treatise on probability*. London/Basingstoke/Macmillan/New York: Cambridge University Press.
Kienpointner, M. (1996). Whorf and Wittgenstein. Language, world view and argumentation. *Argumentation, 10*, 475–494.
Kindi, V. (2017). Wittgenstein and philosophy of science. In Glock & Hyman, 2017, pp. 587–602.
Kitcher, P. (1979). Frege's epistemology. *Philosophical Review, 88*, 235–262.
Kitching, G., & Pleasants, N. (Eds.). (2002). *Marx and Wittgenstein: Knowledge, morality and politics*. London: Routledge.
Klagge, J. C. (2001). *Wittgenstein: Biography and philosophy*. Cambridge: Cambridge University Press.
Klagge, J. C. (2010). *Wittgenstein in exile*. Cambridge: The MIT Press.
Klemke, E. D. (1962). Mr. Warnock on Moore's conception of philosophy. *Philippine Studies, 13*, 81–84.
Kluge, W. H. (2013). *The metaphysics of Gottlob Frege: An essay in ontological reconstruction*. The Hague: Martinus Nijhoff.
Kneale, W. C., & Kneale, M. (1962). *The development of logic*. Bristol: Thoemmes Press.
Kölbel, M., & Weiss, B. (Eds.). (2004). *Wittgenstein's Lasting Significance*. London/New York: Routledge.
Koopman, C. (2010). Bernard Williams on Philosophy's need for history. *The Review of Metaphysics, 64*, 3–30.
Kraft, V. ([1952] 2015). *The Vienna circle: The origin of neo-positivism, a chapter in the history of recent philosophy*. New York: Philosophical Library.

Kreisel, G. (1958). Wittgenstein's remarks on the foundations of mathematics. *British Journal for the Philosophy of Science, 9*, 135–158.
Kreisel, G. (1978). Review of *Wittgenstein's lectures on the foundations of mathematics, Cambridge*, 1939. *Bulletin of the American Mathematical Society, 84*, 79–90.
Kremer, M (2010). Sense and reference: The origins and development of the distinction. In Ricketts & Potter 2010, pp. 220–292.
Kretzmann, N., Kenny, A., & Pinborg, J. (1988). *The Cambridge history of later medieval philosophy: From the rediscovery of Aristotle to the disintegration of Scholasticism*. Cambridge: Cambridge University Press.
Kripke, S. (1972). *Naming and necessity*. Cambridge, MA: Harvard. University Press.
Kripke, S. (1976). *A puzzle about belief*. In Margalit 1976, pp. 239–283.
Kripke, S. (1982). *Wittgenstein on rules and private language: An elementary exposition*. Oxford: Basil Blackwell.
Kripke, S. (2005). Russell's notion of scope. *Mind, 114*, 1005–1037.
Kripke, S. (2008). Frege's theory of sense and reference: Some exegetical notes. *Theoria, 74*, 181–218.
Kuusela, O., & McGinn, M. (Eds.). (2011). *The Oxford handbook of Wittgenstein*. Oxford/New York: Oxford University Press.
Lakatos, I. (1968). Criticism and the methodology of scientific research programmes. *Proceedings of the Aristotelian Society, 69*(1), 149–186.
Lambert, K. (1963). Existential import revisited. *Notre Dame Journal of Formal Logic, IV*, 288–292.
Landini, G. (1998). *Russell's hidden substitutional theory*. New York: Oxford University Press.
Landini, G. (2007). *Wittgenstein's apprenticeship with Russell*. Cambridge/New York: Cambridge University Press.
Landini, G. (2010). *Russell*. London/New York: Routledge.
Landini, G. (2012). *Frege's notations: What they are and how they mean*. Basingstoke: Palgrave Macmillan.
Langford, C. H. (1949). A proof that synthetic a priori propositions exist. *The Journal of Philosophy, 46*, 20–24.
Lehrer, K. J., & Marek, C. (1997). *Austrian philosophy past & present: Essays in Honor of Rudolf Haller*. Dordrecht/Boston: Kluwer Academic.
Leinfellner, W. (1982). Is Wittgenstein a transcendental philosopher? (Wittgenstein and Kant). *Revista Portuguesa de Filosofia*, T. 38, Fasc. 1, *Actas do Colóquio Luso-Austríaco Sobre Ludwig Wittgenstein*: 13–27.
Leinfellner-Rupertsberger, E. (1990). On the purported pragmatico-semantic foundation of linguistics & AI through Wittgenstein's late philosophy. *Journal of Pragmatics, 14*, 853–881.
Lejewski, C. (1954). Logic and existence. *British Journal for the Philosophy of Science, 5*, 104–119.
Lenin, V. I. (1909). Materialism and Empirio-criticism: Critical comments on a reactionary philosophy. Moscow: Zveno Publishers. (2002). (trans: Abraham Fineberg). Honolulu: University Press of the Pacific.
Leonardi, P., & Santambrogio, M. (1995). *On quine: New essays*. Cambridge/New York: Cambridge University Press.
Lepore, E. (Ed.). (1986). *Truth and interpretation: Perspectives on the philosophy of Donald Davidson*. Oxford: Basil Blackwell.
Lerner, B. D. (1994). Wittgenstein's Scapegoat. *Philosophical Investigations, 17*, 604–612.
Lewis, H. D. (Ed.). (1956). *Contemporary British philosophy* (Third series). London: G. Allen & Unwin.
Lewis, D. (1979). Scorekeeping in a language game. *Journal of Philosophical Logic, 8*, 339–359.
Lewis, C. I., & Langford, C. H. (Eds.). (1932). *Symbolic logic*. New York: The Century.
Linsky, L. (Ed.). (1952). *Semantics and the philosophy of language: A collection of readings*. Urbana: University of Illinois Press.

Linsky, L. (1967). *Referring*. New York: Humanities Press.
Linsky, L. (1977). *Names and descriptions*. Chicago/London: The University of Chicago Press.
Lurie, Y. (2012). *Wittgenstein on the human spirit*. Amsterdam: Rodopi.
MacFarlane, J. (2002). Frege, Kant and logicism. *Philosophical Review, 111*, 25–56.
Maimon, S. (1792/3, 2001). *Autobiography* (J. C. Murray, Trans.).
Malcolm, N. (1954). Wittgenstein's philosophical investigations. *The Philosophical Review, 63*, 530–559.
Malcolm, N. (1967). Wittgenstein. In Edwards, 8, pp. 335–337.
Malcolm, N. (1986). *Wittgenstein, nothing is hidden: Wittgenstein's criticism of his early thought*. Oxford: Basil Blackwell.
Malcolm, N. (1989). Wittgenstein on language and rules. *Philosophy, 64*, 5–28.
Malcolm, N. ([1967] 2001). *L. Wittgenstein: A memoir* (2nd ed.). With a Biographical Sketch by G. H. von Wright.
Malcolm, N. ([1993] 2002). *Wittgenstein: A religious point of view?* Edited with Response by Peter Winch.
Malmgren, A.-S. (2011). Rationalism and the content of intuitive judgments. *Mind, 120*, 263–327.
Marchi, P., Agassi J., & Wettersten J. R. (1982). The end of heuristic? (Review of Hintikka & Remes, *The Method of Analysis*). *Philosophia, 11*, 249–76.
Margalit, A. (Ed.). (1976). *Meaning and use*. Boston: Reidel.
Marion, M. (2011). *Wittgenstein on surveyability of proof*. In Kuusela, & McGinn, (Eds.), 2011, Ch. 7.
Markus, G. (2012). *Language and production: A critique of the paradigms*. Dordrecht: Reidel.
Martin, R. M. (1963). On the Frege-Church theory of meaning. *Philosophy and Phenomenological Research, 23*, 605–609.
Matar, A., & Biletzki, A. (2002). *The story of analytic philosophy: Plot and heroes*. London: Routledge.
Mates, B. (1964). On the verification of statements about ordinary language. In Chappell, 1964, pp. 64–74.
Matusova, E. (2015). *The meaning of the letter of Aristeas: In light of biblical interpretation and grammatical tradition, and with reference to its historical context*. Göttingen/Bristol: Vandenhoeck & Ruprecht, cop.
McCutcheon, F. (2001). *Religion within the limits of language alone: Wittgenstein on philosophy and religion*. Florence: Taylor and Francis.
McGuinness, B. (Ed.). (1985). *Moritz Schlick*. Dordrecht: Reidel.
McGuinness, B. (1988). *Wittgenstein: A life: Young Ludwig, 1889–1921*. London: Duckworth.
McGuinness, B. (2002). *Approaches to Wittgenstein*.
McGuinness, B. (2012). *Wittgenstein in Cambridge: Letters and documents 1911–1951* (4th ed.). Malden/Oxford: Wiley-Blackwell.
McGuinness, B., & Haller, R. (Eds.). (1989). *Wittgenstein in focus*. Amsterdam: Rodopi.
McMullin, E. (Ed.). (1963). *The concept of matter in modern philosophy*. Notre Dame: University of Notre Dame Press.
Meinong, A. ([1902] 1983). *On assumptions*. Berkeley: University of California Press
Modrak, D. K. W. (2001). *Aristotle's theory of language and meaning*. Cambridge: Cambridge University Press.
Monk, R. (1990, 2012). *Ludwig Wittgenstein: The duty of genius*. London: Vintage Digital.
Monk, R. (1996). *Bertrand Russell, 1872–1921: The spirit of solitude*. New York: Free Press.
Monk, R. (2000). *Bertrand Russell, 1921–70: The ghost of madness*. London: Vintage.
Monk, R. (2001). Philosophical biography: The very idea. In Klagge 2001, pp. 3–15.
Monk, R. (2005). *How to read Wittgenstein*. New York/London: W.W. Norton & Company.
Moore, G. E. (1953). *Some main problems of philosophy*. London: Allen and Unwin.
Morris, K., & Preti, C. (2015). How to read Moore's "proof of an external world". *Journal for the History of Analytical Philosophy, 4*.

Mumford, S. (2003). *Introduction to Russell on metaphysics: Selections from the writings of Bertrand Russell*. New York: Routledge/Taylor & Francis.
Munitz, M. K. (1965). *The mystery of existence*. New York: Herder and Herder.
Munz, P. (2004). *Beyond Wittgenstein's Poker: New light on Popper and Wittgenstein*. Hampshire/ Aldershot: Ashgate.
Neurath, O. (1973). *Empiricism and sociology*. Dordrecht: Springer.
Neurath, O. (1983). Pseudorationalism of falsification. In Neurath, *Philosophical Papers* 1913–1946, pp. 121–131.
Nieli, R. (1987). *Wittgenstein: From mysticism to ordinary language: A study of Viennese positivism and the thought of Ludwig Wittgenstein*. Albany: State University of New York Press.
Nielsen, K. (2001). Wittgenstein and Wittgensteinians on religion. In Arrington & Addis, 2001.
Nielsen, K., & Phillips, D. Z. (2005). *Wittgensteinian Fideism?* London: SCM Press.
Niiniluoto, I. (1999). *Critical scientific realism*. Oxford/New York: Oxford University Press.
Notturno, M. A. (1984). Frege's justificationism: Truth and the recognition of authority. *Metaphilosophy*, 15210–15224.
Notturno, M. A. (1985). *Objectivity, rationality, and the third realm: Justification and the grounds of psychologism: A study of Frege and Popper*. Dordrecht: Nijhoff.
O'Connor, D. ([1982] 2012). The metaphysics of G. E. Moore. Dordrecht: Reidel, Springer.
Oderberg, D. S. (2011). Essence and properties. *Erkenntnis, 75*, 85–111.
Pap, A. (1950). Logic and the synthetic *a Priori*. *Philosophy and Phenomenological Research, 10*, 500–514.
Parsons, M. (1994). *Touch monkeys: Nonsense strategies for reading twentieth-century poetry*. Toronto: University of Toronto Press.
Pascal, F. (1973). Wittgenstein: A personal memoir. *Encounter*, August 1973, 23–38. Reprinted in Rhees, 1984.
Passmore, J. A. (1954). Professor Ryle's use of 'use' and 'usage'. *Philosophical Review, 63*, 58–64.
Passmore, J. A. (1969). *A hundred years of philosophy* (2nd ed.). Allen and Unwin: London.
Pears, D. F. (Ed.). (1957). *The nature of metaphysics*. London: Macmillan.
Pears, D. F. (1985). Introduction to the 1985 edition of Russell 1918.
Pears, D. F. (Ed.). (1988). *The false prison: A study of the development of Wittgenstein's philosophy. Vol. 2*. Oxford: Clarendon Press.
Pelletier, F. J., & Linsky B. (2009). Russell vs. Frege on definite descriptions as singular terms. In Griffin & Jacquette, 2009, pp. 40–64.
Penco, C. (2010). The influence of Einstein on Wittgenstein's philosophy. *Philosophical Investigations, 33*, 360–379.
Pepper, S. C. (1942). *World hypotheses: A study in evidence*. Berkeley/Los Angeles: University of California Press.
Peterman, J. F. (1992). *Philosophy as therapy: An interpretation and defense of Wittgenstein's later philosophical project*. Albany: State University of New York Press.
Pichler, A., & Säätelä, S. (Eds.). (2005). *Wittgenstein: The philosopher and his works*. Bergen: The Wittgenstein Archives at the University of Bergen.
Pietarinen, A.-V. (2011). Principles and practices of Neurath's picture language. In Symons et al. 2011, pp. 71–82.
Pincock, C. (2005). A reserved reading of Carnap's *Aufbau*. *Pacific Philosophical Quarterly, 86*, 518–543.
Pitcher, G. (1968). *Wittgenstein: The philosophical investigations* [A collection of critical essays]. Notre Dame: University of Notre Dame Press.
Poincaré, H. (1902–1908). *The foundations of science*. Ann Arbor: G. Wahr Pub. Co.
Pole, D. (1958). *The later philosophy of Wittgenstein: A short introduction, with an epilogue on John wisdom*. London: Athlone Press.
Popkin, R., & Stroll, A. (1996). *Philosophy made simple: A complete guide to the World's Most important thinkers and theories*. New York: Doubleday.
Popper, K. ([1935] 1959). *The logic of scientific discovery*. Toronto: University of Toronto Press.

Popper, K. (1945). *The open society and its enemies*. London: Routledge and K. Paul.

Popper, K. (1947a). Logic without assumptions. *Proceedings of the Aristotelian Society, 47*, 251–292.

Popper, K. (1947b). New foundations for logic. *Mind, 56*, 193–235. (Corrections and additions in *Mind, 57*, 1948, 69–70).

Popper, K. (1947c). Functional logic without axioms or primitive rules of inference. *Koninklijke Nederlandse Akademie van Wetenschappen, Proceedings of the Section of Sciences, 50*, 1214–1224.

Popper, K. (1948a). On the theory of deduction, part I. Derivation and its generalizations. *Koninklijke Nederlandse Akademie van Wetenschappen, Proceedings of the Section of Sciences, 51*, 173–183.

Popper, K. (1948b). On the theory of deduction, part II. The definitions of classical and intuitionist negation. *Koninklijke Nederlandse Akademie van Wetenschappen, Proceedings of the Section of Sciences, 51*, 322–331.

Popper, K. (1949). The trivialization of mathematical logic. In E. W. Beth, H. J. Post, & J. H. A. Hollak (Eds.), *Proceedings of the tenth international congress of philosophy,* (vol. 1, pp. 722–727).

Popper, K. (1952). The nature of philosophical problems and their roots in science. *British Journal for the Philosophy of Science, 3*, 124–156 Reprinted in Popper 1963.

Popper, K. (1954). Self-reference and meaning in ordinary language. *Mind, 63*, 162–169 Reprinted in Popper 1963.

Popper, K. (1957). *The poverty of historicism*. London: Routledge & K. Paul.

Popper, K. (1963). *Conjectures and refutations*. New York: Basic Books.

Popper, K. (1986). *Unended quest: An intellectual autobiography*. London: Routledge.

Popper, K. (1995). Hans Hahn – Remembrance of a grateful student. In Hahn, 1995, pp. 1–20.

Preston, A. (Ed.). (2017). *Analytic philosophy: An interpretive history*. Florence: Taylor and Francis.

Preti, C. (2017). *Some problems of Moore interpretation*. (Preston 2017, Chapter 5).

Preyer, G., & Peter, G. (2002). *Logical form and language*. Oxford/New York: Clarendon Press/ Oxford University Press.

Proctor, G. L. (1959). Scientific laws and scientific objects in the *Tractatus*. *British Journal for the Philosophy of Science, 10*, 177–193.

Proops, I. (2001). The new Wittgenstein: A critique. *European Journal of Philosophy, 9*, 375–404.

Proops, I. (2017). Wittgenstein's logical atomism. In *Stanford encyclopedia of philosophy*.

Putnam, H. (2008). *Jewish philosophy as a guide to life: Rosenzweig, Buber, Lévinas, Wittgenstein*. Bloomington: Indiana University Press.

Quine, W. Van O. (1940). *Mathematical logic*. New York: Harper and Row.

Quine, W. Van O. (1943). Notes on existence and necessity. *The Journal of Philosophy, 5*, 113–127.

Quine, W. Van O. (1947). On universals. *The Journal of Symbolic Logic, 12*, 74–84.

Quine, W. Van O. (1950). Two dogmas of empiricism. *Philosophical Review, 60*, 20–43 1943.

Quine, W. Van O. (1960). *Word and object*. Cambridge, MA: MIT Press.

Quine, W. Van O. (1964). Implicit definition sustained. *The Journal of Philosophy, 61*, 71–74.

Quine, W. Van O. (1967). Russell's ontological development. *The Journal of Philosophy, 21*, 657–667.

Quine, W. Van O. (1981). *Theories and Things*. Cambridge, MA: Belknap Press.

Quine, W. Van O. (1988). A comment on Agassi's remarks. *Zeitschrift für allgemeine Wissenschaftstheorie, 19*, 117–118.

Quine, W. Van O. (1990). *Pursuit of truth*. Cambridge, MA: Harvard University Press.

Quine, W. Van O. (2008). *Confessions of a confirmed extensionalist: And other essays*. Cambridge, MA: Harvard University Press.

Rahnfeld, M. (Ed.). (2006). *Is there certain knowledge?/Gibt es sichere erkenntnis? Grundlagenprobleme Unserer Zeit, Vol. V*. Leipzig: Leipziguniversitätverlag.

Ramsey, F. P. (1923). Critical notice of the *Tractatus*. *Mind, 32*, 465–478 Reprinted in Ramsey 1931.
Ramsey, F. P. (1931). *The foundations of mathematics and other logical essays* (R. B. Braithwaite, Ed., Vol. 5). London: Kegan Paul & Co.
Read, R., & Lavely, M. A. (Eds.). (2011). *Beyond the tractatus wars: The new Wittgenstein Debate*. London: Routledge.
Reichenbach, H. (1944). *Philosophic foundations of quantum mechanics*. Berkeley/Los Angeles: University of California Press.
Reimer, M. (2002). Ordinary proper names. In Preyer & Peter, 2002, pp. 444–466.
Reimer, M. (2004). What malapropisms mean: A reply to Donald Davidson. *Erkenntnis, 60*, 317–334.
Rhees, R. (1984). *Recollections of Wittgenstein*. Oxford: Oxford University Press.
Rhees, R. (1998). *Wittgenstein and the possibility of discourse* (D. Z. Phillips, Ed.). Cambridge: Cambridge University Press.
Rhees, R. (2003). *Wittgenstein's on certainty: There – Like our life* (D. Z. Phillips, Ed.). Malden: Blackwell.
Richardson, A., & Uebel, T. (Eds.). (2007). *The Cambridge companion to logical empiricism*. New York: Cambridge University Press.
Richter, D. (2004). *Wittgenstein at his word*. London: Continuum.
Ricketts, T., & Potter, M. (2010). *The Cambridge companion to Frege*. Cambridge: Cambridge University Press.
Robinson, R. (1936). Analysis in greek geometry. *Mind, 45*, 464–473.
Robinson, A. (1966). *Non-standard analysis*. Princeton: Princeton University Press.
Rodych, V. (1999). Wittgenstein's inversion of Gödel's theorem. *Erkenntnis, 51*, 173–206.
Rodych, V. (2018). Wittgenstein's philosophy of mathematics. In *Stanford encyclopedia of philosophy*.
Rorty, R. (Ed.). (1967). *The linguistic turn: Essays in philosophical method*. Chicago: University of Chicago Press.
Rorty, R. (1970). Incorrigibility as the mark of the mental. *The Journal of Philosophy, 67*, 399–424.
Rorty, R. (1971). Verificationism and transcendental arguments. *Noûs, 5*, 3–14.
Rorty, R. (1979). *Philosophy and the mirror of nature*. Princeton: Princeton University Press.
Rorty, R. (2000). Wittgenstein and the linguistic turn. In Ahmed 2000, pp. 129–44.
Royall, N. N. (1952). Logical positivism revisited. *Review of General Semantics, 9*, 139–147.
Russell, B. (1897). *An essay on the foundations of geometry*. Delhi/Kalpaz/New Delhi: Distributed by Gyan Books Pvt. Ltd.
Russell, B. (1900). *A critical exposition of the philosophy of Leibniz*. London: Geo. Allen & Unwin Ltd.
Russell, B. (1902). *Principles of mathematics*. Cambridge: Cambridge University Press.
Russell, B. (1903). *A free man's worship*. Cambridge: Cambridge University Press.
Russell, B. (1905). On denoting. *Mind, 14*, 479–493 reprinted in his 1956a, 39–56.
Russell, B. (1912). *Problems of philosophy*. New York: H. Holt.
Russell, B. (1913). See Russell 1984.
Russell, B. (1914). *Our knowledge of the external world: As a field for scientific method in philosophy*. London: G. Allen & Unwin.
Russell, B. (1917). *Mysticism and logic and other essays*. London: G. Allen & Unwin.
Russell, B. (1918). Philosophy of logical atomism. *The Monist, 28*, 495–527.
Russell, B. (1919). *Introduction to mathematical philosophy*. London: Allen & Unwin.
Russell, B. (1922). Introduction to Ludwig Wittgenstein. In *Tractatus logico-philosophicus*. London: Kegan Paul.
Russell, B. (1923). Vagueness. *The Australasian Journal of Psychology and Philosophy, 1*, 84–92.
Russell, B. (1927). *Introduction to the German edition of his 1912*. Erlangen: Paul Hertz.
Russell, B. (1929). How I came by my creed. *The Realist, 1*, 14–29. Also as What I Believe, *The Forum, 82*, 129–134.

Russell, B. (1940). *An inquiry into meaning & truth*. New York: W. W. Norton & Company.
Russell, B. (1944). My mental development; replies to criticisms. In Schilpp, 1944, pp. 1–20 and 679–741.
Russell, B. (1946). *A history of western philosophy*. London: Allen & Unwin Ltd.
Russell, B. (1948). *Human knowledge, its scope and limits*. New York: Simon and Schuster.
Russell, B. (1950). Logical positivism, in his *Logic and knowledge*, 1956a, pp. 367–382.
Russell, B. (1951). Obituary: Ludwig Wittgenstein. *Mind, 60*, 297–298.
Russell, B. (1953). The cult of common usage. *British Journal for the Philosophy of Science, 3*, 303–307. Reprinted in Russell 1956b, 66–73.
Russell, B. (1956a). *Logic and knowledge*. London: G. Allen & Unwin.
Russell, B. (1956b). *Portraits from memory and other essays*. London: G. Allen & Unwin.
Russell, B. (1957). Mr. Strawson on referring. *Mind, 66*, 385–389.
Russell, B. (1958). Mathematical infinity. *Mind, 67*, 385.
Russell, B. ([1959a] 1993). *My philosophical development*. Basil Blackwell, Oxford
Russell, B. ([1959b] 2005). Foreword to *Gellner's words and things*
Russell, B. (1967, 1968, 1969). *The autobiography of Bertrand Russell*. London: Allen and Unwin.
Russell, B. (1984). *Theory of knowledge*. The 1913 Manuscript. Edited by Elizabeth Eames in collaboration with Kenneth Blackwell.
Russell, B. (1998). *The autobiography of Bertrand Russell (in one volume)*. London: Allen and Unwin.
Ryle, G. ([1949] 2009). *The concept of mind*. New York: Barnes & Noble
Ryle, G. ([1953 and 1957] 1963). Ordinary language; the theory of meaning. In Caton, 1963, pp. 108–127 and 128–153.
Ryle, G. (1957, 1996). The work of an influential but little-known philosopher of science; Ludwig Wittgenstein. *Scientific American, 197*, 251–259. And in Shanker & Shanker 1996, *3*, 121–137.
Ryle, G. (1971). Autobiographical. In O. P. Wood, & G. Pitcher (Eds.), *Ryle* (Modern Studies in Philosophy). London: Macmillan.
Salmon, N. (1993). A problem in the Frege-church theory of sense and denotation. *Noûs, 27*, 158–166.
Sapir, E. (1921, 1949). *Language, an introduction to the study of speech*. New York: Harcourt, Brace and Company.
Sass, L. (2001). Deep disquietudes: Reflections on Wittgenstein as antiphilosopher. In *Wittgenstein: Biography and philosophy* (pp. 98–155). Cambridge/New York: Cambridge University Press.
Savage, C. W., Anthony Anderson, C., & Feigl, H. (Eds.). (1989). *Rereading Russell: Essays on Bertrand Russell's metaphysics and epistemology*. Minneapolis: University of Minnesota Press.
Savickey, B. (2017). *Wittgenstein's investigations: Awakening the imagination*. Cham: Springer.
Scharfstein, B.-A. (1980). *The philosophers: Their lives and the nature of their thought*. New York: Oxford University Press.
Schilpp, P. A. (Ed.). (1944). *The philosophy of Bertrand Russell*. Evanston: Northwestern University.
Schilpp, P. A. (Ed.). (1949). *Albert Einstein, philosopher-scientist*. Evanston: Library of Living Philosophers.
Schilpp, P. A. (Ed.). (1952). *The philosophy of G. E. Moore*. New York: Tudor publishing company.
Schilpp, P. A. (Ed.). (1963). *The philosophy of Rudolf Carnap*. La Salle: Open Court.
Schilpp, P. A. (Ed.). (1974). *The philosophy of Karl Popper*. La Salle: Open Court.
Schlagel, R. H. (1974). Contra Wittgenstein. *Philosophy and Phenomenological Research, 34*, 539–550.
Schlick, M. ([1932–1933] 1959). Positivism and realism. In Ayer, 1959, pp. 82–107.
Schlick, M. (1936). Meaning and verification. *Philosophical Review, 45*, 339–369.
Schlick, M. (1979). *Philosophical papers*, Vol. 2: (1925–1936). H. C. L. Mulder & B. F. B. Velde-Schlick, Editors

Schroeder, S. (2006). *Wittgenstein: The way out of the fly-bottle*. Cambridge: Polity Press.

Schwartz, Y. (2006). The politicization of the mystical in Buber and his contemporaries. In M. Zank (Ed.), *New perspectives on Martin Buber* (pp. 205–218). Tübingen: Mohr Siebeck.

Searle, J. R. (1958). Russell's objections to Frege's theory of sense and reference. *Analysis, 18*, 137–143.

Searle, J. R. (1969). *Speech acts: An essay in the philosophy of language*. Cambridge: Cambridge University Press.

Searle, J. R. (1975). The logical status of fictional discourse. *New Literary History, 6*, 319–332.

Searle, J. R. (1998). *Mind, language, and society: Philosophy in the real world*. New York: Basic Books.

Segre, M. (1994). Peano's axioms in their historical context. *Archive for History of Exact Sciences, 48*, 201–342.

Shanker, S. (1987a). Wittgenstein versus Turing on the nature of church's thesis. *Notre Dame Journal of Formal Logic, 28*, 615–649 (Special Issue on Church's Thesis).

Shanker, S. (1987b). *Wittgenstein and the turning point in the philosophy of mathematics*. London: Croom Helm.

Shanker, S., & Shanker, V. A. (1996). *Ludwig Wittgenstein: Critical assessments*. London: Croom Helm.

Shaw, B. (1903). *Don Juan in hell*. In his *Man and superman*.

Shaw, B. (1908). *The sanity of art: An exposure of the current nonsense about artists being degenerate*. London: New Age Press.

Shaw, B. (1912). *Androcles and the lion*. London: Constable and Company Limited.

Sigwart, C. (1895). *Logic: The judgement, concept and inference*. New York: Macmillan.

Sluga, H. (1987). Frege against the Booleans. *Notre Dame Journal of Formal Logic, 28*, 80–98.

Sluga, H. (1998). What has history to do with me? Wittgenstein and analytic philosophy. *Inquiry, 41*(1), 99–121.

Sluga, H. (2011). The world and its structure. In his *Wittgenstein*.

Sluga, H., & Stern, D. (Eds.). (1996). *The Cambridge companion to Wittgenstein*. Cambridge: Cambridge University Press.

Smit, H. (2015). Popper and Wittgenstein on the metaphysics of experience. *Zeitschrift für allgemeine Wissenschaftstheorie, 46*, 319–336.

Smith, N. J. J. (2009). Frege's judgement stroke and the conception of logic as the study of inference not consequence. *Philosophy Compass, 4*, 639–655.

Stegmüller, W. (2012). *Collected papers on epistemology, philosophy of science and history, volume 2*. Dordrecht: Springer.

Stenius, E. (1960). *Wittgenstein's Tractatus. A critical exposition of its main lines of thought*. Ithaca: Cornell University Press.

Stern, D. (Ed.). (1995). *Wittgenstein: Mind and language*. New York: Oxford University Press.

Stern, D. (2001). Was Wittgenstein a Jew? In Klagge 2001, pp. 237–272.

Stern, D. (2004). *Wittgenstein's philosophical investigations: An introduction*. Cambridge: Cambridge University Press.

Stern, D. (2007). Wittgenstein, the Vienna circle and physicalism: A reassessment. In Richardson & Uebel, 2007, pp. 305–331.

Stokhof, M. (2011). The quest for purity: Another look at the new Wittgenstein. *Croatian Journal of Philosophy, 11*, 275–294.

Strawson, P. F. (1950). On referring. *Mind, 59*, 320–344.

Strawson, P. F. (1959). *Individuals. An essay in descriptive metaphysics*. London: Methuen.

Stroll, A. (1994). *Moore and Wittgenstein on certainty*. New York/Oxford: Oxford University Press.

Stroud, B. (1965). Wittgenstein and logical necessity. *The Philosophical Review, 74*, 504–518.

Stroud, B. (1995). Quine on exile and acquiescense. In Leonardi & Santambrogio 1995, pp. 37–52.

Sweeney, L. (1997). *Christian philosophy: Greek, medieval, contemporary reflections*. New York: P. Lang.
Symons, J., Pombo, O., & Torres, J. M. (Eds.). (2011). *Otto Neurath and the unity of science*. Dordrecht: Springer.
Tait, W. W. (1997a). *Early analytic philosophy: Frege, Russell, Wittgenstein: Essays in honor of Leonard Linsky*. Chicago: Open Court.
Tait, W. W. (1997b). Frege versus cantor and Dedekind: On the concept of number. In Tait 1997, pp. 70–113.
Tanney, J. (2009). Rethinking Ryle: A critical discussion of the concept of mind. In Ryle [1949] 2009.
Tarski, A. ([1936] 1941). *Introduction to logic and to the methodology of deductive sciences*. New York: Oxford University Press.
Tejedor, C. (2011). *Starting with Wittgenstein*. London: Continuum.
Toulmin, S. E. (1958). *The uses of argument*. Cambridge: Cambridge University Press.
Toulmin, S. E. (1975). *Introduction to Engel 1975*. The Hague: Martinus Nijhoff.
Tranøy, K. E. (1988). *The moral import of science. Essays on normative theory, scientific activity and Wittgenstein*. Søreidgrend: Sigma Forlag.
Uschanov, T. P. (2002). Ernest Gellner's criticisms of Wittgenstein and ordinary language philosophy. In Kitching & Pleasants, 2002, pp. 23–35.
Van Bendegem, J. P. (n.d.). *A true(r) history of strict finitism*.
van Brakel, J. (1994). The *Ignis Fatuus* of semantic *Universalia*: The case of colours. *British Journal for the Philosophy of Science, 45*, 770–783.
van Heijenoort, J. (1967). Logic as calculus and logic as language. *Proceedings of the Boston Colloquium for the Philosophy of Science, 1964/1966*, 440–446.
van Peursen, C. A. (1970). *Ludwig Wittgenstein: An introduction to his philosophy*. E.P. Dutton & Co: New York.
Vesey, G. (Ed.). (1974). *Understanding Wittgenstein*. New York: Cornell University Press.
Vinten, R. (2015). Was Wittgenstein a conservative philosopher? *Revista Estudos Hum(e)anos*. http://revista.estudoshumeanos.com/was-wittgenstein-a-conservative-philosopher-por-robert-vinten/
von Hayek, F. A. (1977). Remembering My Cousin Ludwig Wittgenstein. *Encounter*, August 19, 1977, 20–21. Reprinted in his 1992, 176–181.
von Hayek, F. A. (1992). *The fortunes of liberalism: Essays on Austrian economics and the ideal of freedom* (The Collected Works of F. A. Hayek, IV). London: Routledge
von Wright, G. H. (1969). Special supplement: The Wittgenstein papers. *Philosophical Review, 78*, 483–503.
von Wright, G. H. (2006). Remarks on Wittgenstein's use of the terms 'Sinn', 'sinnlos', 'unsinnig', 'wahr', and 'Gedanke' in the *Tractatus*. In Pichler & Säätelä, pp. 98–106.
Waismann, F. (1968). *How i see philosophy*. Also in H. D. Lewis 1956, 447–490 and in Ayer, 1959, 345–380.
Wang, H. (1962). *A survey of mathematical logic*. Peking: Science Press. Reprinted as *Logic, computers, & sets*, 1970.
Wang, H. (1991). To and from philosophy: Discussions with Gödel and Wittgenstein. *Synthese, 88*, 229–277.
Warnock, G. J. (1950). Metaphysics in logic. *Proceedings of the Aristotelian Society, 51*, 197–222.
Warnock, G. J. (1958). *English philosophy since 1900*. New York: Oxford University Press.
Warnock, G. J. (2013). John Langshaw Austin, a biographical sketch. In Fann 2013, pp. 3–21.
Watkins, J. (1957a). Between analytic and empirical. *Philosophy, 32*, 112–131.
Watkins, J. (1957b). Farewell to the paradigm case argument. *Analysis, 18*, 25–33.
Weaver, N. (2014). *Forcing for mathematicians*. Singapore: World Scientific Pub. Co.
Weiler, G. (1961). Beliefs and attributes. *Philosophy, 36*, 196–210.

Weiler, G. (1970). *Mauthner's critique of language*. Cambridge: Cambridge University Press.

Weiss, B. (2014). *Michael Dummett*. London/New York: Routledge.

Wettersten, J. R. (1995). Preliminary report on efforts of Psychologism to gain influence in proper epistemological, methodological and psychological societies. In Jarvie & Laor, 1995, pp. 129–152.

Wettersten, J. R. (2005). *Whewell's critics: Have they prevented him from doing good?* Amsterdam: Rodopi.

Whately, R. (1826). *Elements of logic*.

Wheeler, S. C. (2000). *Deconstruction as analytic philosophy*. Stanford: Stanford University Press.

Whewell, W. (1857). Spedding's complete edition of the works of Bacon. *Edinburgh Review, 106*, 287–322.

Whitehead, A. N., & Russell B. (1910, 1912, and 1913). *Principia Mathematica*. Cambridge: Cambridge University Press

Whittaker, E. T. (1949). *From Euclid to Eddington: A study of conceptions of the external world*. Cambridge: Cambridge University Press.

Will, F. L. (1974). *Induction and justification: Investigation of Cartesian procedures in the philosophy of knowledge*. Ithaca: Cornell University Press.

Williams, B. (1968). Knowledge and meaning in the philosophy of mind. *Philosophical Review, 77*, 216–228.

Williams, B. (1974). Wittgenstein and idealism. In Vesey, 1974, pp. 76–95.

Williams, M. (2004). Nonsense and cosmic exile. An Austere reading of the *Tractatus*. In Kölbel & Weiss, 2004, pp. 1–27.

Williamson, T. (2002). *Vagueness*. Aldershot: Ashgate/Dartmouth.

Williamson, T. (2014). How did we get here from there? The transformation of analytic philosophy. *Belgrade Philosophical Annual, 27*, 7–37.

Winch, P. (1958). *The idea of a social science and its relation to philosophy*. London: Routledge.

Wisdom, J. T. D. ([1934] 1963). *Problems of mind and matter*. Cambridge: Cambridge University Press.

Wisdom, J. O. (1939). The analyst controversy: Berkeley's influence on the development of mathematics. *Hermathena, 29*, 3–29.

Wisdom, J. O. (1953). Berkeley's criticism of the infinitesimal. *British Journal for the Philosophy of Science, 4*, 22–25.

Wisdom, J. O. (1958). Review: The sceptic at bay: The problem of knowledge by A. J. Ayer. *British Journal for the Philosophy of Science, 9*, 159–163.

Wisdom, J. O., ([1947] 1980). *The metamorphosis of philosophy*. Ann Arbor: University Microfilms International.

Witherspoon, E. (2000). *Conceptions of nonsense in Carnap and Wittgenstein*. Abingdon: Taylor & Francis.

Wittgenstein, L. (1921–1922). *Logisch-philosophische Abhandlung*. *Annalen der Naturphilosophie*, 14, 185–262; English translation by C. K. Ogden, *Tractatus Logico-Philosophicus*; New translation by D. F. Pears & B. F. McGuinness. (Here labelled TLP).

Wittgenstein, L. (1929). Some remarks on logical form. *Proceedings of the Aristotelian Society, Supplementary, 9*, 162–171.

Wittgenstein, L. (1938). *Lectures on religious belief*. http://www.depa.univ-paris8.fr/IMG/pdf/Lectures_on_Religious_Belief.pdf

Wittgenstein, L. (1953). *Philosophical investigations* (G. E. M. Anscombe Trans.). (Here labelled PI).

Wittgenstein, L. (1956). *Remarks on the foundations of mathematics* (G. H. von Wright, R. Rhees, & G. E. M. Anscombe, Eds., and G. E. M. Anscombe, Trans.). Cambridge, MA: MIT Press.

Wittgenstein, L. (1958). *The blue and Brown books*. Pittsboro: InteLex Corp.

Wittgenstein, L. (1961). *Notebooks 1914–1916*. Oxford, Basil Blackwell.

Wittgenstein, L. ([1964] 1975). *Philosophical remarks*. Edited by Rush Rhees (1964).

Wittgenstein, L ([1967] 2007). *Lectures and conversations on aesthetics, psychology and religious beliefs*. Compiled from notes taken down at the time by three of the students: Rush Rhees, Yorick Smythies, & James Taylor; edited by Cyril Barrett.

Wittgenstein, L. (1968). Notes for lectures on 'private experience' and 'sense data'. *The Philosophical Review, 77*, 275–320.
Wittgenstein, L. ([1969] 1975). *On certainty* (D. Paul, & G. E. M. Anscombe, Trans.).
Wittgenstein, L. (1974). *Letters to Russell, Keynes, and Moore*. Oxford: Blackwell.
Wittgenstein, L. (1975). *Lectures on the foundations of mathematics, Cambridge, 1939*. Edited by Cora diamond from the notes of R. G. Bosanquet, Norman Malcolm, Rush Rhees, & Yorick Smythies.
Wittgenstein, L. (1976). *Wittgenstein's Lectures on the foundations of mathematics, Cambridge 1939*. Chicago: University of Chicago Press.
Wittgenstein, L. (1977). *Vermischte bemerkungen* (H. Nyman, Ed.). Frankfurt: Suhrakamp.
Wittgenstein, L. (1980a). *Culture and value* (P. Winch, Trans., G. H. von Wright Ed., in collaboration with Heikki Nyman.
Wittgenstein, L. (1980b). *Wittgenstein's lectures, Cambridge: 1930–1932*. From the Notes of John King & Desmond Lee.
Wittgenstein, L. (1983). *Remarks on the foundations of mathematics* (R. Rhees, G. H. von Wright & G. E. M. Anscombe Eds.; G. E. M. Anscombe, Trans.; rev ed.)
Wittgenstein, H. (1984). My brother Ludwig. In Rhees, 1984, pp. 1–11.
Wittgenstein, L. ([1969] 1991). *Philosophical grammar*. Ed. Rush Rhees.
Wittgenstein, L. (2012). *The big typescript: TS213: German-English Scholars' Edition* (C. Grant Luckhardt, & A. E. Maximilian, Ed. and Trans.). Oxford: Wylie-Blackwell.
Wittgenstein, Z., Anscombe, G.E.M., & von Wright, G.H. (Eds.). (1967). (trans: Anscombe, G.E.M.). Oxford: Blackwell.
Wright, C. (1980). *Wittgenstein on the foundations of mathematics*. London: Duckworth.
Zemach, E. M. (1989). Wittgenstein on meaning. In McGuinness & Haller 1989, pp. 415–435.
Ziff, P. (1960). *Semantic analysis*. Ithaca: Cornell University Press.
Ziff, P. (1984). *Epistemic analysis; a coherence theory of knowledge*. Dordrecht/Boston/Lancaster: Reidel.
Zilsel, E. (2003). *The social origins of modern science*. Dordrecht: Springer.

Index

J. Agassi, *Ludwig Wittgenstein's Philosophical Investigations*, Synthese Library 401,
https://doi.org/10.1007/978-3-030-00117-9

Kolibri 360 GmbH
Ferdinand-Jühlke-Straße 7
99095 Erfurt, Deutschland
produktsicherheit@kolibri360.de